AF172766

Communications
in Computer and Information Science 2274

Series Editors

Gang Li⊕, *School of Information Technology, Deakin University, Burwood, VIC,*
Australia
Joaquim Filipe⊕, *Polytechnic Institute of Setúbal, Setúbal, Portugal*
Ashish Ghosh⊕, *Indian Statistical Institute, Kolkata, West Bengal, India*
Zhiwei Xu, *Chinese Academy of Sciences, Beijing, China*

Rationale

The CCIS series is devoted to the publication of proceedings of computer science conferences. Its aim is to efficiently disseminate original research results in informatics in printed and electronic form. While the focus is on publication of peer-reviewed full papers presenting mature work, inclusion of reviewed short papers reporting on work in progress is welcome, too. Besides globally relevant meetings with internationally representative program committees guaranteeing a strict peer-reviewing and paper selection process, conferences run by societies or of high regional or national relevance are also considered for publication.

Topics

The topical scope of CCIS spans the entire spectrum of informatics ranging from foundational topics in the theory of computing to information and communications science and technology and a broad variety of interdisciplinary application fields.

Information for Volume Editors and Authors

Publication in CCIS is free of charge. No royalties are paid, however, we offer registered conference participants temporary free access to the online version of the conference proceedings on SpringerLink (http://link.springer.com) by means of an http referrer from the conference website and/or a number of complimentary printed copies, as specified in the official acceptance email of the event.

CCIS proceedings can be published in time for distribution at conferences or as postproceedings, and delivered in the form of printed books and/or electronically as USBs and/or e-content licenses for accessing proceedings at SpringerLink. Furthermore, CCIS proceedings are included in the CCIS electronic book series hosted in the SpringerLink digital library at http://link.springer.com/bookseries/7899. Conferences publishing in CCIS are allowed to use Online Conference Service (OCS) for managing the whole proceedings lifecycle (from submission and reviewing to preparing for publication) free of charge.

Publication process

The language of publication is exclusively English. Authors publishing in CCIS have to sign the Springer CCIS copyright transfer form, however, they are free to use their material published in CCIS for substantially changed, more elaborate subsequent publications elsewhere. For the preparation of the camera-ready papers/files, authors have to strictly adhere to the Springer CCIS Authors' Instructions and are strongly encouraged to use the CCIS LaTeX style files or templates.

Abstracting/Indexing

CCIS is abstracted/indexed in DBLP, Google Scholar, EI-Compendex, Mathematical Reviews, SCImago, Scopus. CCIS volumes are also submitted for the inclusion in ISI Proceedings.

How to start

To start the evaluation of your proposal for inclusion in the CCIS series, please send an e-mail to ccis@springer.com.

Haipeng Yu · Chengtao Cai · Lan Huang ·
Weipeng Jing · Xuebin Chen · Xianhua Song ·
Zeguang Lu
Editors

Computer Applications

39th CCF National Conference of Computer Applications, CCF NCCA 2024
Harbin, China, July 15–18, 2024
Proceedings, Part I

 Springer

Editors
Haipeng Yu
Northeast Forestry University
Harbin, China

Chengtao Cai
Harbin Engineering University
Harbin, China

Lan Huang
Jilin University
Changchun, China

Weipeng Jing
Northeast Forestry University
Harbin, China

Xuebin Chen
North China University of Science
and Technology
Tangshan, China

Xianhua Song
Harbin University of Science and Technology
Harbin, China

Zeguang Lu
National Academy of Guo Ding Institute
of Data Science
Beijing, China

ISSN 1865-0929   ISSN 1865-0937 (electronic)
Communications in Computer and Information Science
ISBN 978-981-97-9670-0   ISBN 978-981-97-9671-7 (eBook)
https://doi.org/10.1007/978-981-97-9671-7

This Springer imprint is published by the registered company Springer Nature Singapore Pte Ltd.
The registered company address is: 152 Beach Road, #21-01/04 Gateway East, Singapore 189721, Singapore

If disposing of this product, please recycle the paper.

Preface

As the program committee chairs of the 39th CCF National Conference on Computer Applications (CCF NCCA 2024), it is our great pleasure to welcome you to the conference proceedings. CCF NCCA 2024 was held in Harbin, China on July 15–18, 2024, hosted by China Computer Federation (CCF), organized by CCF Computer Application Professional Committee, Northeast Forestry University, and Heilongjiang Computer Society, and co-organized by Harbin Institute of Technology, Harbin Engineering University, Harbin Medical University, Northeast Agricultural University, Heilongjiang University, Northeast Petroleum University, Harbin Normal University, Harbin University of Science and Technology, etc. The goal of this conference series is to provide a forum for computer scientists, engineers, and educators.

This year's conference attracted 238 paper submissions. After the hard work of the Program Committee, 48 papers were accepted to appear in the conference proceedings, with an acceptance rate of 20.2%. The main topic of this conference was to empower the digital economy to open up a new highland to the north, and promote the application of production and research to cultivate new quality productivity. The accepted papers cover a wide range of fields such as Artificial Intelligence and Applications, Pattern Recognition & Machine Learning, Data Science and Technology, Network Communication and Security, Frontier and Comprehensive Applications, including autonomous collaborative control and optimal decision-making, robotics and intelligent systems, intelligent machine perception and pattern recognition, neural networks and deep learning, basic theories and methods of data science, big data mining and knowledge management technology, future grid architecture, network services and management, smart education, intelligent manufacturing, etc.

We would like to thank all the Program Committee members, a total of 288 people from 121 different institutes or companies, for their hard work in completing the review tasks. There were at least 3 reviewers for each article, and each reviewer reviewed no more than 5 articles. Their collective efforts made it possible to attain quality reviews for all the submissions within a few weeks. Their diverse expertise in each research area helped us to create an exciting program for the conference. Their comments and advice helped the authors to improve the quality of their papers and gain deeper insights.

We thank the team at Springer, whose professional assistance was invaluable in the production of the proceedings. A big thanks also to the authors and participants for their tremendous support in making the conference a success.

Besides the technical program, this year CCF NCCA offered different experiences to the participants. We hope you enjoyed the conference.

September 2024

Weipeng Jing
Xuebin Chen

Organization

Honorary Chairs

Hai Jin — Huazhong University of Science and Technology, China

Bin Xu — Tsinghua University, China

General Chairs

Haipeng Yu — Northeast Forestry University, China

Chengtao Cai — Harbin Engineering University, China

Lan Huang — Jilin University, China

Secretary-General

Zeguang Lu — China National Institute of Data Science, China

Deputy Secretary-General

Bing Xia — Zhongyuan University of Technology, China

Program Committee Chairs

Weipeng Jing — Northeast Forestry University, China

Xuebin Chen — North China University of Science and Technology, China

Organizing Committee Chairs

Ming Yu — Northeast Forestry University, China

Jing Liu — Hebei University of Technology, China

Organizing Committee

Rongli Gai	Dalian University, China
Pengtao Zhang	Northeast Forestry University, China
Chao Li	Northeast Forestry University, China
Linhui Li	Northeast Forestry University, China
Wenchao Li	Northeast Forestry University, China
Quan Hu	Northeast Forestry University, China
Yang Li	Northeast Forestry University, China
Peng Liu	Northeast Forestry University, China
Na Niu	Northeast Forestry University, China
Xia Li	Northeast Forestry University, China

Forum Committee

Junyu Lin	Institute of Information Engineering, Chinese Academy of Sciences, China
Rongli Gai	Dalian University, China

Thematic Forum Chairs

Lan Huang	Jilin University, China
Guanghui Yan	Lanzhou Jiaotong University, China
Wei Li	Harbin Engineering University, China
Shiyu Yang	Guangzhou University, China
Hong Yu	Dalian Ocean University, China
Shengke Wang	Ocean University of China, China
Manning Wang	Fudan University, China
Jinwu Wang	The Ninth People's Hospital Affiliated to Shanghai Jiao Tong University School of Medicine, China
Bing Xia	Zhongyuan University of Technology, China
Yong Ding	Guilin University of Electronic Technology, China
Helong Yu	Jilin Agricultural University, China
Qiaolin Ye	Nanjing Forestry University, China
Hongming Cai	Shanghai Jiao Tong University, China
Xianbao Wang	Zhejiang University of Technology, China

Workshop Chairs

Junyu Lin   Institute of Information Engineering, Chinese
                             Academy of Sciences, China
Yun He   Tsinghua AI TIME, China

Publication Chairs

Guangjie Han   Hohai University, China
Xianhua Song   Harbin University of Science and Technology,
                             China

Awards Committee Chair

Hong Yu   Dalian Ocean University, China

Awards Committee Secretary-General

Kui Xiao   Hubei University, China

Awards Committee

Guanghui Yan   Lanzhou Jiaotong University, China
Yun He   Tsinghua AI TIME, China
Zhaowen Qiu   Heilongjiang Tuomeng Technology Co., Ltd.,
                             China
Xiaoyu Chen   China Construction Bank, China
Aibin Chen   Central South University of Forestry and
                             Technology, China
Junhui Zhao   Beijing Jiaotong University, China
Ruiquan Jiang   Shanghai Jinshida Software Technology Co., Ltd.,
                             China

Competition Committee Chair

Bin Xu   Tsinghua University, China

Competition Committee

Zeguang Lu   China National Institute of Data Science, China
Jing Li   Tianjin University of Technology, China
Ke Xu   Weiye Xuanran Education Technology (Beijing)
                          Co., Ltd., China
Lifeng Ren   Universal Selnate (Beijing) Technology Co., Ltd.,
                          China
Xuedong Wang   Test Bar (Beijing) Technology Co., Ltd., China

Publicity Committee Chair

Junyu Lin   Institute of Information Engineering, Chinese
                          Academy of Sciences, China

Publicity Committee

Dalin Li   Zhuhai University of Science and Technology,
                          China
Tian Bai   Jilin University, China
Peng Liu   Changchun University of Science and
                          Technology, China

Exhibition Committee

Yongde Wu   Heilongjiang Provincial Computer Federation
                          Enterprise Working Committee, China
Long Zhao   Neusoft Education & Technology Group, China
Lijun Peng   Harbin Jinchen Weiye Technology Development
                          Co., Ltd., China
Jingyuan Yan   Harbin Jinchen Weiye Technology Development
                          Co., Ltd., China

Contents – Part I

Data Science and Technology

Contents – Part II

Pattern Recognition and Machine Learning

Network Communication and Security

Frontier and Comprehensive Applications

Data Science and Technology

Artificial Intelligence and Applications

On the Behaviors of Fuzzy Knowledge Graphs

Yu Ye[✉]

College of Information Engineering, Yangzhou University, Yangzhou 225127, China
yy4706418@gmail.com

Abstract. A knowledge graph is a common artificial intelligence technology, but we may meet some fuzzy semantics that knowledge graphs cannot effectively solve. Inspired by the fuzzy resource description framework (RDF) proposed by Ma and Li [7], we continue studying knowledge graphs in the framework of the fuzzy setting. Specifically, we first refine fuzzy knowledge graphs and their behaviors. An algorithm for calculating the path in fuzzy knowledge graphs is then provided. Next, the closure properties of the collection of the behaviors of fuzzy knowledge graphs are concentrated on under some familiar operations. Second, the commutative property and associative property of fuzzy knowledge graphs are introduced. Finally, the graph patterns, fuzzy selection operations, and fuzzy projection operations are introduced, demonstrating that these two operations also satisfy the associative and distributive properties. The behavior of fuzzy knowledge graphs provides theoretical support for the practical application of fuzzy knowledge graphs and more application values can be discovered on the basis of the properties of fuzzy knowledge graphs in the future.

Keywords: Fuzzy theory · Fuzzy knowledge graph · Behavior · Closure property

1 Introduction

Currently, with the explosive growth of data on the internet, these large amounts of data will contain valuable information, which needs to be discovered and studied by technical personnel. Therefore, knowledge graphs were invented to address these issues. The knowledge graph can be considered a database and used to manage relationships (connections) between various things. We can use the knowledge graph to retrieve the relationships between desired entities (e.g., the relationship between cars and roads). Additionally, the reasoning function of the knowledge graph enables us to discover new knowledge. Various functions of knowledge graphs are widely used in medicine [1], agriculture [2], etc., and knowledge graphs such as YAGO [8–10], DBpedia [11], Freebase [12], and NELL [13] are widely used. A knowledge graph is a graph model where entities are represented as vertices and relationships are represented as edges. Each vertex is connected by edges. The core data model of the knowledge graph is the

© The Author(s), under exclusive license to Springer Nature Singapore Pte Ltd. 2024
H. Yu et al. (Eds.): CCF NCCA 2024, CCIS 2274, pp. 3–21, 2024.
https://doi.org/10.1007/978-981-97-9671-7_1

resource description framework schema (RDFS) [14]. We call the encoding in the form of a subject-predicate-object a triplet (also called an RDF triplet), such as $<Yangzhou, locatedIn, Jiangsu>$. In everyday life, we might use expressions such as 'very tall' to describe height or 'a bit painful' to convey sensations. However, these semantics are rather fuzzy (how tall is 'very tall'? how painful is 'a bit'?), making it challenging to obtain precise information. Therefore, when such semantics are represented via RDF triples, there is also inherent fuzziness. For triples with fuzzy semantics, classic knowledge graphs cannot effectively describe fuzzy semantic information. Therefore, the knowledge graph is extended into a fuzzy knowledge graph to solve these situations.

The core part of the fuzzy knowledge graph is the fuzzy RDF triple, which combines RDF triples with fuzzy sets. M. Mazzieri [15] and others pioneered the integration of RDF triples with fuzzy theory. They mapped the elements of RDF triples through a function to the interval [0,1]. The resulting values represent the membership degree of the corresponding elements. Consequently, this type of RDF triple is referred to as a fuzzy RDF triple. Another form of fuzzy RDF was discussed in [16]. Li pu [17] et al. studied fuzzy semantic reasoning on knowledge graphs and built a corresponding reasoning model. Bai [19] and others have focused on the task of representing the membership degrees of spatiotemporal information. While fuzzy knowledge graphs are gaining traction in practical applications, few theoretical studies on this subject exist. Our work on algebraic operations in fuzzy RDFs was inspired by the work given in [7]. This work provides relevant definitions of knowledge graphs and integrates Zadeh's [18] fuzzy set theory with RDF to define this fuzzy knowledge graph. In [7], the modeling, query and subgraph matching of fuzzy knowledge graphs were studied. However, not much explanation was provided for the given properties, so we studied the closure properties of the behavior set of fuzzy knowledge graphs, as well as the associativity and commutativity of fuzzy knowledge graphs. In this paper we define the behaviors of fuzzy knowledge graphs and graph patterns. Then the closure and other related properties of the fuzzy knowledge graph are introduced on the basis of the behavior of the fuzzy knowledge graph and the behavior of the graph pattern. The goal of this step is to understand the properties of fuzzy knowledge graphs and develop new application values on the basis of these properties.

The remainder of this article is organized as follows. Section 2 introduces some basic notions of RDF triples, knowledge graphs and fuzzy sets. Section 3 introduces fuzzy RDF triples, fuzzy knowledge graphs and graph patterns; defines the behavior of fuzzy knowledge graphs and graph patterns, and illustrates the effect of behaviors with examples. Section 4 introduces the relevant properties and proofs of the fuzzy knowledge graph. Section 5 provides a summary and highlights further research directions.

2   Preliminaries

In this section, we introduce the relevant definitions and concepts of resource description frameworks(RDF [3]), knowledge graphs and fuzzy sets.

An RDF is a markup language used to describe web resources and composed of three sets: $\mathcal{U}$, $\mathcal{B}$, and $\mathcal{L}$. $\mathcal{U}$ represents the set of all unique identifiers. $\mathcal{B}$ represents the set of blank nodes. $\mathcal{L}$ represents the set of all text sets. Text sets are used to describe the relationships between entities. For mutually disjoint sets $\mathcal{U}$, $\mathcal{B}$, and $\mathcal{L}$, we can denote them as: $\mathcal{U} \cap \mathcal{B} \cap \mathcal{L} = \emptyset$. An RDF triple can be expressed as $<s, p, o>$. In this RDF triple, s represents the subject, o represents the object, and p represents the predicate(p describes the relationship between s and o). Therefore, each triplet can be interpreted as a relationship p between object s and object o.

A knowledge graph is composed of many nodes, edges, and characteristic labels. However by introducing RDF triples, we can also consider a knowledge graph as being composed of many RDF triples.

Through the relationship between RDF triples and knowledge graphs, we can define knowledge graphs as follows.

Definition 2.1. (*See* [7]). A knowledge graph (KG) is a four-tuple (V, E, Σ, L), where

(i) V is a finite set of vertices,

(ii) E is a finite set of edges,

(iii) Σ represents a set of strings. These strings are composed of labels from nodes and edges within the knowledge graph,

(iv) L: $V \cup E \rightarrow \Sigma$, called the label function.

If we have an RDF triple $<s, p, o>$, we need to convert this RDF triple into a knowledge graph. First, we need to store s and o in the knowledge graph. Assuming that we have two nodes v_s and v_o. Therefore a vertex set has two nodes v_s, v_o and $V = \{v_s, v_o\}$. Then, for strings s and o, we can use the label function L to associate nodes with labels. Therefore, $L(v_s) = s, L(v_o) = o$ and $\Sigma = \{s, o\}$. Second, we need to store p in the knowledge graph. The method is the same as above.

Example 2.2. We have some RDF triples $<Jack, Has, Car>$, $<Jack, Has, Meeting>$, $<Tom, Is, Student>$, $<Tom, Like, Football>$, $<American, Is, Country>$, $<Jack, Friend, Tom>$, $<Tom, BornIn, American>$.Through the above conversion method, we can obtain the following knowledge graph in Fig. 1. For this KG, we can obtain:

$$V = \{v_{Car}, v_{Jack}, v_{Meeting}, v_{Student}, v_{Tom}, v_{Fooball}, v_{American}, v_{Country}\},$$

$$E = \{(v_{Jack}, v_{Car}), (v_{Jack}, v_{Meeting}), (v_{Tom}, v_{Student}),$$

$$(v_{Tom}, v_{Football}), (v_{American}, v_{Country}), (v_{Jack}, v_{Tom}), (v_{Tom}, v_{American})\},$$

$$\Sigma = \{Car, Jack, Meeting, Tom, Student, Football, American, Country,$$

$$Has, Friend, Is, Like, BornIn\}.$$

Next, we introduce fuzzy sets [4].

Classical sets are basic mathematical concepts used to describe deterministic phenomena. However, classical sets cannot resolve some fuzzy phenomena

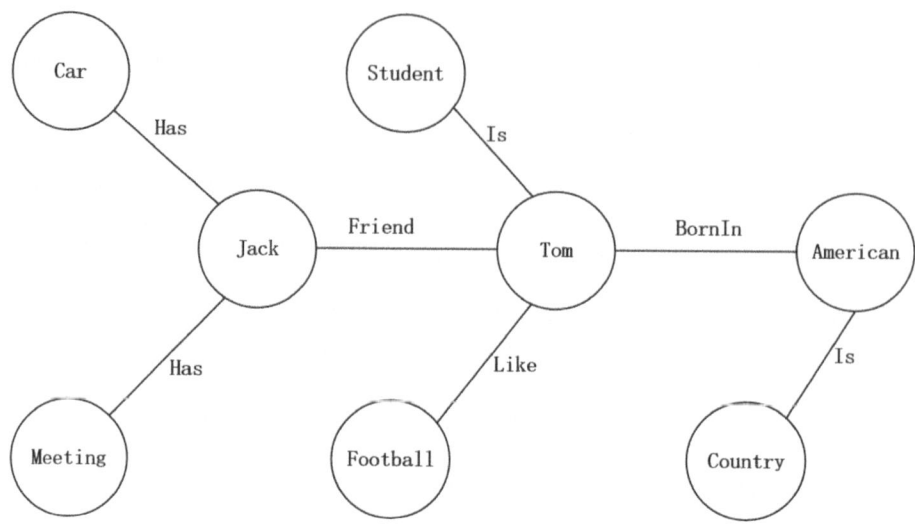

Fig. 1. Knowledge graph

(such as: temperature, speed). Therefore, to address fuzzy phenomena (concepts), Zadeh [18] introduced the concept of fuzzy sets.

Definition 2.3. A fuzzy set F is an ordered pair (X, F), where:
    (i) X represents a nonempty set,
    (ii) $F\colon X \to [0,1]$,called the membership function(commonly denoted as μ_F), which describes the membership degree of each element in the fuzzy set.
    Below, we introduce the operations of the intersection, union, and complement of fuzzy sets.$T(X)$ is used to represent all fuzzy sets on X, that is: $T(X) = \{A|A : X \to [0,1]\}$.

Definition 2.4. (*See* [20]). Let $A, B \in T(X)$, the union of A and B($A \cup B$), the intersection of A and B($A \cap B$), and the complement of A(A^c) be defined as:
    $(A \cup B)(x) = A(x) \vee B(x) = \max\{A(x), B(x)\}$
    $(A \cap B)(x) = A(x) \wedge B(x) = \min\{A(x), B(x)\}$
    $A^c(x) = 1 - A(x)$

Theorem 2.5. (*See* [20]). Let X set, $A, B, C \in F(X)$, then: (i) Commutative Law
    $A \cup B = B \cup A, A \cap B = B \cap A$
(ii) Associative Law
    $(A \cup B) \cup C = A \cup (B \cup C)$
    $(A \cap B) \cap C = A \cap (B \cap C)$
(iii) Distributive Law
    $A \cap (B \cup C) = (A \cap B) \cup (A \cap C)$
    $A \cup (B \cap C) = (A \cup B) \cap (A \cup C)$
The other properties of fuzzy sets can be found in [5,6,20].

3 Fuzzy Knowledge Graphs

In this section, RDF triples and knowledge graphs are expanded. They are combined with fuzzy sets to form fuzzy RDF triples and fuzzy knowledge graphs. Subsequently, we define the behaviors of fuzzy knowledge graphs and provid examples. We introduce graph patterns and establish their behaviors in the framework of fuzzy circumstances.

Assigning membership degrees to each element of the RDF triples through membership functions. We call these new RDF triples fuzzy RDF triples [7]. In this article, the form of fuzzy RDF triples is as follows: $< \mu_s/s, \mu_p/p, \mu_o/o >$.

$< \mu_s/s, \mu_p/p, \mu_o/o >$ is called an element-level fuzzy RDF triple. μ_s, μ_p, μ_o represent the membership degrees of $s, p, o(\mu_i \in [0,1], i = s, p, o)$ respectively. This fuzzy RDF triple assigns membership degrees to s, p, and o so that the semantics of these fuzzy RDF triples are more precise. For example, we use the $[0,1]$ interval to show how long a person has lived in a city. 0 represents never lived in this city, and 1 represents always lived here without leaving. Then, a fuzzy RDF triple $< 1/Jack, 0.8/LiveIn, 1/Yangzhou >$ represents Jack lives in Yangzhou but because $\mu_{LiveIn} = 0.8$. Therefore, we can consider that Jack has been residing in Yangzhou for a long time, but he has also lived in other cities. At this point, $Jack$ and $Yangzhou$ are associated through the relationship of $liveIn$. If $Jack$ was born in $Yangzhou$, the triple representation of fuzzy RDF would be $< 1/Jack, 1/BornIn, 1/Yangzhou >$(For elements with a membership degree of 1, in order to avoid confusion, we need to write "1" on them.). At this moment, $Jack$ is associated with $Yangzhou$ through the $BornIn$ relationship. So in real life, for the same subject (s) and object (o), there may exist multiple different predicates (p).

Similar to a knowledge graph, a fuzzy knowledge graph is composed of numerous fuzzy RDF triples. In a fuzzy RDF triple, the subject s and object o act as two nodes, and the predicate p serves as the edge connecting these two nodes. Therefore, p can also be represented as (s, o). Therefore, for ease of explanation, we establish a membership function ρ, such that $\rho(s, o) = \mu_p$. If there are multiple predicates $p^i(i = 1, 2, ..., n)$ between s and o, then $\rho(s, o) = \bigwedge_{i=1,2,...,n} \mu_{p^i}$.

Definition 3.1. (*See* [7]). A fuzzy knowledge graph (FKG) is a seven-tuple $(V, E, \Sigma, L_V, L_E, \mu, \rho)$, for the functions ρ and μ, we only need to use a random function to limit the membership range to $[0, 1]$, where

(*i*) V is a finite set of vertices,

(*ii*) E is a finite set of edges,

(*iii*) Σ represents a set of strings. These strings are composed of labels from nodes and edges within the fuzzy knowledge graph,

(*iv*) $L_V: V \to \Sigma$ is a function assigning labels to vertices,

(*v*) $L_E: E \to \Sigma$ is a function assigning labels to edges,

(*vi*) $\mu: V \to [0, 1]$, called the node membership function,

(*vii*) $\rho: E \to [0, 1]$, called the edge membership function.

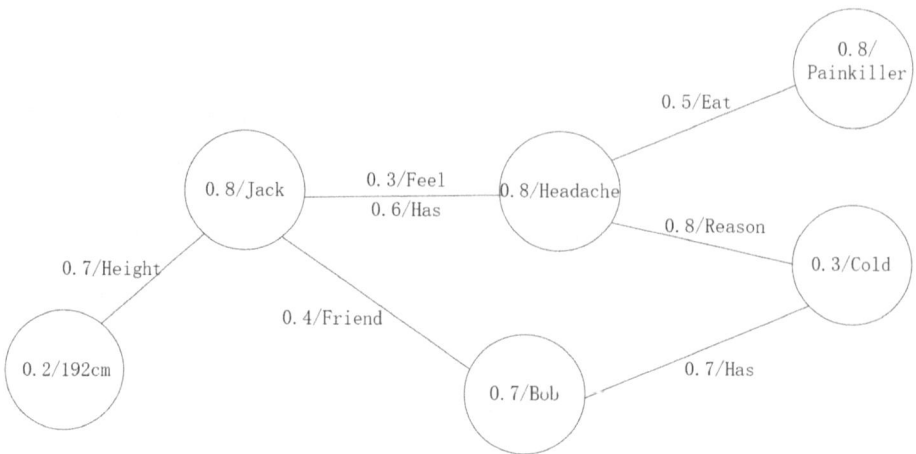

Fig. 2. Fuzzy knowledge graph

Example 3.2. We can see a fuzzy knowledge graph, as shown in Fig. 2. For this FKG, we can obtain:

$$V = \{v_{192cm}, v_{Jack}, v_{Bob}, v_{Headache}, v_{Painkiller}, v_{Cold}\},$$
$$E = \{(v_{Jack}, v_{192cm}), (v_{Jack}, v_{Headache}), (v_{Jack}, v_{Bob}), (v_{Bob}, v_{Cold}),$$
$$(v_{Headache}, v_{Painkiller}), (v_{Headache}, v_{Cold})\},$$
$$\Sigma = \{192cm, Height, Jack, Friend, Bob, Has, Cold, Reason,$$
$$Headache, Feel, Eat, Painkiller\}.$$

In the fuzzy knowledge graph, $(v_m, v_{m+1})(v_{m+1}, v_{m+2})...(v_{n-1}, v_n)$. These sequentially connected edges are called paths (a single edge can also be called a path). As shown in Fig. 2, we refer to $(v_{Jack}, v_{Bob})(v_{Bob}, v_{Cold})$ as a path, then (v_{Jack}, v_{192cm}) can also be referred to as a path. We denote these paths as d_i ($i = 1, 2, ..., n$, where n is the total number of paths in the FKG). We define $\mathcal{D}$ as the set of all paths in the fuzzy knowledge graph($d_i \in \mathcal{D}$). For a path d, we define a function α, where $\alpha(d) = \rho(v_m, v_{m+1}) \wedge \rho(v_{m+1}, v_{m+2}) \wedge ... \wedge \rho(v_{n-1}, v_n)$. By using the fuzzy operation $\wedge$, we can express the semantics of the KFG more accurately. In a FKG, as there may be multiple paths from node v_i to node v_j, we define $P(v_i, v_j)$ to represent all paths starting from node v_i and ending at node $v_j(P(v_i, v_j) \subseteq \mathcal{D})$. For example, in Fig. 2, $P(v_{Jack}, v_{Cold}) = \{d_1, d_2\}$ where $d_1 = (v_{Jack}, v_{Headache})(v_{Headache}, v_{Cold})$, $d_2 = (v_{Jack}, v_{Bob})(v_{Bob}, v_{Cold})$.

Fuzzy knowledge graphs have a greater semantic advantage over traditional knowledge graphs. For example, the statement "Tom is slightly taller than Jack" is abstract. By adding membership degrees to their heights, such as $\mu_{Tom}(Tall) = 0.9$ and $\mu_{Jack}(Tall) = 0.7$, we make their height comparison more concrete. In everyday life, many concepts are fuzzy. Adding membership degrees makes these fuzzy concepts clearer. Therefore, compared with traditional knowledge graphs. Our fuzzy knowledge graph can handle fuzzy semantics (such as "very tall" or "extremely good") better, allowing readers to understand relationships more intuitively through the membership degrees.

Definition 3.3. In a fuzzy knowledge graph, there is a path from vertex v_i to $v_j(P(v_i, v_j))$. We can express this path as follows,

$$\theta(d) = \bigvee_{m=1,2,\ldots,n} \{\alpha(d_m)|d_m \in P(v_i, v_j)\}$$

In FKG, when there are multiple paths between two nodes, the semantics between the nodes can be interpreted differently. We define the θ function to make the semantics between two nodes more easily understandable. When faced with multiple paths between two nodes, we can use the θ function to obtain the path with the maximum membership degree (highest influence) among these paths. In this way, we can gain a more accurate understanding of the relationship between the two nodes. For the aforementioned $P(v_{Jack}, v_{Cold})$, we use the θ function to obtain:$\theta(P(v_{Jack}, v_{Cold})) = \alpha(d_1) \vee \alpha(d_2) = \alpha(d_2)$. Therefore, by using the θ function, we can consider that the influence of Jack's friend Bob on Jack having a cold is more significant. We can write $\theta(d)$ as $\theta_{d_{i,j}}$ (i and j represent the head and tail nodes of path P).

$2^{\mathcal{D}}$ is the power set of $\mathcal{D}$. Suppose that $L(2^{\mathcal{D}}) = \{f : 2^{\mathcal{D}} \to [0,1]|\ f$ is a function that maps $2^{\mathcal{D}}$ to the interval $[0, 1]\}$ is the collection of all FKG behaviors in $2^{\mathcal{D}}$.

Next, we can define the behavior of FKG.

Definition 3.4. For any FKG, suppose that it has two nodes: the head node **H** and the tail node **T**. These two nodes are both n-dimensional. After that, we define the behavior(denoted as f_{FKG}) recognized by FKG as $f_{FKG} \in L(2^{\mathcal{D}})$. $\circ$ represents $\wedge, \vee, \complement, \times$ compositional operation. Therefore, for $\forall P(v_h, v_t) \in 2^{\mathcal{D}}$

$$f_{FKG}(d) = \mu_h \circ \theta_{d_{h,t}} \circ \mu_t = \bigvee_{v_h \in \mathbf{H}, v_t \in \mathbf{T}} \{\mu(v_h) \wedge \theta(d) \wedge \mu(v_t)|d \in P(v_h, v_t)\}$$

Within the realm of a fuzzy knowledge graph, nodes v_h and v_t symbolize two distinct entities. These nodes encapsulate the attributes or features of both the head node and the tail node. For nodes v_h and v_t, we need to consider only the membership of these two nodes and the membership of the paths. Because we are investigating the degree of relationship between these two nodes, we do not consider the relationships between these two nodes and other nodes. By performing fuzzy operations on the membership of two nodes and the membership of paths between them, we can determine which type of relationship is stronger between the two nodes. This makes it easier to understand the relationship between the two nodes. By defining the behavior of fuzzy knowledge graphs, we can effectively elucidate the pertinent characteristics inherent to them. In fact, we can leverage the behavior of fuzzy knowledge graphs to assess the validity of node membership through reasoned analysis.

Algorithm 1. Path calculation in fuzzy knowledge graph (FKG)

Input: an FKG, v_h, v_t
Output: Fuzzy value f

1: Assign membership degrees to v_s and v_t through the function μ
2: $\mu(v_h)$ and $\mu(v_t)$ are random functions on the interval $[0,1]$
3: $P(v_h, v_t)$ represents all the paths between v_h and v_t
4: $P(v_h, v_t) = \{d_1, d_2, \ldots, d_n\}$
5: d_i is one of the paths between v_h and v_t
6: **for** d_i in $P(v_h, v_t)$ **do**
7:      $\theta(d_i) = \alpha(d_i)$
8: **end for**
9: **if** only one path d **then**
10:      $\theta(d) = \theta(d_i)$
11:      $f = f(d)$
12: **else**
13:      **for** d_i in $P(v_h, v_t)$ **do**
14:          $\theta(d) = \bigvee \theta(d_i)$
15:          $f = \bigvee f(d_i)$
16:      **end for**
17: **end if**
18: **return** f

Next, we present an example of the behavior of a fuzzy knowledge graph.

Example 3.5. . We have an FKG, as shown in Fig. 3. We denote the behavior of this FKG as $f_1(f_1 \in L(2^D))$.

Let us discuss the membership degree between vertex v_{Jack} and v_{Book}. First, there are multiple paths(d_1, d_2) between v_{Jack} and v_{Book}.

$$d_1 = (v_{Jack}, v_{John})(v_{John}, v_{Book}), d_2 = (v_{Jack}, v_{Book})$$

Therefore,$P(v_{Jack}, v_{Book}) = \{d_1, d_2\}$. For $d \in P(v_{Jack}, v_{Book})$, we can obtain

$$f_1(d) = 0.6$$

From the above equation, we know that Jack's degree of purchase of the book is 0.6. Therefore, Jack has the idea of buying books, therefore, we can recommend books to Jack on shopping apps.

For the above example, we can obtain more accurate semantics in the FKG by leveraging the behavior of the FKG. Through the functions and behavior defined above, we design Algorithm 1.

Algorithm 1 consists of three parts. In the first part (Lines 1–6), the membership degrees of the head node and tail node are assigned via the function μ, where $\mu(v_h), \mu(v_t) \in [0,1]$. Then, all paths between the head node and tail node are stored in the set P, with d_i representing one of these paths. In the second part (Lines 6–8), we traverse the set P and assign a membership degree to each path. In the third part (Lines 9–18), we evaluate the number of paths between

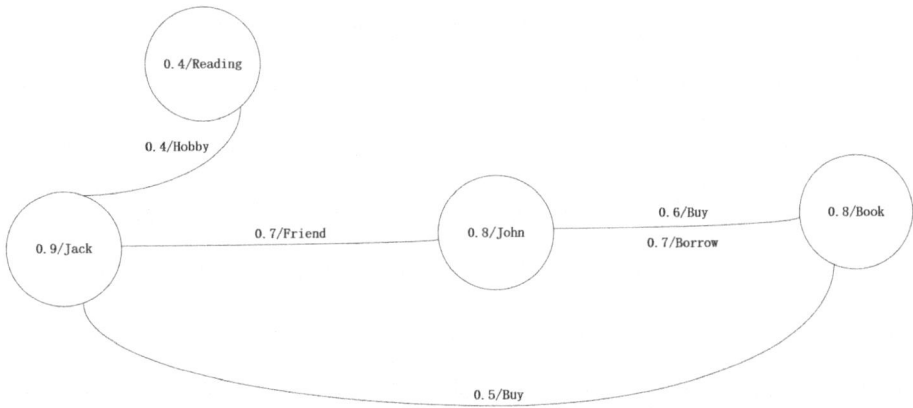

Fig. 3. Fuzzy knowledge graph of people

the head node and the tail node. If there is only one path, then $\theta(d)$ is directly substituted into the fuzzy knowledge graph's behavior function f (Lines 9–11). Otherwise (Lines 12–18), we perform $\bigvee$ operation on the membership degrees of multiple paths, similarly substituting each $\theta(d)$ into the function f and finally performing $\bigvee$ operation. Using Algorithm 1, we can obtain the fuzzy value f of the paths between the two nodes v_h and v_t.

In an FKG, we query the desired results. At this time, we need a graph pattern to match the fuzzy knowledge graph, a graph pattern is a graph with query information. We can match subgraphs that satisfy the graph pattern in the fuzzy knowledge graph.

Definition 3.6. (*See* [7]). A graph pattern (GP) is a six-tuple $(V_P, E_P, R_E, \Sigma_{PV}, \Sigma_{PE}, F_P)$, where

(i) V_P is a finite set of vertices,

(ii) E_P is a finite set of edges,

(iii) R_E represents a mapping. The corresponding query requirements are mapped to regular expressions.

(iv) Σ_{PV} represents a set of strings. These strings are composed of labels from nodes within the graph pattern.

(v) Σ_{PE} represents a set of strings. These strings are composed of labels from edges within the graph pattern.

(vi) $F_P: V \cup E \rightarrow \Sigma_P(\Sigma_P = \Sigma_{PV} \cup \Sigma_{PE})$, called the label function,

Example 3.7. We can see the graph pattern, as shown in Fig. 4.

This means that there is some connection between a person and a number. Therefore we match Fig. 4 with Fig. 2, and obtain the subgraph shown in Fig. 5.

Finally, we obtain the corresponding results by matching the graph pattern with the fuzzy knowledge graph. This is very meaningful for searching fuzzy knowledge graphs.

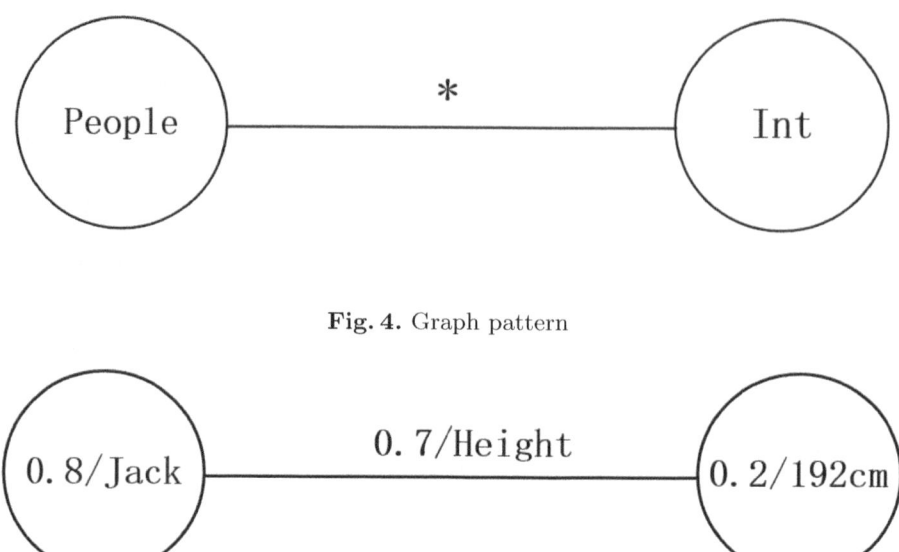

Fig. 4. Graph pattern

Fig. 5. Matching results between Fig. 4 and Fig. 2

We next need to define the behavior of the graph pattern. The behavior of graph patterns and fuzzy knowledge graphs facilitates a better understanding of the properties of fuzzy knowledge graphs.

Definition 3.8. $L(GP)$ is the collection of all the GP behaviors. We have a $GP = (V_P, E_P, R_E, \Sigma_{PV}, \Sigma_{PE}, F_P)$ then the behavior of the GP is recorded as $\beta_{GP}(\beta_{GP} \in L(GP))$. So,

$$\beta_{GP}(s) = \bigvee_{s_{v_i} \in \Sigma_{PV}, s_{e_j} \in \Sigma_{PE}} \{T(s_{v_i}) \wedge T(s_{e_j}) | i = 1, 2, ..., m. j = 1, 2, ..., n\}$$

where we denote the elements in the set Σ_P as $s_c(s_c \in \Sigma_P, c = v_1, ..., v_m, e_1, ..., e_n.v_m \in V_P, e_n \in E_P)$. Therefore, for $F_P(v_i)$, let $F_P(v_i) = s_{v_i} \in \Sigma_{PV}$; similarly, $F_P(e_j) = s_{e_j} \in \Sigma_{PE}$. s_{v_i} describes the character of node v_i, and s_{e_j} describes the character of edge e_j. $T : \Sigma_P \rightarrow [0, 1]$, map the characters in the Σ_P set to [0,1]. Because the behavior of the graph pattern is similar to that of the fuzzy knowledge graph, we do not provide examples in this section.

4 The Properties of Fuzzy Knowledge Graphs

In this section, first, we use the behavior of fuzzy knowledge graphs to show that the sets formed by fuzzy knowledge graphs are closed under operations such as union, intersection, complement, and Cartesian products. Through these behaviors, we elucidate the relevant properties of fuzzy knowledge graphs. Finally, we introduce two operations(fuzzy selection and fuzzy projection) to illustrate the relevant properties of the graph patterns and fuzzy knowledge graphs.

Theorem 4.1. *The set consisting of FKGs is closed under union.*

Proof. Given any two fuzzy knowledge graphs FKG_1 and FKG_2, let their accepted behaviors be f_1 and f_2 respectively.

First, let $FKG_i = (V_i, E_i, \Sigma_i, L_{V_i}, L_{E_i}, \mu_i, \rho_i)(i = 1, 2)$. Let us assume $V_1 \cap V_2 = \emptyset$, $E_1 \cap E_2 = \emptyset$, $\Sigma_1 \cap \Sigma_2 = \emptyset$. We construct a fuzzy knowledge graph $FKG = (V, E, \Sigma, L_V, L_E, \mu, \rho)$. Here, $V = V_1 \cup V_2, E = E_1 \cup E_2$.

$$\mu(v_i) = \mu_1(v_{1i}) \vee \mu_2(v_{2i}), v_{1i} \in V_1, v_{2i} \in V_2$$
$$\rho(v_i, v_j) = \rho_1(v_{1i}, v_{1j}) \vee \rho_2(v_{2i}, v_{2j}), (v_{1i}, v_{1j}) \in E_1, (v_{2i}, v_{2j}) \in E_2$$
$$\theta(d) = \theta_1(d) \vee \theta_2(d)$$

Then,

$$\theta_1(d) = \bigvee_{i=1,2,\ldots,n} \{\alpha_1(d_i)|d_i \in P(v_{1h}, v_{1t})\}$$

$$\theta_2(d) = \bigvee_{i=1,2,\ldots,n} \{\alpha_2(d_i)|d_i \in P(v_{2h}, v_{2t})\}$$

Therefore, we have,

$$f_{FKG}(d) = (\mu_{1h} \circ \theta_{1d_{h,t}} \circ \mu_{1t}) \vee (\mu_{2h} \circ \theta_{2d_{h,t}} \circ \mu_{2t}) = f_1(d) \vee f_2(d)$$

Thus, *The set consisting of FKGs is closed under union.*

Theorem 4.2. *The set consisting of FKGs is closed under intersection.*

Proof. Given any two fuzzy knowledge graphs FKG_1 and FKG_2, let their accepted behaviors be f_1 and f_2 respectively.

First, let $FKG_i = (V_i, E_i, \Sigma_i, L_{V_i}, L_{E_i}, \mu_i, \rho_i)(i = 1, 2)$. We construct a fuzzy knowledge graph $FKG = (V, E, \Sigma, L_V, L_E, \mu, \rho)$. Here, $V = V_1 \cap V_2, E = E_1 \cap E_2$.

Because it is an intersection operation, the vertices and edges in the graph do not need to be distinguished.

$$\mu(v) = \mu_1(v) \wedge \mu_2(v), v \in V$$
$$\rho(v_i, v_j) = \rho_1(v_i, v_j) \wedge \rho_2(v_i, v_j), (v_i, v_j) \in E$$
$$\theta(d) = \theta_1(d) \wedge \theta_2(d)$$

Then,

$$\theta_1(d) = \bigvee_{i=1,2,\ldots,n} \{\alpha_1(d_i)|d_i \in P(v_h, v_t)\}$$

$$\theta_2(d) = \bigvee_{i=1,2,\ldots,n} \{\alpha_2(d_i)|d_i \in P(v_h, v_t)\}$$

Therefore, we have,

$$f_{FKG}(d) = (\mu_{1h} \circ \theta_{1d_{h,t}} \circ \mu_{1t}) \wedge (\mu_{2h} \circ \theta_{2d_{h,t}} \circ \mu_{2t}) = f_1(d) \wedge f_2(d)$$

Thus, *The set consisting of FKGs is closed under intersection.*

Theorem 4.3. *The set consisting of FKGs is closed under complement.*

Proof. Given any two fuzzy knowledge graphs FKG_1 and FKG_2, let their acceptance behaviors be f_1 and f_2 respectively.

First, let $FKG_i = (V_i, E_i, \Sigma_i, L_{V_i}, L_{E_i}, \mu_i, \rho_i)(i = 1, 2)$. Then, we denote the complement of the fuzzy knowledge graph as $(FKG_1 \cap FKG_2)^c$. Therefore, we can obtain a fuzzy knowledge graph $FKG_3 = (V_3, E_3, \Sigma_3, L_{V_3}, L_{E_3}, \mu_3, \rho_3)$. $FKG_3 = FKG_1 \cap FKG_2$, so $V_3 = V_1 \cap V_2, E_3 = E_1 \cap E_2, \Sigma_3 = \Sigma_1 \cap \Sigma_2, \mu_3 = \mu_1 \wedge \mu_2, \rho_3 = \rho_1 \wedge \rho_2$. Let the behavior accepted by FKG_3 be f_3.

We construct a fuzzy knowledge graph $FKG = (V, E, \Sigma, L_V, L_E, \mu, \rho)$. Here $V = V_3^c$, $E = E_3^c$.

$$\mu(v_i) = 1 - \mu_3(v_i), v_i \in V,$$
$$\rho(v_i, v_j) = 1 - \rho_3(v_i, v_j), (v_i, v_j) \in E,$$
$$\theta(d) = 1 - \theta_3(d)$$

Then,

$$\theta_3(d) = \bigvee_{i=1,2,\ldots,n} \{\alpha_3(d_i) | d_i \in P(v_{3h}, v_{3t})\}$$

Therefore, we have,

$$f_{FKG}(d) = 1 - (\mu_{3h} \circ \theta_{3d_{h,t}} \circ \mu_{3t}) = 1 - f_3(d)$$

Thus, *The set consisting of FKGs is closed under complement.*

Next, we will define the Cartesian product operation for the FKG($*$ represents the multiplication operation). For any two $FKG_i = (V_i, E_i, \Sigma_i, L_{V_i}, L_{E_i}, \mu_i, \rho_i)$ and $FKG_j = (V_j, E_j, \Sigma_j, L_{V_j}, L_{E_j}, \mu_j, \rho_j)$, their Cartesian product is denoted as $FKG_i \times FKG_j = FKG = (V, E, \Sigma, L_V, L_E, \mu, \rho)$, where $V = \{(v_i, v_j) | v_i \in V_i, v_j \in V_j\}$, $E = \{[(v_i, v_{i+1}), (v_j, v_{j+1})] | (v_i, v_{i+1}) \in E_i, (v_j, v_{j+1}) \in E_j\}$, $\mu(v_i, v_j) = \mu_i(v_i) * \mu_j(v_j), \rho[(v_i, v_{i+1}), (v_j, v_{j+1})] = \rho_i(v_i, v_{i+1}) * \rho_j(v_j, v_{j+1})$.

Therefore, we can have the following theorem.

Theorem 4.4. *The set consisting of FKGs is closed under cartesian product.*

Proof. Given any two fuzzy knowledge graphs FKG_1 and FKG_2, let their accepted behaviors be f_1 and f_2 respectively.

First, let $FKG_i = (V_i, E_i, \Sigma_i, L_{V_i}, L_{E_i}, \mu_i, \rho_i)(i = 1, 2)$. We construct a fuzzy knowledge graph $FKG = (V, E, \Sigma, L_V, L_E, \mu, \rho)$. Here, $V = V_1 \times V_2, E = E_1 \times$

E_2.

$$\mu(v_{1i}, v_{2i}) = \mu_1(v_{1i}) * \mu_2(v_{2i}), v_{1i} \in V_1, v_{2i} \in V_2$$
$$\rho[(v_{1i}, v_{1j}), (v_{2i}, v_{2j})] = \rho_1(v_{1i}, v_{1j}) * \rho_2(v_{2i}, v_{2j}), (v_{1i}, v_{1j}) \in E_1, (v_{2i}, v_{2j}) \in E_2$$
$$\theta(d) = \theta_1(d) \times \theta_2(d)$$

Then,

$$\theta_1(d) = \bigvee_{i=1,2,\ldots,n} \{\alpha_1(d_i)|d_i \in P(v_{1h}, v_{1t})\}$$

$$\theta_2(d) = \bigvee_{i=1,2,\ldots,n} \{\alpha_2(d_i)|d_i \in P(v_{2h}, v_{2t})\}$$

Therefore, we have,

$$f_{FKG}(d) = (\mu_{1h} \circ \theta_{1d_{h,t}} \circ \mu_{1t}) \times (\mu_{2h} \circ \theta_{2d_{h,t}} \circ \mu_{2t}) = f_1(d) \times f_2(d)$$

Thus, *The set consisting of FKGs is closed under cartesian product.*

Through the above four theorems, we find that the fuzzy knowledge graph has closed properties in the intersection, union, difference and Cartesian product operations. Next, we introduce other properties of fuzzy knowledge graphs.

First, let us look at other operations. For $a \in [0, 1]$, $f \in L(2^{\mathcal{D}})$, the quantity product $af \in L(2^{\mathcal{D}})$ of a and f is defined as: $\forall d \in \mathcal{D}, af(d) = a \wedge f(d)$. Then, we can define the connection operation on $L(2^{\mathcal{D}})$:

Let $f, g \in L(2^{\mathcal{D}})$, the connection operation of f and g be recorded as fg. Then, the connection operation is defined as:

$$\forall d \in \mathcal{D}, fg(d) = \bigvee_{d_1 d_2 = d} (f(d_1) \wedge g(d_2))$$

Proposition 4.5. *The connection operation satisfies the associative law.*

Proof. *For fuzzy behaviors $f, g, h \in L(2^{\mathcal{D}})$, we have*

$$(fg)h = f(gh)$$

For the above formula, on the left side there is

$$(d) = \bigvee_{d' d_3 = d} [(fg)(d') \wedge h(d_3)]$$
$$= \bigvee_{d' d_3 = d} [\bigvee_{d_1 d_2 = d'} [f(d_1) \wedge g(d_2)] \wedge h(d_3)]$$
$$= \bigvee_{d_1 d_2 d_3 = d} [f(d_1) \wedge g(d_2) \wedge h(d_3)]$$

For the above formula, on the right side,

$$(d) = \bigvee_{d_1 d' = d} [f(d_1) \wedge (gh)(d')]$$
$$= \bigvee_{d_1 d' = d} [f(d_1) \wedge \bigvee_{d_2 d_3 = d'} [g(d_2) \wedge h(d_3)]]$$
$$= \bigvee_{d_1 d_2 d_3 = d} [f(d_1) \wedge g(d_2) \wedge h(d_3)]$$

From the above proof, it can be concluded that the left formula is equal to the right formula: $(fg)h = f(gh)$.
Thus The connection operation satisfies the associative law.

Proposition 4.6. *Given three fuzzy knowledge graphs* FKG_1, FKG_2, *and* FKG_3, *then,*

(i) $FKG_1 \cup FKG_2 = FKG_2 \cup FKG_1$

(ii) $FKG_1 \cap FKG_2 = FKG_2 \cap FKG_1$

(iii) $(FKG_1 \cup FKG_2) \cup FKG_3 = FKG_1 \cup (FKG_2 \cup FKG_3)$

(iv) $(FKG_1 \cap FKG_2) \cap FKG_3 = FKG_1 \cap (FKG_2 \cap FKG_3)$

The above concerns the relevant properties of fuzzy knowledge graphs. Next, we introduce the relevant properties of the combination of fuzzy knowledge graphs and graph patterns. The graph pattern has two main forms of operations on the fuzzy knowledge graph: fuzzy selection and fuzzy projection.

Definition 4.7. *(See* [7]*).* Given a fuzzy knowledge graph $FKG = (V, E, \Sigma, L_V, L_E, \mu, \rho)$ and a graph pattern $GP = (V_P, E_P, R_E, \Sigma_{PV}, \Sigma_{PE}, F_P)$. The expression for fuzzy selection is as follows:

$$\sigma_{GP}(FKG) = \{\langle g, \delta_{GP}(g) \rangle \, | g = \varepsilon(GP, FKG), \delta_{GP}(g) > 0\}$$

The fuzzy knowledge graph FKG and the graph pattern GP are matched through the ε function to obtain the corresponding subgraph g. $\delta_{GP}(g) > 0$ is used to limit the membership degree to be greater than 0.

Definition 4.8. *(See* [7]*).* Given a fuzzy knowledge graph $FKG = (V, E, \Sigma, L_V, L_E, \mu, \rho)$ and graph pattern $GP = (V_P, E_P, R_E, \Sigma_{PV}, \Sigma_{PE}, F_P)$. The expression for fuzzy projection is as follows:

$$\pi_{GP,PL}(FKG) = \{\langle g, \delta_{GP}(g) \rangle \, | g = \varpi(GP, PL, FKG), \delta_{GP}(g) > 0\}$$

The fuzzy knowledge graph FKG and the fuzzy graph pattern GP are matched through the ϖ function to obtain the corresponding subgraph g. Then, subgraph g selects subgraphs that meet the conditions specified in the condition list PL. $\delta_{GP}(g) > 0$ is used to limit the membership degree to be greater than 0.

The functions of fuzzy selection and fuzzy projection are to select the corresponding subgraph in the fuzzy knowledge graph on the basis of the graph pattern. However, fuzzy selection does not have a condition list PL, so the subgraph selected according to the graph pattern is not unique. Unlike fuzzy selection, fuzzy projection has a condition list PL, so the subgraph selected according to the graph pattern is unique. Therefore, when we do not consider the list of conditions, the operation forms of the two are the same. Because both fuzzy selection and fuzzy projection query subgraphs that satisfy the graph pattern using the fuzzy knowledge graph, we can consider fuzzy selection and fuzzy projection as intersection operations between the fuzzy knowledge graph and the graph pattern. The resulting subset is the subgraph that satisfies the graph pattern. Therefore, we can consider fuzzy selection $\sigma_{GP}(FKG) = FKG \cap GP$. For fuzzy projection $\pi_{GP,PL}(FKG)$, this article considers only the case of $PL = \emptyset$. Therefore, $\pi_{GP,PL}(FKG)$ is written as $\pi_{GP}(FKG)$ and $\pi_{GP}(FKG) = FKG \cap GP$. When $PL = \emptyset$, the operations of fuzzy selection $\sigma_{GP}(FKG)$ and fuzzy projection $\pi_{GP}(FKG)$ are the same.

Through the above introduction, we can obtain the following properties:

Proposition 4.9. $\pi_{GP}(FKG_1 \cup FKG_2) = \pi_{GP}(FKG_1) \cup \pi_{GP}(FKG_2)$

Proof. Given any two fuzzy knowledge graphs FKG_1, FKG_2 and a graph pattern GP, let their accepted behaviors be f_1, f_2 and β respectively. To the left,

$$\pi_{GP}(FKG_1) \cup \pi_{GP}(FKG_2) = (GP \cap FKG_1) \cup (GP \cap FKG_2)$$

Therefore, we can express it as,

$$(\beta \wedge f_1) \vee (\beta \wedge f_2) = (f_1 \vee f_2) \wedge \beta$$

To the right,

$$\pi_{GP}(FKG_1 \cup FKG_2) = (FKG_1 \cup FKG_2) \cap GP$$

Then, we can express it as,

$$(f_1 \vee f_2) \wedge \beta$$

Because the left side is equal to the right side, we have $\pi_{GP}(FKG_1 \cup FKG_2) = \pi_{GP}(FKG_1) \cup \pi_{GP}(FKG_2)$.

Proposition 4.10. $\pi_{GP_1}(\pi_{GP_2}(FKG)) = \pi_{GP_1 \cap GP_2}(FKG)$

Proof. Given a fuzzy knowledge graph FKG and two graph patterns GP_1 and GP_2, let their accepted behaviors be f, β_1 and β_2 respectively. To the left,

$$\pi_{GP_1}(\pi_{GP_2}(FKG)) = \pi_{GP_1}(GP_2 \cap FKG) = GP_1 \cap GP_2 \cap FKG$$

Therefore, we can express it as,

$$\beta_1 \wedge \beta_2 \wedge f$$

To the right,

$$\pi_{GP_1 \cap GP_2}(FKG) = GP_1 \cap GP_2 \cap FKG$$

Then, we can express it as,

$$\beta_1 \wedge \beta_2 \wedge f$$

Because the left side is equal to the right side, we have $\pi_{GP_1}(\pi_{GP_2}(FKG)) = \pi_{GP_1 \cap GP_2}(FKG)$.

Proposition 4.11. $\sigma_{GP_1 \cap GP_2}(FKG) = \sigma_{GP_1}(\sigma_{GP_2}(FKG))$

Proof. Because the condition list in fuzzy projection is empty, the operations of fuzzy projection and fuzzy selection are the same. Therefore, the proof method is the same as that of Proposition 4.10.

Proposition 4.12. $\sigma_{GP_1 \cup GP_2}(FKG) = \sigma_{GP_1}(FKG) \cup \sigma_{GP_2}(FKG)$

Proof. Given a fuzzy knowledge graph FKG and two graph patterns GP_1 and GP_2, let their accepted behaviors be f, β_1 and β_2 respectively.
To the left,

$$\sigma_{GP_1 \cup GP_2}(FKG) = (GP_1 \cup GP_2) \cap FKG$$

Therefore, we can express it as,

$$(\beta_1 \vee \beta_2) \wedge f = (\beta_1 \wedge f) \vee (\beta_2 \wedge f)$$

To the right,

$$\sigma_{GP_1}(FKG) \cup \sigma_{GP_2}(FKG) = (GP_1 \cap FKG) \cup (GP_2 \cap FKG)$$

Then, we can express it as,

$$(\beta_1 \wedge f) \vee (\beta_2 \wedge f)$$

Because the left side is equal to the right side, we have $\sigma_{GP_1 \cup GP_2}(FKG) = \sigma_{GP_1}(FKG) \cup \sigma_{GP_2}(FKG)$.

Proposition 4.13. $\sigma_{GP_1}(\sigma_{GP_2}(FKG)) = \sigma_{GP_2}(\sigma_{GP_1}(FKG))$ *Proof. Given a fuzzy knowledge graph FKG and two graph patterns GP_1 and GP_2, let their accepted behaviors be f, β_1 and β_2 respectively.*
To the left,

$$\sigma_{GP_1}(\sigma_{GP_2}(FKG)) = GP_1 \cap GP_2 \cap FKG$$

Therefore, we can express it as,

$$\beta_1 \wedge \beta_2 \wedge f$$

To the right,

$$\sigma_{GP_2}(\sigma_{GP_1}(FKG)) = GP_2 \cap GP_1 \cap FKG$$

Then, we can express it as,

$$\beta_2 \wedge \beta_1 \wedge f$$

Because the left side is equal to the right side, we have $\sigma_{GP_1}(\sigma_{GP_2}(FKG)) = \sigma_{GP_2}(\sigma_{GP_1}(FKG))$.

Proposition 4.14. $\sigma_{GP}(FKG_1 \cup FKG_2) = \sigma_{GP}(FKG_1) \cup \sigma_{GP}(FKG_2)$

Proof. Given any two fuzzy knowledge graphs FKG_1, FKG_2 and a graph pattern GP, let their accepted behaviors be f_1, f_2 and β respectively.
To the left,

$$\sigma_{GP}(FKG_1 \cup FKG_2) = GP \cap (FKG_1 \cup FKG_2)$$

Therefore, we can express it as,

$$\beta \wedge (f_1 \vee f_2) = (\beta \wedge f_1) \vee (\beta \wedge f_2)$$

To the right,

$$\sigma_{GP}(FKG_1) \cup \sigma_{GP}(FKG_2) = (GP \cap FKG_1) \cup (GP \cap FKG_2)$$

Then, we can express it as,

$$(\beta \wedge f_1) \vee (\beta \wedge f_2)$$

Because the left side is equal to the right side, we have $\sigma_{GP}(FKG_1 \cup FKG_2) = \sigma_{GP}(FKG_1) \cup \sigma_{GP}(FKG_2)$.

We introduced the common properties of fuzzy knowledge graphs, and in [7], introduced other properties of fuzzy knowledge graphs. We provide theoretical support for fuzzy knowledge graphs by defining their behavior.

5   Conclusion

Faced with a multitude of fuzzy semantics that arise in real-life scenarios, the traditional knowledge graph is unable to effectively address this issue. As a result, fuzzy knowledge graphs have been developed. Although the application of fuzzy knowledge graphs is gradually becoming increasingly popular, the theory of fuzzy knowledge graphs is still in its early stages. In response to this situation, we have initiated research on the theoretical aspects of fuzzy knowledge graphs. The main focus of this paper is to study the properties of fuzzy knowledge graphs by defining the behaviors of fuzzy knowledge graphs. First, we briefly introduce the fundamental concepts of fuzzy knowledge graphs, gaining an understanding of RDF triples, knowledge graphs, and fuzzy sets. In conjunction with fuzzy sets, we introduce the relevant definitions of fuzzy RDF triples, fuzzy knowledge graphs, and graph patterns. Next, we define the behaviors of fuzzy knowledge graphs and graph patterns. Then, on the basis of these behaviors, we provide an algorithm for calculating the path in a fuzzy knowledge graph. Through this algorithm, we can obtain the membership degree between two nodes in the fuzzy knowledge graph, thereby clarifying the relationship between these two nodes. In conjunction with real-world scenarios, we provide examples illustrating the practical implications of fuzzy knowledge graph behaviors. At the end of this paper, we discovered that the set formed by fuzzy knowledge graphs is closed under intersection, union, complement, and Cartesian operations on the basis of the behaviors of fuzzy knowledge graphs. Additionally, we discuss the relevant properties of fuzzy knowledge graphs and graph patterns. This article is the first attempt to study the properties of fuzzy knowledge graphs. On the basis of our current research on the properties and algorithm of fuzzy knowledge graphs, in future research, we will continue to study the basic operations of fuzzy knowledge graphs through the closure properties discussed in this paper. In terms of applications, we need to consider how to incorporate membership degrees into knowledge graphs and apply algorithms to fuzzy knowledge graphs to calculate paths. We will conduct comprehensive research on the properties of fuzzy knowledge graphs, which will provide stronger theoretical support for their application.

References

1. Xiang, X., Wang, Z., Jia, Y., Fang, B.: Knowledge graph-based clinical decision support system reasoning: a survey. In: 2019 IEEE Fourth International Conference on Data Science in Cyberspace (DSC), pp. 373–380 (2019). https://doi.org/10.1109/DSC.2019.00063
2. Chenglin, Q., Qing, S., Pengzhou, Z., Hui, Y.: Cn-MAKG: China meteorology and agriculture knowledge graph construction based on semi-structured data. In: 2018 IEEE/ACIS 17th International Conference on Computer and Information Science (ICIS), pp. 692–696 (2018). https://doi.org/10.1109/ICIS.2018.8466485
3. Manola, F., Miller, E.: RDF Primer. W3C Recommendation 10 February 2004. https://www.w3.org/TR/2004/REC-rdf-primer-20040210/. Accessed 17 Aug 2023
4. Klir, G., Yuan, B.: Fuzzy Sets and Fuzzy Logic. Prentice Hall, New Jersey (1995)
5. Dubois, D.J.: Fuzzy Sets and Systems: Theory and Applications. Academic Press (1980)
6. Zadeh, L.A.: Fuzzy sets. Inf. Control $8(3)$, 338–353 (1965)
7. Ma, Z.M., Li, G., Yan, L.: Fuzzy data modeling and algebraic operations in RDF. Fuzzy Sets Syst. $\mathbf{351}$, 41–63 (2018). https://doi.org/10.1016/j.fss.2017.11.013
8. Suchanek, F.M., Kasneci, G., Weikum, G.: Yago: a core of semantic knowledge. In: 16th International Worldwide Web Conference, WWW 2007, pp. 697–706 (2007). https://doi.org/10.1145/1242572.1242667
9. Hoffart, J., Suchanek, F.M., Berberich, K., Weikum, G.: YAGO2: a spatially and temporally enhanced knowledge base from Wikipedia. Artif. Intell. $\mathbf{194}$, 28–61 (2013). https://doi.org/10.1016/j.artint.2012.06.001
10. Mahdisoltani, F., Biega, J., Suchanek, F.M.: YAGO3: a knowledge base from multilingual wikipedias. In: CIDR 2015 - 7th Biennial Conference on Innovative Data Systems Research (2015)
11. Auer, S., Bizer, C., Kobilarov, G., Lehmann, J., Cyganiak, R., Ives, Z.: DBpedia: a nucleus for a web of open data. In: Aberer, K., et al. (eds.) ASWC/ISWC -2007. LNCS, vol. 4825, pp. 722–735. Springer, Heidelberg (2007). https://doi.org/10.1007/978-3-540-76298-0_52
12. Bollacker, K., Evans, C., Paritosh, P., Sturge, T., Taylor, J.: Freebase: a collaboratively created graph database for structuring human knowledge. In: Proceedings of the ACM SIGMOD International Conference on Management of Data, pp. 1247–1249 (2008). https://doi.org/10.1145/1376616.1376746
13. Carlson, A., Betteridge, J., Kisiel, B., Settles, B., Hruschka, E.R., Mitchell, T.M.: Toward an architecture for never-ending language learning. In: Proceedings of the 24th AAAI Conference on Artificial Intelligence, AAAI 2010, pp. 1306–1313 (2010)
14. Brickley, D., Guha, R.V., McBride, B.: RDF schema 1.1. W3C Recommendation 25 February 2014. https://www.w3.org/TR/2014/REC-rdf-schema-20140225/. Accessed 25 Aug 2023
15. Mazzieri, M., Dragoni, A.F.: A fuzzy semantics for the resource description framework. Lecture Notes in Computer Science (Including Subseries Lecture Notes in Artificial Intelligence and Lecture Notes in Bioinformatics), LNAI, vol. 5327, pp. 244–261 (2008). https://doi.org/10.1007/978-3-540-89765-1-15

16. Lv, Y., Ma, Z.M., Yan, L.: Fuzzy RDF: a data model to represent fuzzy metadata. In: IEEE International Conference on Fuzzy Systems, pp. 1439–1445 (2008). https://doi.org/10.1109/FUZZY.2008.4630561
17. Li, P., et al.: A fuzzy semantic representation and reasoning model for multiple associative predicates in knowledge graph. Inf. Sci. **599**, 208–230 (2022). https://doi.org/10.1016/j.ins.2022.03.079
18. Zadeh, L.A., Fu, K.S., Tanaka, K.: Fuzzy Sets and their Applications to Cognitive and Decision Processes. Academic press (1974)
19. Bai, L., Wang, J., Di, X., Li, N.: Fixing the inconsistencies in fuzzy spatiotemporal RDF graph. Inf. Sci. **578**, 166–180 (2021). https://doi.org/10.1016/j.ins.2021.07.038
20. Li, Y.M., Li, P.: Fuzzy Computing Theory. Academic Press, Beijing (2016)

Research and Development of a Voice Interaction Platform for Medical Assistive Robots Based on the ROS

Haiyan Wang[✉], Guiyuan Gao, Zhan Shi, and Zhiwei Xu

College of Computer Science and Technology, Key Laboratory of Intelligent Rehabilitation and Barrier-Free for the Disabled (Changchun University), Ministry of Education, Changchun University, Changchun 130022, China
wanghy80@ccu.edu.cn

Abstract. With respect to the voice interaction problem of the ROS medical auxiliary robot, the design and development method of the ROS-based medical auxiliary robot voice interaction system is described in detail, which has certain reference value and promotion significance for the development of complex intelligent service robot projects. For this reason, this paper develops an ROS-based medical assistant robot voice interaction platform to provide convenience for patients and medical staff.

Keyword: Voice interaction · ROS · Medical auxiliary robot

1 Introduction

The development process of the voice system can be traced back to a number of key historical nodes and technological advances. Research on speech recognition began in the 1950s [1]. At that time, AT & T Bell Labs successfully implemented the first speech recognition system that can recognize ten English number-Audry systems [2]. This lays the foundation for subsequent speech recognition technology. In the 1960s, the application of computers promoted the development of speech recognition. Important technical achievements include dynamic programming (DP) [3] and linear predictive analysis (LP) [4]. LP technology solves the problem of the speech signal generation model and has a profound impact on the development of speech recognition. The 1970s was a breakthrough period in the field of speech recognition. During this period, LP technology was further developed, and dynamic time correction (DTW) technology was basically mature [5]. Moreover, vector quantization (VQ) [6] and the hidden Markov model (HMM) [7] theory have also been proposed, which provide strong theoretical support for the accuracy and efficiency of speech recognition. With the advancement of technology, voice systems will play a more important role in the future, increasing convenience and benefits to human life and work.

The technology of speech system application involves many key fields, which together support the processing, recognition, synthesis and interaction of speech signals. Signal processing technology [8] involves preliminary analysis and processing of

H. Yu et al. (Eds.): CCF NCCA 2024, CCIS 2274, pp. 22–31, 2024.
https://doi.org/10.1007/978-981-97-9671-7_2

speech signals. Multiple processing, noise processing, speech enhancement and other methods eliminate the influence of the external environment on speech and remove redundant information to improve the quality and recognizability of speech signals. After signal processing, the acoustic feature extraction technology receives the effective signal and extracts the acoustic features of the waveform. This feature information is the key information, which is helpful for subsequent speech recognition and natural language understanding. Speech recognition technology is the key technology for converting human speech into machine-recognizable text or commands [9]. It is based on an acoustic model, a language model, deep neural network technology, and pattern recognition of speech signals to achieve voice-to-text conversion. Speech synthesis technology generates artificial speech through mechanical and electronic methods [10]. It can convert text information into comprehensible and fluent speech outputs so that the machine can speak like a human. Natural language processing technology is used to understand and analyze text information after speech conversion [11]. It involves syntactic analysis, semantic understanding, dialog management and other aspects so that the robot can accurately understand the user's intentions and problems. These technologies enable the voice system to be continuously optimized and improved. Through a large amount of data training, the speech system can improve its recognition accuracy, reduce the false recognition rate, and better adapt to different accents, speech rates and noise environments. The acoustic model is the most critical part of speech recognition technology [12]. It can extract the speech feature vector sequence, calculate the distance of the pronunciation template, and generate the acoustic model score to improve the accuracy of speech recognition. Speech models are also very important for speech recognition technology. It usually uses a statistical grammar speech model and a semantic and grammatical structure command language model to analyze grammar and semantics, reduce the search space and improve the system recognition rate. These technologies cooperate with each other to form the core technology system of the speech system, which enables robots to interact with humans efficiently and naturally. With the continuous advancement and innovation of technology, the application of voice systems will be more extensive, increasing the convenience and fun of human life.

2   Correlation Technique

Traditional human–computer interactions generally involve strong vision and weak logic through graphics, also known as the graphic user interface (GUIs). The voice user interface (VUI) is a human–computer interaction through voice, strong logic without vision (or weak vision) [13]. The VUI delivers all sufficient information through voice, carrying all the elements of cognition, logic, emotion and so on, which is the core of intelligent voice interaction. The voice interaction system is one of the core modules of the service robot. This allows the robot to have the interactive ability of 'listening, speaking, asking and answering', similar to a human. It is the key link for realizing communication between a service robot and a human [14]. First, the voice input device (such as a microphone array module) loaded on the service robot collects the user's voice and completes the input, sampling and coding of the audio. The speech recognition link completes the conversion of speech information to machine-recognizable text information. The language processing step completes the corresponding operation processing according to the logical judgment of the text characters or commands after speech recognition conversion. Finally, the speech synthesis link completes the conversion of text information to sound information and feeds back the speech signal output to humans. During the working process of the service robot, when humans give voice commands to the robot, the robot will first understand the human voice. After the recognition and processing of the voice signal, the robot answers human questions and completes a task according to human instructions. This is the function of the voice interaction control system. A block diagram of the voice interaction system is shown in Fig. 1.

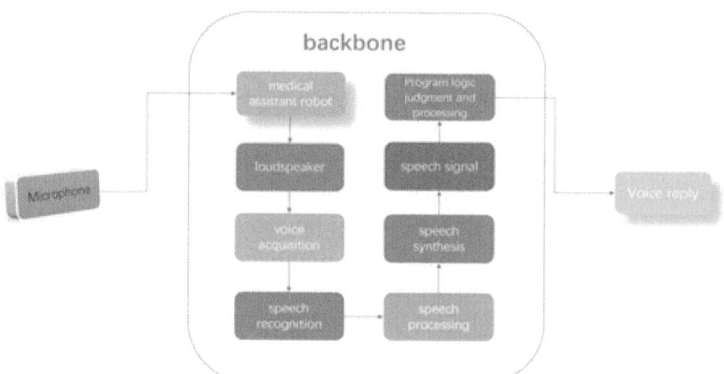

Fig. 1. Voice interaction system block diagram

The robot operating system (ROS) is an open source meta-operating system for robots that complies with the BSD open source license. The ROS is a type of secondary operating system that is based on the traditional operating system and provides a structured communication layer. It provides a series of functions similar to traditional operating systems, such as hardware abstraction, device drivers, library functions, visualization, messaging, and package management, to help software developers create robot applications. The powerful features of the ROS enable developers to easily deploy voice interaction engineering, realize communication between service robots and humans, and complete a task according to human commands. The ROS has the characteristics of point-to-point design, multilanguage support, a streamlined architecture, high integration, rich toolkits, and free and open source.

The software development kit (SDK) launched by IFLYTEK provides developers with rich voice interaction functions. These SDKs usually contain core functions such as speech recognition, speech synthesis, and speech wake-up, providing powerful speech processing capabilities for intelligent devices such as robots. We need to integrate the IFLYTEK SDK into the ROS robot. To this end, this part develops a speech control function package based on the ROS and IFLYTEK, which includes speech recognition, speech synthesis and speech wake-up functions.

For speech synthesis, the open platform of IFLYTEK is used to realize the speech output function of the ROS. In the terminal, a string is issued to the/voiceWords topic through the terminal to turn on the function of text to speech. The function of text to speech is then realized by modifying tts_subscribe.cpp, and the converted speech is broadcast in real time. Its workflow is shown in Fig. 2.

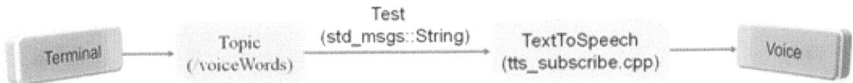

Fig. 2. Speech synthesis

For speech recognition, the IFLYTEK open platform is used to achieve the ROS voice input function. In the terminal, a wake-up word is issued through the/voiceWakeup topic to turn on the function of voice to text. Then, by modifying iat_publish.cpp, the function of voice to text is realized, and the converted text is sent to the/voiceWords topic. Its workflow is shown in Fig. 3.

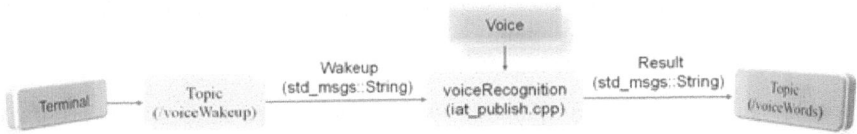

Fig. 3. Speech recognition

For speech recognition and synthesis output, the IFLYTEK open platform is used to achieve simultaneous ROS input and output speech functions. First, in the terminal, the wake-up word is released to the/voiceWakeup topic to open the voice-to-text function. Then, the text-to-speech function is turned on to obtain text from the/voiceWords topic into voice playback. Its workflow is shown in Fig. 4.

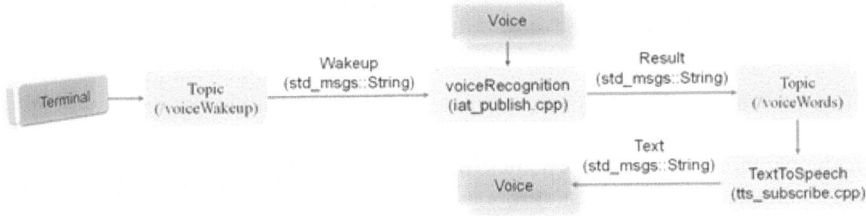

Fig. 4. Speech recognition and synthesis

Voice interaction occurs through the wake-up word to start the interaction function after the voice inputs an audio, according to the input audio, to take the corresponding action, such as returning to the other end of the audio or publishing a specified topic. A voice interaction process is performed, as shown in Fig. 5.

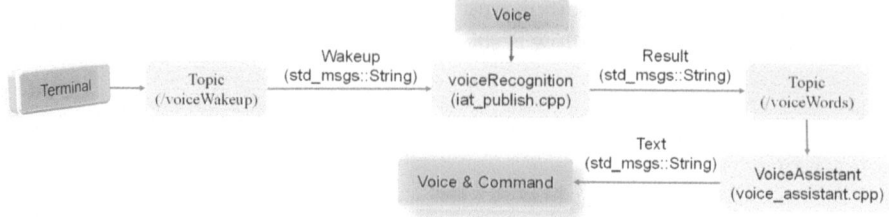

Fig. 5. Voice interaction

3 System Service Object

The service objects of medical auxiliary robots include people who need medical care and assistance. These populations may include disabled patients, autistic patients and medical staff. What they all have in common is the need for some form of medical support or assistance to improve quality of life or meet the challenges of everyday life. The main service object is oriented for the following situations.

1. Patients with disabilities: For patients with hearing impairment, we can present the results that the patient wants to know in the form of text according to his search; for patients with limited vision, we can present the search results in voice form according to his search. To better serve patients and reduce the burden on medical staff.
2. Patients with autism rehabilitation: For patients with autism rehabilitation who do not want to talk to outsiders, medical-assisted robots can solve problems in the form of one question and one answer. Medical-assisted robots can provide continuous emotional value and psychological comfort, giving them the companionship they need in their hearts.
3. Medical staff: Although medical assistant robots mainly serve patients, they also provide important support for medical staff. Through automatic and intelligent operation, medical assistant robots can directly answer what patients want to know, reduce the workload of medical staff, save time for medical staff, and improve the quality and efficiency of medical services.

The medical auxiliary robot can provide patients with daily nursing knowledge and precautions that they want to know. By combining multidisciplinary knowledge such as medicine, biomechanics, mechanics, materials science, computer vision and mathematical analysis, medical-assisted robots aim to improve the quality and efficiency of medical services while reducing the workload of medical staff. In addition, medical-assisted robots can adapt to the needs of different patients. Through good interaction with patients, medical-assisted robots can enhance patients' treatment experience and increase their confidence and comfort.

In summary, the service objects of medical assistant robots are mainly those who need medical care and assistance. They bring good news to these people and promote the upgrading and development of modern medical care by providing accurate, safe, adaptable and interactive services.

4   Experiment

4.1   Experimental Environment

The experimental environment is ubuntu18.04, and the ROS version is melodic. The ROS environment is installed via the following commands, including related libraries and core components:

- sudo apt-get install ros-melodic-desktop-full
- rosdep init
- rosdep update

Environment variable configuration: Open the terminal to run sudo gedit.bashrc, input at the end of the text.

- source/opt/ros/melodic/setup.bash
- source ~ /catkin_ws/develop/setup.bash

4.2   Experimental Procedure

First, they log into the open platform; after registration, they click on the console to enter. The application can then download its own exclusive APPID Linux sdk.

Enter the workspace catkin_ws and compile, and enter the following commands to install libasound2-dev and the mplayer player:

- sudo apt-get install libasound2-dev mplayer

Enter the command to modify the mplayer player configuration file:

- echo ' lirc = no ' > ~ /.mplayer/config

The terminal input command roslaunch robot_voice iat_publish.launch is opened to start the voice recognition function. When the node is successfully started, the interface shown in Fig. 6 below will appear.

At this time, it is necessary to wake up. The new terminal inputs the rostopic pub/voiceWakeup std_msgs/String 'data: ' input any words ', ' to wake up. All the current active services and topics can be viewed through the instructions rosservice list and rostopic list. At this time, the microphone is used to input the voice, and the implementation result is shown in Fig. 7.

The voice synthesis function converts text into a realistic voice service, allowing the machine to generate natural voice output and interact with users. The terminal input command roslaunch robot_voice tts_subscribe.launch is opened to start the speech synthesis node, and at the same time, the new terminal publishes the topic and inputs rostopic pub/voiceWords std_msgs/String 'data: ', ", where the single quotation mark position is set as the target content. At this time, the speaker starts the voice broadcast, and the result is shown in Fig. 8.

Speech recognition and synthesis combine the above two functions to obtain audio streams in real time into text and convert text into realistic speech services so that the machine can generate natural speech outputs and interact with users. Open the terminal

```
(base) sz@ubuntu:~$ roslaunch robot_voice iat_publish.launch
... logging to /home/sz/.ros/log/f8b41652-e5cd-11ee-b4b4-000c29880d69/roslaunch-ubuntu-28
10.log
Checking log directory for disk usage. This may take a while.
Press Ctrl-C to interrupt
Done checking log file disk usage. Usage is <1GB.

started roslaunch server http://192.168.133.128:34481/

SUMMARY
========

PARAMETERS
 * /rosdistro: melodic
 * /rosversion: 1.14.12

NODES
 /
    iat_publish (robot_voice/iat_publish)

auto-starting new master
process[master]: started with pid [2826]
ROS_MASTER_URI=http://192.168.133.128:11311

setting /run_id to f8b41652-e5cd-11ee-b4b4-000c29880d69
process[rosout-1]: started with pid [2837]
started core service [/rosout]
process[iat_publish-2]: started with pid [2840]
[ INFO] [1710838336.328062418]: Sleeping...
```

Fig. 6. Speech recognition process

Fig. 7. Recognition results

input roslaunch robot_voice repeat_voice.launch, colleagues wake up the speech recognition and speech synthesis node, and at the same time, the new terminal releases the topic, rostopic pub/voiceWakeup std_msgs/String 'data: 'any words'. At this time, the speech recognition function is awakened, and the microphone is used to input the speech. After the recognition is successful, speech synthesis is performed on the obtained results. At this time, the speaker begins to broadcast the content of the speech recognition. The experimental results are shown in Fig. 9.

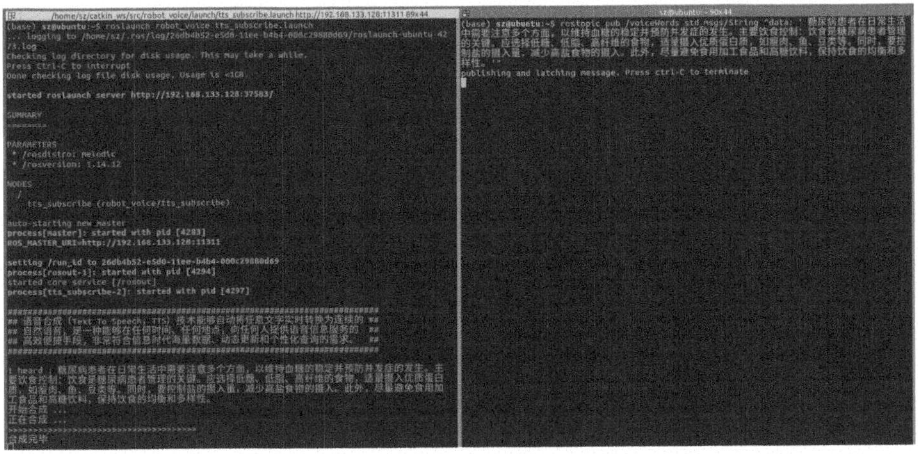

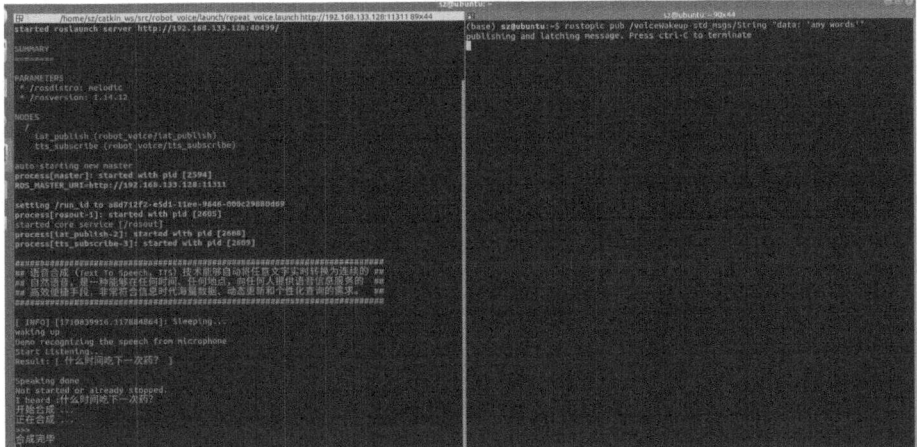

Fig. 8. Speech synthesis

Fig. 9. Speech recognition and synthesis

Voice interaction: Voice wake-up is a key prestep of voice interaction. Before voice interaction, the device needs to be awakened first and enter the working state from the sleeping state to process the user's instructions normally. Voice wake-up switches the device from the sleep state to the working state via voice. Common wake-up methods include 'one call and one answer ' and ' wake-up word + command word '. The interactive function starts with the wake-up word. After the voice inputs an audio item, the response action is taken according to the input audio. Open the terminal input roslaunch robot_voice assistant.launch, create a new terminal input wake-up node, first perform speech recognition, and perform the corresponding function wake-up according to the recognition content, as shown in Fig. 10.

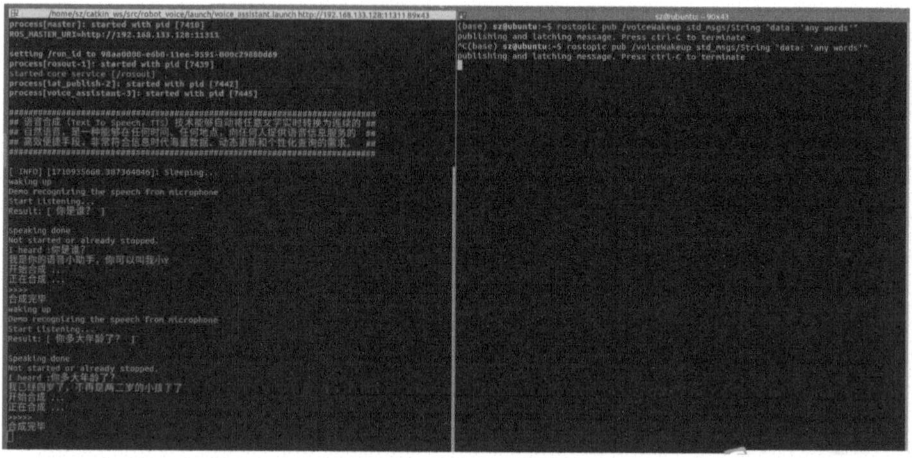

Fig. 10. Voice interaction

5 Summary

The voice interaction system of the medical assistant robot based on the ROS designed in this paper has a good voice interaction function, which can communicate with the service object smoothly and can complete the corresponding tasks according to the voice instructions of the service object. The ROS can be used to conveniently and quickly deploy the voice interaction engineering of medical assistant robots. This design method is typical and practical and can be extended to other complex project developments of robots, which has certain reference value and broad application prospects.

Acknowledgments. This work was supported in part by the Jilin Province Science and Technology Development Plan Project (Grant No. YDZJ202201ZYTS549) and the Science and Technology Research Project of the Education Department of Jilin Province (Grant No. JJKH20220597KJ).

References

1. Nan, J., Pang, Y., Gao, S.: Speech recognition based on attention mechanism spectrogram feature extraction. J. Jilin Univ. Sci. Ed. **62**(2), 320–330 (2024). https://doi.org/10.13413/j.cnki.jdxblxb.2023080
2. Ling, H., Zhou, Y., Huang, Y.: Design of smart home systems based on speech recognition technology. Comput. Knowl. Technol. **19**(31), 38–40 (2023). https://doi.org/10.14004/j.cnki.ckt.2023.1636
3. Xiuping, G., Yusha, Z.: Using multiobjective dynamic programming to solve Pareto optimal frontier for disassembly sequences. J. Syst. Manag. **32**(06), 1205–1212 (2023)
4. Yongheng, L., Jiaming, Y., Feng, J.: Research on linear predictive analysis in conjunction speech recognition. Comput. Simul. **27**(11), 340–344 (2010)
5. Li, P., Yang, W., Li, J.: Implementation of speech recognition system based on DTW/SVM in DSP. Electroacoust. Technol. **9**, 40–44 (2006). https://doi.org/10.16311/j.audioe2006.09.013

6. He, Z., Wang, K., Ji, L.: Hardware implementation of speaker recognition system based on vector quantization. Mod. Electron. Technol. **45**(01), 171–175 (2022). https://doi.org/10.16652/j.issn.1004-373x.2022.01.032

7. Su, Y., Hong, R., Liu, T.: Research on precise reconstruction method of robot demonstration trajectory based on feedforward hidden Markov model. J. Instrum. **44**(12), 199–207 (2023). https://doi.org/10.19650/j.cnki.cjsiJ2311989

8. Zuhong, Z.: Speech signal processing technology and its application analysis. Electron. Technol. **51**(12), 151–153 (2022)

9. Sha, Y., Xiao, F., Cao, D., et al.: The application of AI intelligent voice technology in intelligent scheduling platforms. Chin. Inf. Technol. **8**, 55–56 (2023)

10. Wang, Y.: Design of machine translation robot based on speech synthesis. Autom. Instrum. **4**, 185–190 (2023). https://doi.org/10.14016/j.cnki.1001-9227.2023.04.185

11. Yonghong, B.: Application and optimization of natural language processing technology in intelligent customer service system. Internet Wkly **02**, 21–23 (2024)

12. Wang, X., Sun, Z.: Chinese speech recognition acoustic model based on QRNN-CTC. Comput. Appl. Softw. **40**(12), 184–188+262 (2023)

13. Liao, C.: Research and design of voice interaction system for welcoming robots based on ROS. Electromechanical Inf. **23**, 58–61 (2023). https://doi.org/10.19514/j.cnki.cn32-1628/tm.2023.23.015

14. Xu, X.: Research on intelligent restaurant service robot based on ROS. South. Met. **2**, 17–19+31 (2022)

A Rule-Based Multidimensional Axiomatic Fuzzy Set Knowledge Graph Question-Answering Model

Jin Du[(✉)]

School of Information Engineering, Yangzhou University, Yangzhou 225127, China
mz220220338@stu.yzu.edu.cn

Abstract. The pretrained large language model has made significant progress in intelligent question-answering, reading comprehension, and other aspects. This paper refers to a rule-based mathematical model-the multidimensional axiomatic fuzzy set (AFS) model and extends it by using logic inverse and fuzzy implication operations to increase the interpretability of this model. For open-book question answering (QA), AFS theory can generate a knowledge graph of passages, and retrieve the context most closely related to the question on the basis of the graph retrieval algorithm. We utilize the reading comprehension ability of the large language model to analyze the retrieved context on the basis of the question and obtain the final answer. The experimental results show that the proposed model can perform better on several datasets.

Keywords: Axiomatic fuzzy set · logical inverse · fuzzy implication · large language model · question answering · interpretability

1   Introduction

The emergence of large language models has revolutionized the application of natural language processing in the real world, such as Intelligent question answering (QA), fact-checking (FC), and arithmetic reasoning (AR) [1–6]. In recent years, large language models have gained widespread attention in open-book question-answering (open-book QA). This method requires large language models combined with the question and a large amount of context to reason and arrive at the final answer. Some scholars [7] have also reported that large models reading the context of multiple paragraphs can improve the accuracy of question-answering. This is because open-book QA can combine more external knowledge to complete the QA process without being limited by the knowledge reserve of the model itself. Generally, an open-book QA model consists of two parts: a retriever and a reader. The retriever is responsible for retrieving the context related to the question from the knowledge base, whereas the reader is responsible for jointly analyzing the question and context to obtain the final answer. As shown in Fig. 1, if we want the open-book question-answering model to answer

© The Author(s), under exclusive license to Springer Nature Singapore Pte Ltd. 2024
H. Yu et al. (Eds.): CCF NCCA 2024, CCIS 2274, pp. 32–52, 2024.
https://doi.org/10.1007/978-981-97-9671-7_3

the question "Who is older, Annie Morton or Terry Richardson?", we need to set a "retriever model" to retrieve some key passages such as "Annie Morton" and "Terry Richardson" from the entire Wikipedia. Moreover, this also tests the retrieval accuracy of the retriever model. With the above key information and the ability of the reader model to understand language for analysis and reasoning, we can obtain the final answer "Terry Richardson". In the field of large models, large language models can be chosen as readers for open-book QA.

Question： Who is older, Annie Morton or Terry Richardson?
Answer： Terry Richardson

"Annie Morton" : Annie Morton (born October 8, 1970) is an American model・・・・・・

"Decoding Annie Parker" : Decoding Annie Parker is a 2013 drama film written・・・・・・

"Alicia Morton" : Alicia Morton (born April 29, 1987) is a former American actress and singer・・・・・・

"Lady Gaga x Terry Richardson" : "Lady Gaga x Terry Richardson is a photo-book・・・・・・

"Terry Richardson" : Terrence Richardson (born August 14, 1965) is an American fashion and portrait photographer who has shot advertising campaigns for Marc Jacobs, Aldo, Supreme, Sisley, Tom Ford, and Yves Saint Laurent among others. He has also done work for magazines such as "Rolling Stone", "GQ", "Vogue", "Vanity Fair", "Harper's Bazaar", "i-D", and "Vice".

Fig. 1. Open-book Q A example. The passages with blue entities are key passages, which is the task of the retriever. The red entities are key information in the passage, which needs to be fully captured by the large language model. (Color figure online)

As the retriever in open book QA, there are two ways to achieve this goal. One way is based on deep learning. These methods have been confirmed by many experts and scholars to achieve good results in terms of information extraction and knowledge reasoning [8–11]. In addition, some models set up special inference mechanisms [12,13] to extract key information. Among them, the attention mechanism plays an important role, and it can pay extra attention to key information in a wide range of information. Transformer [14] is a typical representative of the attention mechanism and embodies the great potential of the attention mechanism in the domain of deep learning. However, the interpretability of deep learning models is still a difficult problem.

The second way is rule-based methods [15–19]. These models usually have strong interpretability and good inference capabilities but are limited by feature representation. As a tool to embody knowledge structure, the knowledge graph can be applied to these two ways of methods [20,21]. This is because first, knowledge graphs can integrate more knowledge structure information to enhance the embedded representation of each data in the knowledge graph; second, many mature graph algorithms can enhance the interpretability of the model. Lang et al. [20] used AFS theory to construct an AFS knowledge graph, which provides an interpretable semantic basis for passage retrieval and the construction of reasoning chains. The AFS graph model was built by Lang et al. [20] is still the best rule-based retriever model thus far. However, some parts of the model are still not sufficiently rigorous and rely on human intuition. Moreover, the construction of semantics is not concise enough and affects the retrieval efficiency to a certain extent.

Inspired by the AFS graph model [20] and combined with the current mainstream trend of LLM in the field of intelligent QA, we propose an open-book question-answering model with the AFS graph retrieval model + large language model. The model consists of two parts: a retriever and a reader. We use the AFS graph model as our retriever and the large language model as our reader. In the AFS graph model part, we apply logic inverse and fuzzy implication to improve the algorithm. Through such improvements, the AFS graph model has become more rigorous and more interpretable, enhancing the interpretability of AFS theory [22], and the semantic description of the context has become more concise and clear. Finally, we use some prompt words to encapsulate the context we extracted and question as the input to the large language model. By utilizing the language understanding and reasoning capabilities of the large language model, the final answer can be obtained after the large language model reasoning and analyzing the context on the basis of the known question.

The main contributions of this paper are as follows.

1) We refer to the AFS question-answering model, propose the idea of the AFS graph retrieval model + large language model, and further develop the advantages of AFS theory in the domain of large language models.
2) We use fuzzy implication and logical inverse to improve the AFS theory, further enhancing the interpretability of the AFS theory and improving the retrieving effect of the AFS retrieval model.
3) Using a special graph retrieval strategy, the context most relevant to the question can be found in the AFS semantic knowledge graph, improving the effect of information extraction and filtering.

2   Related Works

In this section, three related parts are introduced: open-book QA models, multidimensional axiomatic fuzzy set (AFS), and the mainstream Large Language Models (LLMs) in the field of intelligent question-answering.

2.1   Open-Book QA Models

Open-book QA models first retrieve supporting passages from a large corpus such as Wikipedia and then apply a more expensive reader to predict answers via the passages [23]. Unlike closed-book QA models, open-book QA models need to go through a passage retrieval module in advance. Figure 1 is an example of open-book QA, where each passage is composed of a title entity and several sentences. The entities marked in blue and their corresponding context are the supporting passages for answer prediction. There are currently multiple methods to accomplish this task, with the TFIDF [1] algorithm being the most common.

Currently, passage retrieval methods based on graph neural networks combined with attention strategies have been widely proposed. EmbedKGQA [8] uses different KG embedding strategies for the question and passages and then calculates the similarity score between the question and the passages to obtain passages that are more relevant to the question. Deep Question Generation (DQG) [10] uses a natural language processing parsing tool (SpaCy) to generate passage knowledge graphs and then uses a gated graph neural network to encode the knowledge graph so that the embedded representation of passages integrates more knowledge base information. The graph-level representation of the passages serves as input for the context selection.

Moreover, some methods enhance the interpretability of the model by constructing unique inference paths. Yu et al. [24] constructed reasoning paths by extracting satisfying logic related to the question from the given text. L. Qiu et al. [12] constructed a semantic graph by using a dynamically fused graph network to simulate the process of human reasoning, and the entity graph mask achieves the explainability of the reasoning process.

2.2   Axiomatic Fuzzy Set

The AFS theory, proposed by Liu [22], aims to simulate human relational reasoning ability by transforming human fuzzy semantics into computable language, achieving explainability at the data level. The general process of AFS theory is as follows: First, fuzzy concepts are added to data attributes to create simple concepts. These simple concepts are then processed via EI algebra [25] and AFS natural language axioms [26], forming complex concepts with richer semantics. Moreover, membership functions [27] can be used to calculate the degree of membership between the data and these complex concepts. After filtering through a threshold, the final semantic description of each piece of data is obtained. Additionally, EI algebra [25] and membership functions [27] can be utilized to determine the semantic similarity between different pieces of data. On the basis of the semantic similarity between data, we can further achieve the purpose of data clustering [28, 29] classification [30], and retrieval [20].

2.3 Logical Inverse and Fuzzy Implication:

In this paper, we use the logical inverse and the fuzzy implication to improve the AFS theory and enhance the theoretical validity and explainability of the AFS theory. The logical inverse of a proposition refers to the negative form of the proposition, and its definition is as follows:

Suppose p is a proposition. The compound proposition "not p" is called the inverse of p, denoted as ¬P. It is stipulated that ¬P is true if and only if p is false.

The implication is a mathematical term used to express whether two propositions have a certain conditional relationship. It is usually expressed in the form of p→q, where p is the antecedent of the implication, q is the consequent of the implication, and → is called the implication connective. Usually, in discrete mathematics, the implication values of p and q are both Boolean-type values, so the implications in discrete mathematics are usually Boolean-type implications, but sometimes they cannot handle some fuzzy and uncertain situations in real life. If we want to determine the degree of correlation between two fuzzy values located in the interval [0, 1], we need to use fuzzy implications. We considered several implication rules and compared them briefly. We find that the values obtained by the Lukasiewicz implication are more realistic and discriminative, and the definition of the Lukasiewicz implication is as follows:

$$a \rightarrow_L b = min\{a, 1 - a + b\}, \forall a, b \in [0, 1] \tag{1}$$

The fuzzy implications are used to determine the degree of correlation between two scores and construct multidimensional features, which will be described later.

2.4 LLM: Large Language Model

Large language models are a popular type of model in the domain of artificial intelligence today. They are characterized by large numbers of parameters (often composed of billions or more parameters) and use semisupervised methods to train on large numbers of unlabeled text, images, etc. In the domain of NLP, large language models have excellent language learning and understanding abilities and can be adapted to many natural language tasks.

3 Method

Our question-answering model is an open-book QA model consisting of a retriever and a reader. In the retriever part. We improve the AFS theory via logical inversion and fuzzy implications to build an AFS semantic knowledge graph and then use the graph retrieval algorithm to retrieve key passages related to the question. In the reader part, we encapsulate the retrieved passages and the question through prompt learning techniques and send them to the large language model. The large language model analyzes the retrieved passages on the basis of the known question to obtain the answer. The key to the model is the construction process of the entire AFS semantic knowledge graph, which is an important part of this paper. The entire process of constructing the AFS semantic knowledge graph is shown in Fig. 2. First, in Sect. 3.1, we set up five scores to evaluate the degree of relevance between passages and the given question. Then, we use fuzzy implication rules to divide the five scores into three types of features, this part of the content corresponds to Sect. 3.2 of this paper. In Sect. 3.3, we use logical inverse to create two fuzzy concepts, "large" and "not large", and combine them with three features to obtain 10 simple concepts. In Sect. 3.4, EI algebra is mentioned in this section, serving to manipulate simple concepts and combine them into more intricate, semantically rich complex concepts. In Sect. 3.5, we construct the AFS structure as a basis for establishing membership functions. In Sect. 3.6 we use the membership function to calculate the membership degree of a passage to each complex concept. Through the filtering of hyperparameter ε, the semantic description of each passage can be obtained. Finally, Sect. 3.7 introduces the computation of similarity between pairs of passages on the basis of their respective semantic representations. In this way, the entire AFS semantic knowledge graph can be established, in which nodes represent passages, each node has a corresponding semantic description, and each edge represents the degree of semantic similarity between two passages. This is also the main reason why it is called a semantic knowledge graph.

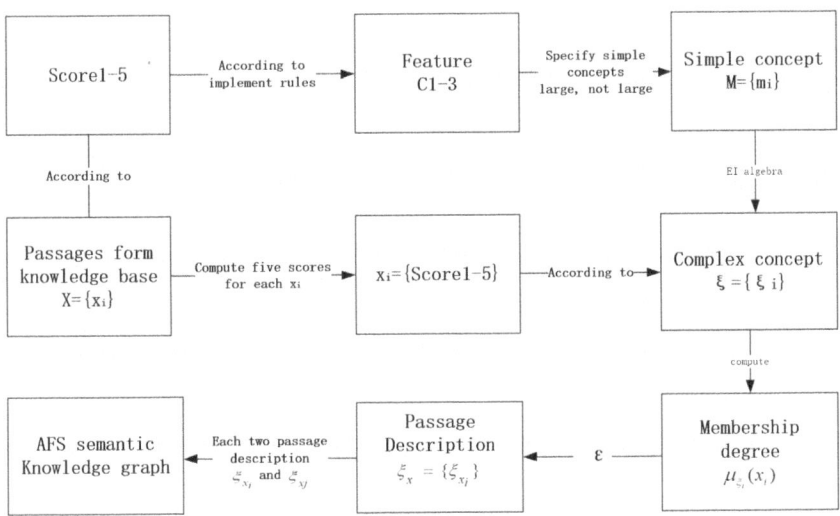

Fig. 2. AFS semantic knowledge graph construction process

3.1   Data Preprocessing

The key of our model is the retrieval part, in which we use logical inversion and fuzzy implications to improve the AFS theory [22] and construct an AFS semantic knowledge graph. Question-related passages can be obtained from the knowledge graph via a special graph retrieval algorithm. These passages ultimately serve the large language model as context for the question. The entire process of our model is shown in Fig. 3.

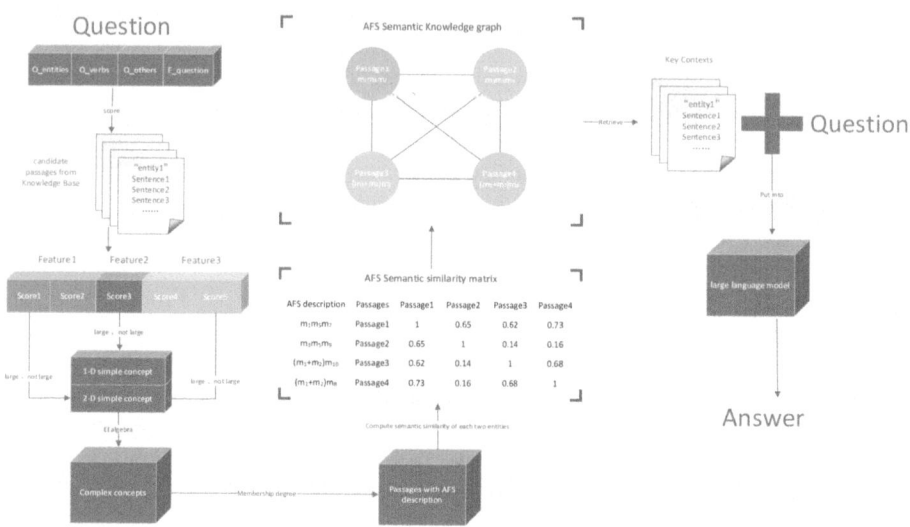

Fig. 3. The whole process of the AFS retriever + large language model open book question-answering model

Figure 3 shows the entire process of our model. First, we leverage the TFIDF algorithm to retrieve relevant passages from the knowledge base (e.g., Wikipedia), which are pertinent to the given question, serving as candidate passages of the question. As shown in Fig. 1, each passage consists of a title entity and several sentences. Next, we preprocess the data in the same way as in [20]. We used the Stanfnlp [31] tool to identify the keywords in the known question and used the Flair [32] tool to encode the entire question. We then set five scores: $Score_1$ to $Score_5$ measure the relevance between the question and the candidate passages from different perspectives. For the specific implementation and meaning of the five scores, please refer to [20]. We use (2) to scale these five scores to the [0, 1] interval.

$$\tilde{s} = \frac{s - min}{max - min} \tag{2}$$

In formula (2), s represents any score from $Score_1$ to $Score_5$ corresponding to the passage. Min and max represent the maximum and minimum values of all passages on this score class respectively. The entire data preprocessing algorithm is shown in Algorithm 1.

Algorithm 1: Data preprocessing
___

Input: candidate passages; question;

Output: Five scores for each passage

1 Use the Stanfnlp tool to identify keywords in question

2 Use the Flair tool to encode the entire question

3 **for** *passage x in passages* **do**

4   Calculate the relevance of the question to x using the different formulas from [20] as $Score_1$ to $Score_5$ Scale fives Scores to the range 0 to 1 using formula (2)

5 **end**

3.2 Build Multidimensional Features

Multidimensional features are a major feature of AFS theory. They are usually formed by bundling multiple features. In the original AFS theory, feature bundling is performed in a way that relies on human intuition. In this paper, we use fuzzy implication rules to bundle scores with high similarity to form high-dimensional features. This is also a major innovation of this paper, which provides an explainable basis for feature bundling and further enhances the theoretical validity of AFS theory. With respect to the role of feature bundling, from a semantic perspective, high-dimensional features help to measure a thing comprehensively. For example, if we want to describe whether a person is obese or thin, we cannot measure it by using just one feature such as weight, but need to rely on a two-dimensional feature such as height and weight to measure it comprehensively. Moreover, compared with one-dimensional features, multidimensional features can greatly reduce the number of complex concepts and reduce semantic redundancy, making the semantic description of a passage clearer and more concise and helping to reduce computational complexity.

This paper uses the following formula as the basis for feature bundling:

$$S_{ij}(x) = (Score_i \rightarrow_L Score_j) \wedge (Score_j \rightarrow_L Score_i) \tag{3}$$

where $\rightarrow_L$ is the Lukasiewicz implication and x represents a candidate passage. $S_{ij}(x)$ represents the implication similarity score between $Score_i$ and $Score_j$ for passage x. For the five scores, 10 classes of implication similarity scores can be ultimately divided. We used a large amount of data to conduct statistics on these ten classes of implication similarity scores. We bundle features on the basis of the ranking of ten classes of similarity scores. Finally, according to the similarity ranking, $Score_1$ and $Score_2$ are bundled to form a two-dimensional feature C_1, $Score_4$ and $Score_5$ are bundled to form another two-dimensional feature C_3, and the remaining $Score_3$ is used as a one-dimensional feature C_2. The detailed experimental datasets and statistical histograms are described in Sect. 4.

Moreover, through this data statistics method, some potential feature correlations can be discovered. Because the correlation between some features is not so obvious. For example, in the story of beer and diapers, on the surface, there is no correlation between these two things. However, in the 1990s, among American families with infants, many fathers in supermarkets bought a bottle of beer when they bought diapers. This shows that two seemingly unrelated things may have a certain implicit relationship in a specific scenario. It is the same in this paper, refer to [20], $Score_1$ measures the similarity between the entities in the question and the entity in the candidate passage title, and $Score_2$ measures the number of entities in the question contained in the candidate passage. Although there is no relationship between the two scores on the surface, their implicit relationship may be seen from their implication similarity.

3.3 Specify Simple Concepts

A simple concept is a computable fuzzy semantic description, which is the basis of complex concepts and has relatively simple semantics. To form simple concepts, we must first define fuzzy concepts. Fuzzy concepts are usually vague words that describe degrees in daily life, such as "tall" and "not tall" to describe a person's height, "old" and "young" to describe age, or "more" and "less" to describe quantity. In this paper, we use logical inverse to define our fuzzy concepts. We adopted "large" and the inverse of large "not large" as fuzzy concepts. These two fuzzy concepts can be used to describe the level of a score, where "not large" is the opposite and the negative form of "large". Of course, more than two fuzzy words can be used to describe the level of a score, and they may not always appear in pairs. More fuzzy concepts can also be set up to make the passage description richer in semantics, but this also greatly increases the number of calculations. We only set two fuzzy concepts because we do not care whether a passage's relevance score to the question is small, medium, upper-medium, or lower-medium. We focus only on whether this passage is relevant to the question, that is, whether the score is large or not large. Moreover, this setting approach can also make the final semantic description of passages more concise and clearer.

With fuzzy concepts, we can define simple concepts on the basis of three features: F_1, F_2, and F_3. Simple concepts are semantic combinations of fuzzy concepts and one-dimensional or multidimensional features. Let us take feature F_1 as an example. Since feature F_1 is related to $Score_1$ and $Score_2$, we can define four simple concepts in this feature: "$Score_1$ is large and $Score_2$ is large", "$Score_1$ is large and $Score_2$ is not large", "$Score_1$ is not large and $Score_2$ is large", and "$Score_1$ is not large and $Score_2$ is not large". Thus, we can define all the simple concepts, as shown in Table 1.

Table 1. Simple concepts of corresponding features and their semantic descriptions

Feature	Simple Concept	Semantic
F_1	c_1	*Score1 is large and Score2 is large*
	c_2	*Score1 is large and Score2 is not large*
	c_3	*Score1 is not large and Score2 is large*
	c_4	*Score1 is not large and Score2 is not large*
F_2	c_5	*Score3 is large*
	c_6	*Score3 is not large*
F_3	c_7	*Score4 is large and Score5 is large*
	c_8	*Score4 is large and Score5 is not large*
	c_9	*Score4 is not large and Score5 is large*
	c_{10}	*Score4 is not large and Score5 is not large*

3.4   Generate Complex Concepts

As shown in Table 1, the set $C = \{c_1, ..., c_{10}\}$ is our simple concept set defined on F_1, F_2, and F_3. Then, we use the EI algebra shown in (4) to combine the simple concepts to form complex concepts:

$$EM^* = \{\sum_{i \in I}(\prod_{c \in A_i} c)|A_i \subset C, i \in I\} \tag{4}$$

where $\prod_{c \in A_i} c$ represents the conjunction of simple concepts in A_i, and where $\sum_{i \in I}(\prod_{c \in A_i} c)$ represents the disjunction of some $\prod_{c \in A_i} c$. In addition, we stipulate that simple concepts within the same feature cannot perform conjunction operations; otherwise, semantic ambiguity will occur. EI algebra also satisfies the following natural language axioms [26]:

1) *Absorption Law*:

$$c_1 c_5 + c_1 c_5 c_9 + c_3 = c_1 c_5 + c_3 \tag{5}$$

2) *Commutative Law*:

$$c_1 c_7 + c_2 c_7 = c_2 c_7 + c_1 c_7 \tag{6}$$

$$c_1 c_7 = c_7 c_1 \tag{7}$$

3) *Distributive Law*:

$$c_1 c_7 + c_2 c_7 = (c_1 + c_2)c_7 \tag{8}$$

Complex concepts derived from EI algebra can serve as a semantic description of a passage. For example, through EI algebra, we obtain this complex concept $\delta = c_1 c_5 c_{10} + c_2 c_5 c_{10} = (c_1 + c_2)c_5 c_{10}$. Its semantic description is as follows: there is a passage, among its five relevance scores to the question, its $Score_1$ is large, its

$Score_2$ is large or not large(uncertain), its $Score_3$ is large, and its $Score_4$ and $Score_5$ are not large. Moreover, this complex concept also satisfies the natural language axioms [26] and the requirement that simple concepts within the same feature cannot perform conjunction.

We can also build semantic equivalence classes to identify whether multiple complex concepts have consistent semantics. To establish semantic equivalence classes, we must first define the semantic equivalence relationship. The definition rules of the semantic equivalence relation R on EM^* are as follows: for $\alpha = \sum_{i \in I}(\prod_{c \in A_i} c)$ and $\beta = \sum_{j \in J}(\prod_{c \in A_j} c) \in EM^*$:

$$\alpha\ R\ \beta \iff \forall A_i(i \in I), \exists B_h(h \in J) \text{ such that } A_i \subseteq B_h$$
$$\text{and } \forall B_j(j \in J), \exists A_k(k \in I) \text{ such that } B_j \subseteq A_k$$

The quotient set EM^*/R is also called the semantic equivalence class EM. If two complex concepts belong to the same semantic equivalence class, then we consider these two complex concepts to have completely consistent semantics. If two complex concepts have a consistent form after being simplified or changed by natural language axioms, then they are equivalent. However, because this proof process is not the main part of this paper, we do not prove it in this paper.

3.5   Constructing the AFS Structure

The purpose of constructing the AFS structure [28] is to establish membership functions. The AFS structure is defined as follows: let X and C be nonempty sets, and 2^C is the power set of C. If $\tau: X \times X \to 2^C$ satisfies Axiom 1 and Axiom 2, then $(C, \tau, \text{and } X)$ is called the AFS structure [20].

Axiom 1: $\forall (x_1, x_2) \in X \times X, \tau(x_1, x_2) \subseteq \tau(x_1, x_2)$

Axiom 2: $\forall (x_1, x_2), (x_2, x_3) \in X \times X, \tau(x_1, x_2) \cap \tau(x_2, x_3) \subseteq \tau(x_1, x_3)$

In this paper, structure τ appears in the form of the following linear order relation:

$$\tau(x, y) = \{c | c \in C, x \geq_c y \subseteq 2^C\} \tag{9}$$

where $x \geq_c y$ means that the degree to which x belongs to c is greater than or equal to the degree to which y belongs to c.

To make the "degree" computable, we set different cut points on different simple concepts according to their semantics. Since we have scaled all five scores to the range 0 to 1 previously, we use 0 and 1 as the standard values ??of "not large" and "large", respectively. For example, since the semantic description of c_2 is "$Score_1$ is large and $Score_2$ is not large", its cut point is (1,0). Table 2 shows the cut points of some simple concepts and the semantics corresponding to the cut points. Moreover, if a simple concept is one-dimensional, then it has only one cut point such as c_5 or c_6.

With the cut points corresponding to each simple concept, we can further use the Gaussian kernel function shown in (10) to calculate the degree to which a passage belongs to a certain simple concept:

$$\rho_{c_i}(x) = e^{-\frac{\|s_i - cp_i\|^2}{2\sigma^2}} \tag{10}$$

Table 2. Cut points and semantic descriptions of some simple concepts

Concept	Cutpoints	Semantic of corresponding cutpoints
c_1	$(1,1)$	*Score1 and Score2 are all large, closer to 1*
c_4	$(0,0)$	*Score1 and Score2 are all not large, closer to 0*
c_5	1	*Score3 is large, closer to 1*
c_6	0	*Score3 is not large, closer to 0*
c_7	$(1,1)$	*Score4 and Score5 are all large, closer to 1*
c_{10}	$(0,0)$	*Score4 and Score5 are all not large, closer to 0*

where $\rho_{m_i}(x)$ represents the degree to which passage x belongs to the simple concept c_i, s_i represents the scores involved in the simple concept c_i, and cp_i represents the cut point corresponding to the simple concept c_i. σ is an adjustable hyperparameter that determines the effect of the model.

3.6 Calculating the Membership Degree

With the AFS structure, we can further define the membership function as shown in (11), which can be used to calculate the degree to which a passage belongs to a complex concept.

$$\mu_\xi(x) = \sup_{i \in I}\left\{ \prod_{\gamma \in A_i} \left(\frac{\sum_{u \in A_i^\tau} \rho_\gamma(u)}{\sum_{x \in X} \rho_\gamma(x)} \right) \right\} \tag{11}$$

$$A_i^\tau(x) = \{y \in X | c \in \tau(x,y), \forall c \in A_i\} = \{y \in X | A_i \in \tau(x,y)\} \tag{12}$$

In the above formula, $\mu_\xi(x)$ represents the membership degree of passage x to complex concept $\xi = \sum_{i \in I}(\prod_{c \in A_i} c) \in EM$, where A_i is the ith set of simple concepts and the calculation method of $\rho_\gamma(\cdot)$ is shown in (10). We set the threshold ε to filter out those complex concepts with $\mu_\xi(x)$ greater than ε, and add them to the semantic description of the passage x.

3.7 Generate the AFS Semantic Knowledge Graph

After obtaining the description of candidate passages, we can calculate the semantic similarity between each two passages through the following formula (13):

$$f_{ij} = min\{\mu_{\xi_{x_i} \wedge \xi_{x_j}}(x_i), \mu_{\xi_{x_i} \wedge \xi_{x_j}}(x_j)\} \tag{13}$$

In this formula, f_{ij} represents the semantic similarity value between passage x_i and passage x_j. ξ_{x_i} and ξ_{x_j} represent the semantic descriptions of x_i and x_j respectively; of course, they must also be complex concepts. The calculation rules for $\vee$ and $\wedge$ satisfy the operation rules of the fuzzy logic system proposed by Liu [33]. After the semantic similarity between each pair of passages is determined, our AFS semantic knowledge graph is determined. The nodes are each passage, each with a semantic description that matches it. There are edges between any two passages in the graph, and each edge has a semantic similarity value between two passages. The construction algorithm of the entire AFS semantic knowledge graph is shown in Algorithm 2.

Algorithm 2: Generating the AFS Semantic Knowledge Graph

Input: Candidate passages set X; Five scores for passages; threshold ε
Output: Semantic description for each passage; Semantic similarity value f_{xy};

1 According to the five scores as well as the fuzzy concepts, simple concepts set $C=\{c_1,c_2,\cdots,c_{10}\}$ are determined as shown in Table 1.
2 **for** $x, y \in X$ **do**
3 $\quad$ construct the order relation $\tau(x,y)$ by (9)
4 $\quad$ **for** c *in* C **do**
5 $\quad\quad$ compute the weights $\rho_{c_i}(x)$ by (10)
6 $\quad\quad$ generate the AFS structure $A^\tau(x), A = \{c\}$
7 $\quad$ **end**
8 **end**
9 **for** $x \in X$ **do**
10 $\quad$ **for** c *in* M **do**
11 $\quad\quad$ calculate the membership degree of $\mu_c(x)$ by (11)
12 $\quad\quad$ **if** $\mu_c(x) \geq \varepsilon$ **then**
13 $\quad\quad\quad$ add m to the description of x
14 $\quad\quad$ **end**
15 $\quad$ **end**
16 **end**
17 **for** $x, y \in X$ **do**
18 $\quad$ calculate the similarity f_{xy} between passage x and passage y by (13)
19 **end**

3.8 Extracting Key Contextual Information

After obtaining the AFS semantic knowledge graph, we can further extract the context that is most relevant to the question on the basis of a special graph retrieval strategy. First, we create an information base list, select the passages with the largest scores from $Score_1$ to $Score_5$ among the candidate passages as our basic passages, and enter the first cycle. In the first cycle, we first create an internal subrepository list, and we select a passage from the basic passages as the first element of the internal subrepository list. Then, we create a subinformation base list. The top ten passages with the highest similarity to the first element are retrieved from the knowledge graph through the breadth-first search strategy as the second batch of passages and placed in the subinformation base list. For each passage in the second batch of passages, we again use the breadth-first search algorithm to retrieve the top five passages with the highest similarity to it from the knowledge graph and put them into the subinformation base list again. The passages in the subinformation base list are deduplicated to obtain the final subinformation base set. We then put the passages in the subinformation base set into the information base list and enter the subsequent cycle. In the second and subsequent cycles, we select the second and subsequent passages from the basic passages that do not appear in the internal subrepository list to obtain the second and subsequent subinformation base sets. For the second and subsequent subinformation base sets, we need to make a judgment: if this subinformation base set has more passages than the number of passages in the information base list and if the passages in it exist in the information base list less than 60%, we will put these passages in this subinformation base set into the information base list. After multiple cycles and deduplication of elements in the information base list, we regard the passages in the final information base list as our key passages. These key passages are considered to be most relevant to the question and will also be passed to the large language model as the key context of the question. We call this process information extracting which is the last step of the entire retriever part.

3.9 Reader

In this section, the large language model (LLM) is used as the reader of the entire open-book QA. We use some prompt words to encapsulate the key context and question and pass them to the LLM. By utilizing the reading comprehension capabilities of the LLM, the final answer can be obtained after the LLM reads and understands the context on the basis of the known question. The encapsulation method is as follows:

prompt = f "according to :{context},question is :{question}, what is answer"

Through this method of contextual reading comprehension, the large language model can obtain more accurate answers on the basis of the prompts of the key context. Moreover, through this open-book QA method of the AFS retriever + large language model, large language models eliminate the limitations of their knowledge reserves and can obtain more external knowledge through the retriever.

4   Experiments

In this chapter, we first conduct comparative experiments and ablation experiments to demonstrate the effectiveness of our method. We then studied the effects of different model hyperparameters on model performance and finally determined the best hyperparameters for the model.

4.1 Comparative Experiment

In this section, we refer to our model as the LIFI-AFSGraph question-answering model, which is an AFS graph question-answering model based on logical inverse and fuzzy implication. The purpose of this is to differentiate our model from the AFSGraph question-answering model reported in [20], and the abbreviated names are easier to distinguish in comparative experiments. Here, we performed our experiments on the HotoptQA dataset [34]. It is a classic dataset in the field of open-book QA and includes two main settings: the distractor setting and the full wiki setting. We compared the paragraph retrieval capabilities of our model with several models reported in [2]. The main measurement indicators are answer recall (AR), paragraph recall (PR), and paragraph EM (P EM). The experimental results are shown in Table 3. Our model achieves better performance on these three indicators. On the one hand, this result illustrates the better retrieval performance of the LIFI-AFSGraph retriever model. Second, the application of logical inverse and fuzzy implication further improves the AFS theory.

Table 3. Paragraph retrieval capabilities on the HotpotQA Full Wiki

Models	AR	PR	P EM
TF-IDF [1]	39.7	66.9	10.0
Entity-centric IR [35]	63.4	87.3	34.9
Cognitive Graph [36]	76.0	87.6	57.8
Semantic Retrieval [37]	77.9	93.2	63.9
PathRetriever [2]	87.0	93.3	72.7
AFSGraph [20]	87.8	93.8	74.2
LIFI-AFSGraph	**87.9**	**94.1**	**74.4**

In terms of the question-answering effect of the model, we measured the effect of the retriever + the large language models. We conducted comparative experiments using GPT-3 [38] and LLaMA-7B [39] as readers of open-book QA and using LIFI-AFSGraph and other mainstream retrievers as retrievers. The dataset we use here is the Natural Questions dataset [40], and we use the exact match score (EM) as our measure. As shown in Table 4, we find that open-book QA combined with large language models indeed improves the question-answering effect of large language models to a certain degree. Our LIFI-AFSGraph retrieval model achieved better performance after being combined with large language models for question-answering.

4.2   Ablation Study

Additionally, we verified the efficacy of our LIFI-AFSGraph retriever model through an ablation study. Here, we only use GPT-3 as our reader to conduct this study. As shown in Table 5, to verify the effectiveness of our LIFI-AFS retriever model, we replaced the retriever of our model with the classic TFIDF. We found that after replacement, the effect of the model dropped by 1.5% points, whereas

Table 4. Comparison of the effects of the use of a retriever combined with large language models on the NATURAL QUESTIONS of question-answering performance

Models	EM
GPT-3	14.6
PathRetriever [2]+GPT-3	15.3
MDR [41]+GPT-3	15.8
AFSGraph [20]+GPT-3	16.2
LIFI-AFSGraph+GPT-3	**16.4**
LLaMA-7B	16.8
PathRetriever [2]+LLaMA-7B	17.0
MDR [41]+LLaMA-7B	17.9
AFSGraph [20]+LLaMA-7B	18.2
LIFI-AFSGraph+LLaMA-7B	**18.5**

compared with using only the large language model for question-answering, the effect improved by only 0.3% points. This result shows that the context retrieved by the TFIDF algorithm contains considerable noise, and our model can further filter this noise to ensure that the retrieved context is highly correlated with the question. In addition, we also replaced our LIFI-AFSGraph retriever model with a one-dimensional AFS retriever model, which we call the 1D-AFSGraph model. This is a simple feature processing model. Compared with our model, this model cancels the feature bundling module and uses a combination of one-dimensional features and fuzzy concepts to form simple concepts. We found that the one-dimensional AFS retriever model still has some effectiveness. However, compared with the multidimensional AFS retriever model, the effect is still far behind. This shows that the explainable multidimensional feature bundling method based on fuzzy implications can better exert the effect of AFS theory.

Table 5. Ablation study on natural questions

Models	EM
GPT-3	14.6
TFIDF [1]+GPT-3	14.9
1D-AFSGraph+GPT-3	15.6
LIFI-AFSGraph+GPT-3	**16.4**

4.3 Effect of Hyperparameters

We study the two hyperparameters σ and ε of the model. The former appears in formula (10) and is used to calculate the weight of a passage for a sim-

ple concept and affects the calculation of the membership function. The latter
appears in Sect. 3.6 and is used to determine the semantic description of the passages. Here, we still use the HotoptQA dataset for the experiment. We use the
ratio of retrieved passages containing correct answers to all retrieved passages
as the retrieval accuracy to measure the retrieval effect. As shown in Fig. 4, we
measure our model under the conditions of $\sigma \in [0.1, 1]$ and $\varepsilon \in [0.5, 1]$. We found
that when $\sigma = 0.3$ and $\varepsilon = 0.9$, the model achieved the best effect of 93.4%.

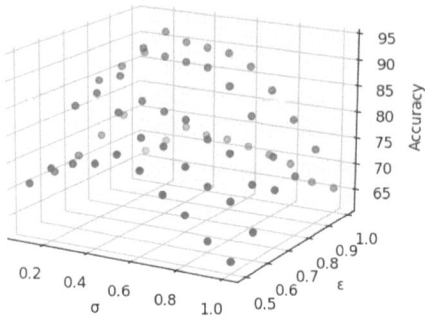

Fig. 4. Retrieval accuracy of the LIFI AFSGraph on HotpotQA at $\sigma \in [0.1, 1]$ and
$\varepsilon \in [0.5, 1]$

We then studied the effect of varying the number of passages retrieved by
the TFIDF on the extraction efficiency of the retriever model. We control the
range of passages retrieved by the TFIDF to 10-400 to observe the accuracy of
our model in terms of passage retrieval. As shown in Fig. 5, when the number
of passages retrieved by the TFIDF reaches fifty or more, the improvement in
accuracy becomes smaller and smaller. Therefore, we usually set the number of
passages retrieved by the TFIDF to 50 or 100.

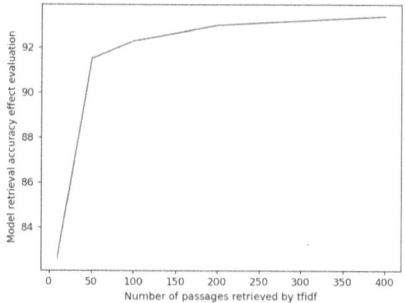

Fig. 5. The effect of varying the number of passages retrieved by the TFIDF on the
extraction efficiency

At the end of this section, we supplement the dataset required for feature bundling mentioned in Sect. 3.2 and the statistical results of the ten classes of implication similarity scores. We experiment by using the first 100 questions in the HotpotQA dataset [34]. For each question, 50 passages are retrieved from Wikipedia via the TFIDF algorithm; thus, each class of the implication similarity score will eventually have 5,000 values. We average these 5000 values to obtain the final implication similarity score. Figure 6 shows the final implication similarity score results.

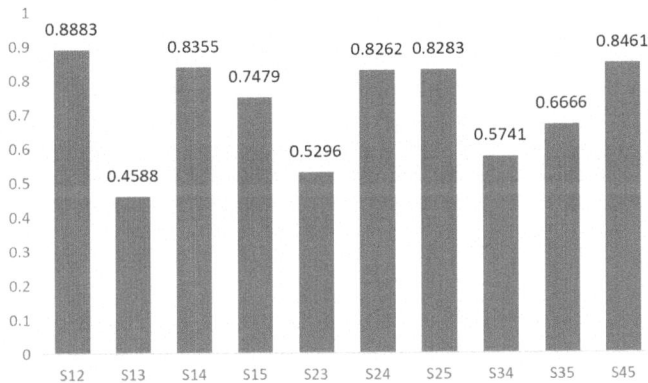

Fig. 6. The final implication similarity score between each score (where S_{ij} represents the implication similarity score between $Score_i$ and $Score_j$)

According to the figure, $Score_1$ and $Score_2$ exhibit the closest similarity, with a score of 0.8883, whereas $Score_3$ and $Score_5$ closely follow, with a similarity score of 0.8461. Therefore, we believe that the similarity between $Score_1$ and $Score_2$ or between $Score_4$ and $Score_5$ is greater than that of other combinations, so we can bundle the features of $Score_1$ with $Score_2$ and those of $Score_4$ with $Score_5$ to form two multidimensional features, F_1 and F_3, and $Score_3$ as a one-dimensional feature, F_2. Compared with the feature combination method which relies on intuition, the multidimensional features obtained via the fuzzy implication-based feature combination method are more explainable. Moreover, this method can also discover potential implicit correlations between different features.

5 Conclusion

We propose an open-book QA model of the AFSGraph retriever + large language model (LLM). This model builds an AFS semantic knowledge graph and applies a graph retrieval algorithm to retrieve context directly related to the question. Then, through prompt learning techniques, the question and context are encapsulated and sent to the LLM, and through the knowledge reasoning and language understanding capabilities of the LLM, the final answer can be predicted. We applied fuzzy computing theory to improve the AFSGraph model, which is also an improvement over the AFS theory. In the construction stage of fuzzy concepts, we introduce logical inverse to make the semantics of the concepts more concise and clear while also reducing the number of calculations. In the multi-dimensional feature combination stage, we apply fuzzy implication to increase the interpretability of the combination of multiple features, while fuzzy implication also has the ability to explore deeper semantic implicit connections between different features. The experimental results show that our model achieves better results on both the hotpotQA dataset and the Natural Questions dataset. However, in the experiment, we also found that although this kind of question-answering method of retriever + large language model can further improve the question-answering effect of the large language model, owing to financial and technical limitations, we have not created a special dataset for supervised fine-tuning, so the improvement in the model effect is not relatively obvious enough. This is also a point we need to solve in the future.

References

1. Chen, D., Fisch, A., Weston, J., Bordes, A.: Reading wikipedia to answer open-domain questions (2017)
2. Asai, A., Hashimoto, K., Hajishirzi, H., Socher, R., Xiong, C.: Learning to retrieve reasoning paths over wikipedia graph for question answering. arXiv preprint arXiv:1911.10470 (2019)
3. Karpukhin, V., et al.: Dense passage retrieval for open-domain question answering. arXiv preprint arXiv:2004.04906 (2020)
4. Thorne, J., Vlachos, A., Christodoulopoulos, C., Mittal, A.: Fever: a large-scale dataset for fact extraction and verification. arXiv preprint arXiv:1803.05355 (2018)
5. Aly, R., et al.: Feverous: fact extraction and verification over unstructured and structured information. arXiv preprint arXiv:2106.05707 (2021)
6. Qin, C., Zhang, A., Zhang, Z., Chen, J., Yasunaga, M., Yang, D.: Is chat-gpt a general-purpose natural language processing task solver? arXiv preprint arXiv:2302.06476 (2023)
7. Yu, Y., et al.: Large language model as attributed training data generator: a tale of diversity and bias. In: Advances in Neural Information Processing Systems, vol. 36 (2024)
8. Saxena, A., Tripathi, A., Talukdar, P.: Improving multi-hop question answering over knowledge graphs using knowledge base embeddings. In: Proceedings of the 58th Annual Meeting of the Association for Computational Linguistics, pp. 4498–4507 (2020)
9. Lee, J., Seo, M., Hajishirzi, H., Kang, J.: Contextualized sparse representations for real-time open-domain question answering. arXiv preprint arXiv:1911.02896 (2019)
10. Pan, L., Xie, Y., Feng, Y., Chua, T.S., Kan, M.Y.: Semantic graphs for generating deep questions. arXiv preprint arXiv:2004.12704 (2020)
11. Tan, C., et al.: Context-aware answer sentence selection with hierarchical gated recurrent neural networks. IEEE/ACM Trans. Audio Speech Lang. Process. **26**(3), 540–549 (2017)

12. Qiu, L., et al.: Dynamically fused graph network for multi-hop reasoning. In: Proceedings of the 57th Annual Meeting of the Association for Computational Linguistics, pp. 6140–6150 (2019)
13. Liu, D., et al.: Rikinet: reading wikipedia pages for natural question answering. arXiv preprint arXiv:2004.14560 (2020)
14. Wolf, T., et al.: Transformers: state-of-the-art natural language processing. In: Proceedings of the 2020 Conference on Empirical Methods in Natural Language Processing: System Demonstrations, pp. 38–45 (2020)
15. Asai, A., Hajishirzi, H.: Logic-guided data augmentation and regularization for consistent question answering. arXiv preprint arXiv:2004.10157 (2020)
16. Pota, M., Esposito, M., De Pietro, G.: Learning to rank answers to closed-domain questions by using fuzzy logic. In: 2017 IEEE International Conference on Fuzzy Systems (FUZZ-IEEE), pp. 1–6. IEEE (2017)
17. Chen, J., Lin, S.t., Durrett, G.: Multi-hop question answering via reasoning chains. arXiv preprint arXiv:1910.02610 (2019)
18. Yadav, V., Bethard, S., Surdeanu, M.: Quick and (not so) dirty: unsupervised selection of justification sentences for multi-hop question answering. arXiv preprint arXiv:1911.07176 (2019)
19. Yadav, V., Bethard, S., Surdeanu, M.: Unsupervised alignment-based iterative evidence retrieval for multi-hop question answering. arXiv preprint arXiv:2005.01218 (2020)
20. Lang, Q., Liu, X., Jia, W.: AFS graph: multidimensional axiomatic fuzzy set knowledge graph for open-domain question answering. IEEE Trans. Neural Netw. Learn. Syst. (2022)
21. Lan, Y., Wang, S., Jiang, J.: Knowledge base question answering with a matching aggregation model and question-specific contextual relations. IEEE/ACM Trans. Audio Speech Lang. Process. **27**(10), 1629–1638 (2019)
22. Xiaodong, L.: The fuzzy sets and systems based on AFS. structure, EI algebra and EII algebra. Fuzzy Sets Syst. **95**(2), 179–188 (1998)
23. Yen, H., Gao, T., Lee, J., Chen, D.: Moqa: benchmarking multi-type open-domain question answering. In: Proceedings of the Third DialDoc Workshop on Document grounded Dialogue and Conversational Question Answering, pp. 8–29 (2023)
24. Yu, J., et al.: Low-resource generation of multi-hop reasoning questions. In: Proceedings of the 58th Annual Meeting of the Association for Computational Linguistics, pp. 6729–6739 (2020)
25. Wang, Y., Duan, X., Liu, X., Wang, C., Li, Z.: A spectral clustering method with semantic interpretation based on axiomatic fuzzy set theory. Appl. Soft Comput. **64**, 59–74 (2018)
26. Liu, X., Jia, W., Wang, Y., Guo, H., Ren, Y., Li, Z.: Knowledge discovery and semantic learning in the framework of axiomatic fuzzy set theory. Wiley Interdisc. Rev. Data Mining Knowl. Discov. **8**(5), e1268 (2018)
27. Liu, X., Chai, T., Wang, W., Liu, W.: Approaches to the representations and logic operations of fuzzy concepts in the framework of axiomatic fuzzy set theory i. Inf. Sci. **177**(4), 1007–1026 (2007)
28. Liu, X., Jia, W., Liu, W., Pedrycz, W.: AFSSE: an interpretable classifier with axiomatic fuzzy set and semantic entropy. IEEE Trans. Fuzzy Syst. **28**(11), 2825–2840 (2019)
29. Liu, X., Wang, W., Chai, T.: The fuzzy clustering analysis based on AFS theory. IEEE Trans. Syst. Man Cybern. Part B (Cybern.) **35**(5), 1013–1027 (2005)

30. Duan, X., Wang, Y., Pedrycz, W., Liu, X., Wang, C., Li, Z.: AFSNN: a classification algorithm using axiomatic fuzzy sets and neural networks. IEEE Trans. Fuzzy Syst. **26**(5), 3151–3163 (2018)

31. Qi, P., Dozat, T., Zhang, Y., Manning, C.D.: Universal dependency parsing from scratch. arXiv preprint arXiv:1901.10457 (2019)

32. Akbik, A., Blythe, D., Vollgraf, R.: Contextual string embeddings for sequence labeling. In: Proceedings of the 27th International Conference on Computational Linguistics, pp. 1638–1649 (2018)

33. Xiaodong, L.: The fuzzy theory based on AFS algebras and AFS structure. J. Math. Anal. Appl. **217**(2), 459–478 (1998)

34. Yang, Z., et al.: Hotpotqa: a dataset for diverse, explainable multi-hop question answering. arXiv preprint arXiv:1809.09600 (2018)

35. Das, R., et al.: Multi-step entity-centric information retrieval for multi-hop question answering. In: Proceedings of the 2nd Workshop on Machine Reading for Question Answering, pp. 113–118 (2019)

36. Ding, M., Zhou, C., Chen, Q., Yang, H., Tang, J.: Cognitive graph for multi-hop reading comprehension at scale. arXiv preprint arXiv:1905.05460 (2019)

37. Nie, Y., Wang, S., Bansal, M.: Revealing the importance of semantic retrieval for machine reading at scale. arXiv preprint arXiv:1909.08041 (2019)

38. Floridi, L., Chiriatti, M.: GPT-3: its nature, scope, limits, and consequences. Mind. Mach. **30**, 681–694 (2020)

39. Touvron, H., et al.: Llama: open and efficient foundation language models. arXiv preprint arXiv:2302.13971 (2023)

40. Kwiatkowski, T., et al.: Natural questions: a benchmark for question answering research. Trans. Assoc. Comput. Linguist. **7**, 453–466 (2019)

41. Xiong, W., et al.: Answering complex open-domain questions with multi-hop dense retrieval. arXiv preprint arXiv:2009.12756 (2020)

Multi-population Evolutionary Computation Based on Lethal Chromosome and Its Application in Path Planning

Minjian Sun[iD], Guosheng Hao[✉][iD], Xia Wang, Xilong Feng[iD], Peng Zhang,
Yi Zhu, and Shijin Ren

Jiangsu Normal University, Xuzhou 221116, Jiangsu, China
`hgskd@jsnu.edu.cn`

Abstract. Evolutionary computation has been widely applied in many
fields. However, there are several disadvantages when evolutionary computation is applied to the field of path planning: Firstly, the fitness
function struggles to discern discontinuous paths. Secondly, in cases of
small population sizes, there is a notable tendency for homogenization
within the population. Lastly, the reliance on hyperparameters becomes
excessive in determining the optimal path. The paper tackles the aforementioned issues by introducing a multi-population genetic algorithm
based on lethal chromosomes. Firstly, discontinuous paths are identified as lethal chromosomes and eliminated via a deletion operator. Secondly, the utilization of multi-population addresses the homogenization
phenomenon within the population. Finally, adding a fitness evaluation process to the mutation operator promotes positive mutation and
reduces reliance on hyperparameters. The paper presents experimental evidence showcasing the algorithm's effectiveness in obstacle avoidance, path smoothness, and path length optimization. This has some
instructive significance for mitigating discontinuous paths in path planning challenges using genetic algorithms.

Keywords: Path planning · Genetic algorithm · Lethal
chromosomes · Multi-population

1 Research Background

Path planning is to determine a collision-free route from a starting point to an
endpoint. It is widely applied in many fields, such as mobile cleaning [1], aircraft [2], and agricultural robots [3], etc. Typically, path planning experiments
are conducted in grid-based environments [4], where the optimal route is delineated by a series of grid points. Despite significant research efforts into leveraging
evolutionary computation for path planning, there are still relatively few methods for handling discontinuous paths.

© The Author(s), under exclusive license to Springer Nature Singapore Pte Ltd. 2024
H. Yu et al. (Eds.): CCF NCCA 2024, CCIS 2274, pp. 53–70, 2024.
https://doi.org/10.1007/978-981-97-9671-7_4

Evolutionary computation like genetic algorithm [5], particle swarm algorithm [6], and ant colony algorithm [7] are commonly used for path planning problems. However, challenges persist when applying these algorithms: (1) Some studies, focus on generating initial and mutated paths that satisfy constraint conditions, often ignoring discontinuous paths [8]; (2) The issue of individual homogenization, still need to be solved [9]; (3) The evolutionary direction induced by mutation sometimes mislead the evolution [10].

To solve these issues, this paper proposes the following methods: (1) Defining discontinuous paths as lethal chromosomes and designing a deletion operator to handle them; (2) Utilizing multi-population to mitigate homogenization phenomenon; (3) Incorporating the fitness evaluation process into the mutation operator to ensure positive variation of individuals and integrating the above methods to reduce reliance on hyperparameters.

The structure of this paper is as follows: Sect. 2 introduces related work and preparatory knowledge, and Sect. 3 details the methods employed, including the definition and deleting method of lethal chromosomes, the multi-population, and the mutation operator design. Section 4 presents experimental results and summarizes the experiment. Finally, Sect. 5 provides the conclusion.

2 Related Work and Background

2.1 Genetic Algorithm in Path Planning

Genetic algorithm have found widespread application in path planning problems. For instance, Chen et al. [8] and Xiao et al. [11] have enhanced path planning effectiveness by employing adaptive crossover and mutation probabilities, respectively. Sarkar et al. [4] improved traditional genetic algorithm by integrating insert-delete operators, fine-tuning operators, target alignment operators, etc., onto conventional genetic operators. Orozco-Rosas et al. [12] proposed a membrane evolutionary artificial potential field algorithm by combining traditional path planning algorithms with genetic algorithm. Charis Ntakolia et al. [13] introduced fuzzy logic into genetic algorithm to enhance algorithm performance. Yibo Li et al. [14] reduced iteration times and improved algorithm effectiveness by integrating the evaluation function of the A* algorithm into genetic algorithm. Lakshman-an et al. [15] combined reinforcement learning methods with genetic algorithm to establish a robust model adaptable to various environments. Tharwat et al. [16] integrated Bezier curves with genetic algorithm to address non-smooth path problems in grid environments. Cheng et al. [17] achieved Pareto-optimal results by combining multi-objective optimization algorithms with genetic algorithm. Ortiz et al. [18] combined genetic algorithm with SLAM to achieve navigation in unknown environments. Lu et al. [19] integrated the Frenet coordinate system with genetic algorithm for adaptive trajectory control.

Despite the significant achievements in path planning on grid maps in the aforementioned studies, several issues exists: (1) All studies addressed the problem under the assumption that all initialized paths and paths during iterations are continuous, without considering how to deal with discontinuous paths; (2) In some studies, with fixed hyperparameters such as mutation and crossover probabilities, the overall characteristics of the population remain relatively constant, leading to severe homogenization and inefficient crossover operations, thereby causing the algorithm with large iteration number; (3) The direction of the mutation process is random, neglecting the positive variation of individuals. This may slow down the convergence rate of iterations and overlook the possibility of generating discontinuous paths during the iteration process, affecting the generation of continuous results.

2.2   Environment Modeling in Path Planning

The space designated for path planning comprises both free space and space obstructed by obstacles, with predetermined start and end points situated within the free space. The objective is to plan an optimal path for the robot from the starting point to the destination. Typically, there exist multiple potential paths between the starting and ending points. Algorithms generally consider objectives such as minimizing path length, ensuring smoothness, and guaranteeing movement safety to identify the optimal path based on these criteria.

This paper presents a path-planning algorithm and utilizes a grid map to represent the environment. Initially, the continuous space is discretized into a 20×20 grid map. The feasible path for the robot is a collection of free grids within the grid map. The points within the path set are referred to as waypoints in the subsequent discussion.

To evaluate the performance of the improved genetic algorithm, a grid-based environment is established. Figure 1 illustrates the environmental characteristics of the map where black denotes obstacles and white signifies free space. This setup enables the evaluation of the proposed enhanced genetic algorithm in terms of fitness function. It is noteworthy that both the grid size and the overall map size are dimensionless.

The subsequent section delineates the process of environment modeling, encompassing the encoding of the aforementioned waypoints. Paths are comprised of waypoints, thereby a path can be denoted as $(p_1, p_2, \ldots, p_n)$, where p_i represents the waypoint number and i denotes the number of waypoints. Alternatively, paths can be represented as $((x_1, y_1), (x_2, y_2), \ldots, (x_n, y_n))$, where (x_i, y_i) signifies the coordinates of waypoints. Equation (1) is utilized to convert grid numbers to grid map coordinates, while Eq. (2) is employed to convert grid map coordinates to grid numbers.

$$\begin{cases} x = \lceil \frac{p_i}{N} \rceil \\ y = \lfloor \frac{p_i}{N} \rfloor + 1 \end{cases} \tag{1}$$

$$p_i = (x - 1) + (y - 1) * N \tag{2}$$

where x and y represent the horizontal and vertical coordinates of the grid in the grid map, i is the waypoint p_i grid number, and N is the number of rows and columns in the grid map.

Fig. 1. Grid Map Environment

2.3   Initial Path Generation

To deal with the continuous paths, this paper adopts the approach introduced
in [20] which generates continuous initial paths through interpolation. The idea
of the algorithm is to insert waypoint between non-adjacent waypoints in the
grid. Initially, the algorithm scans through the list of waypoints and identifies all
gaps between their positions. Upon detecting a gap, the midpoint is computed,
and a new waypoint is inserted at that location. To ensure that the inserted
waypoint does not coincide with the obstacle, collision checks with obstacles in
the grid map are conducted. If a suitable insertion point cannot be found nearby,
alternative directions such as left, right, up, or down are explored until a valid
position is identified. This iterative process continues until all waypoints in the
path are interconnected, resulting in a continuous path.

However, in practical scenarios, discontinuous paths may emerge due to various factors, for example, the issues with raw data (such as resolution, noise,
or missing values), inaccuracies in parameter settings during the computation
process, or limitations of the algorithm itself when dealing with complex environments, etc.

To maintain path continuity, nodes are continuously inserted until all gaps
between adjacent nodes are eliminated. This approach ensures that the generated
path offers a continuous and obstacle-free trajectory in the grid environment.
After a certain number of iterations, if the insertion point is not found the
algorithm will forego the insertion operation and directly preserve this path.

2.4   Fitness Function

The fitness function is used to measure the quality of a path. As shown in Eq.
(3), fitness is a linear combination of path length and path smoothness [21].
Higher fitness indicates better quality for an individual.

$$fitness(X) = \alpha \cdot f_1(X) + \beta \cdot f_2(X) \tag{3}$$

here, X represents the entire path $((x_1, y_1), (x_2, y_2), \ldots, (x_n, y_n))$, $fitness(X)$ denotes the fitness function, $f_1(X)$ is the reciprocal of the path length, $f_2(X)$ is the reciprocal of the path smoothness, and α and β are the weights for $f_1(X)$ and $f_2(X)$ respectively. The definitions of $f_1(X)$ and $f_2(X)$ are given by Eq. (4):

$$\begin{cases} f_1(X) &= \frac{1}{pathvalue(X)} \\ f_2(X) &= \frac{1}{smoothness(X)} \end{cases} \tag{4}$$

where $pathvalue$ represents the total length of the path, and $smoothness$ represents the total smoothness of the path. Assuming there are n waypoints in the path, and the coordinate of the ith waypoint are (x_i, y_i), $pathvalue$ and $smoothness$ can be defined as follows in Eqs. (5) and (6):

$$pathvalue(X) = \sum_{i=1}^{n-1} \sqrt{(x_{i+1} - x_i)^2 + (y_{i+1} - y_i)^2} \tag{5}$$

$$smoothness(X) = \sum_{i=2}^{n-1} smoothness_i \tag{6}$$

where the definition of $smoothness_i$ is given in Eq. (7), and its values are determined by experiments and empirical knowledge:

$$smoothness_i = \begin{cases} 5, & \text{if } \theta_i = \frac{\pi}{4} \\ 30, & \text{if } \theta_i = \frac{\pi}{2} \\ 5000, & \text{if } \theta_i = \frac{3\pi}{4} \\ 10000, & \text{if } \theta_i = \pi \end{cases} \tag{7}$$

Equation (8) demonstrates the calculation of θ_i based on the cosine theorem as depicted in Fig. 2.

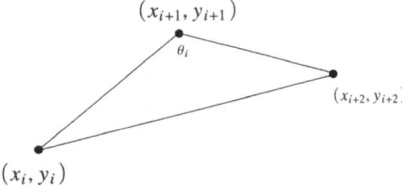

Fig. 2. Diagram for computing θ_i

$$\theta_i = \cos^{-1}\Big(((x_{i+1} - x_i)^2 + (y_{i+1} - y_i)^2 + (x_{i+2} - x_{i+1})^2 + (y_{i+2} - y_{i+1})^2$$
$$- (x_{i+2} - x_i)^2 - (y_{i+2} - y_i)^2) / \tag{8}$$
$$\Big(2\sqrt{(x_{i+1} - x_i)^2 + (y_{i+1} - y_i)^2} \sqrt{(x_{i+2} - x_{i+1})^2 + (y_{i+2} - y_{i+1})^2} \Big) \Big)$$

2.5   Crossover Operator

The method employed in this paper adopts the crossover technique as described in [20]. In addition to facilitating crossover between two parents with the same number of waypoint and those with different numbers, it also enhances population diversity and endeavors to explore the entire free space as extensively as possible. The crossover takes place at positions corresponding to the same waypoints in the two parents, following which all subsequent waypoints are exchanged to generate two offspring individuals. The crossover process is illustrated in Fig. 3. The same waypoint position is 20, and then all waypoints after 20 are swapped, and the length of each path changes after the two parents cross, increasing the diversity of the population and may result in individuals with higher fitness.

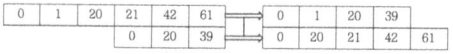

Fig. 3. Crossover Operator

The content of this chapter is mainly the common method of using genetic algorithm to solve the path planning problem. The next chapter is the main problem solved in this paper and the detailed content of the proposed algorithm.

3   Multi-population Path Planning Algorithm Based on Lethal Chromosomes

The primary aim of the algorithm in this paper is to expedite convergence and attain optimal continuous paths by eliminating discontinuous paths. Additionally, it tackles the issue of diminished population diversity resulting from path deletion.

The flow chart of the algorithm is shown in Fig. 4. Initially, two populations are initialized, each comprising both continuous and discontinuous paths. The adoption of the multi-population approach aims to enhance population diversity through population migration. Subsequently, evolutionary operations such as selection, crossover, and mutation are executed on each population to refine the initial solutions. The mutation process will also generate discontinuous paths, which cannot be identified by fitness function. Following this, discontinuous paths are eliminated using a deletion operator, and individuals are randomly migrated between the two populations. This iterative process continues until a termination condition is met.

The key features of the improved genetic algorithm are as follows: (1) a customized mutation operator tailored for optimal path exploration, offering positive mutation for faster convergence; (2) a deletion operator to weed out discontinuous paths and expedite convergence; (3) Multi-population to ensure population diversity, achieve faster convergence, evade local optima, reduce reliance on

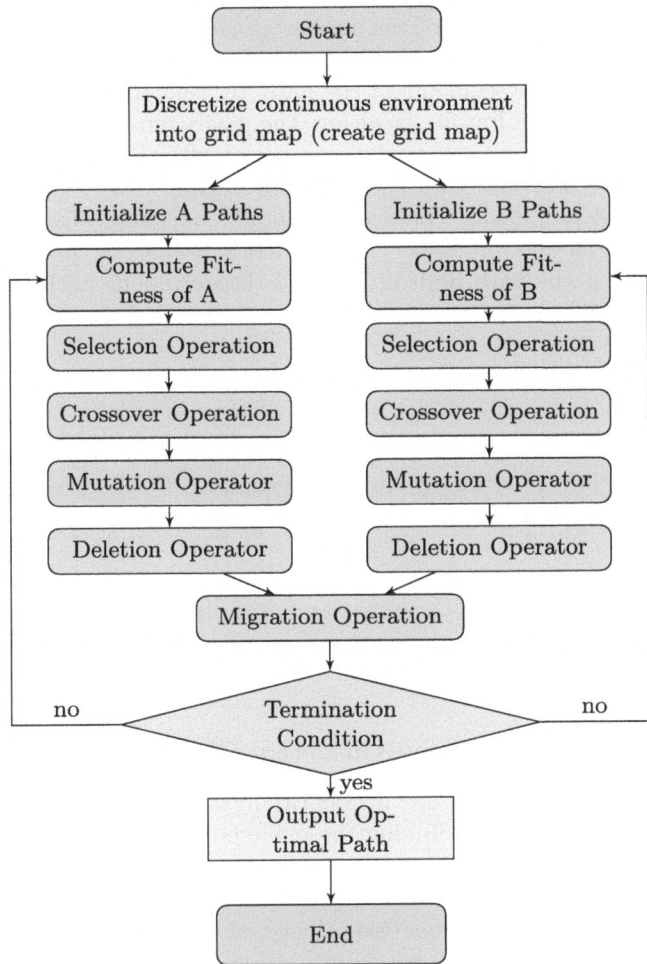

Fig. 4. Improved Genetic Algorithm Flowchart

hyperparameters, and enhance the optimal solution. Each of these features is elaborated upon below.

3.1 Mutation Operator

The mutation operator plays a crucial role in enhancing the exploration capability of genetic algorithms by increasing the diversity of the population and avoiding local optima. For path planning problems, variation is realized by changing the position of waypoints in the path.

This paper introduces a novel mutation method, wherein two random waypoints are selected along the path, thereby dividing it into three segments. Subsequently, the sub-path between these two waypoints is regenerated using the initial path generation method. This method of generating variation paths aims to address the approach proposed in this paper for handling discontinuous paths. Following this, the three segments of the path are concatenated to form a complete path as depicted in Fig. 5. To ensure the positive mutation of the chromosome, the fitness of the mutated path is compared with that of the original path. If the fitness increases, the mutated path is preserved; otherwise, the mutation process is carried out again. However, it should be noted that the fitness of discontinuous paths may be higher. Consequently, following the mutation step, both continuous paths with high fitness and discontinuous paths may coexist. The methods for handling such discontinuous paths are discussed in Sect. 3.2.

Fig. 5. Mutation Operator

This approach not only contributes to expanding the search space and fostering exploration but also addresses the issue of potentially discontinuous paths arising from mutation. By evaluating the fitness of mutated paths, this method ensures the positive variation of the population and the diversity of the population.

3.2 Deletion Operator Based On Lethal Chromosomes

The essence of path planning lies in identifying a set of waypoints that meet specified constraints and then linking these waypoints together to form a path. It sounds that any waypoint can be connected to others, but connecting non-adjacent waypoints is discontinuous. Unfortunately, discontinuous paths cannot be filtered out by fitness function. Hence, this paper adopts a strategy to remove discontinuous paths before each iteration, thereby preventing computational resource wastage, reducing iterations number, and obtaining continuous optimal paths.

This paper leverages the concept of "lethal chromosomes" proposed to craft a deletion operator [22]. In a grid environment of 20×20, a path is deemed a "lethal chromosome" if its base points fail to adhere to the criteria outlined in Eq. (9). For example, 1,20,42,... 379,399 is an discontinuous path because the difference between the second and third waypoints is 22, which is not in the range 1,19,20,21.

$$\forall i = 1, 2, \ldots, 399, \quad |p_{i+1} - p_i| \notin \{1, 19, 20, 21\} \tag{9}$$

Through the iterative process, the deletion operator removes the discontinuous path in the population and leaves the continuous path. Lethal chromosomes can impact the optimal path since, as per the above definition, they may acquire the highest fitness among discontinuous paths during selection, crossover, and mutation processes. Due to the use of delete operator, the search space contracts, leading to a decrease in population diversity. The reasons for the reduction in search space can be explained as follows.

The path's definition is $(p_1, p_2, \ldots, p_n)$, each of these waypoints is treated as a set. By applying Eq. (10), the waypoints are gathered into a set S, and the cardinality of the set is computed using Eq. (11). With a total of 400 grids, the proportion of the search space to the entire space R is given by Eq. (12). Upon employing the deletion operator, the value of Q decreases, meaning the proportion R of the search space to the whole space decreases. The reduction of the search space implies an increased likelihood of local optima.

$$S = \bigcup_{i=1}^{n} p_i \tag{10}$$

$$Q = |S| \tag{11}$$

$$R = \frac{|\bigcup_{i=1}^{n} p_i|}{400} \tag{12}$$

3.3   Multi-population

The deletion operator, as described above, often leads to the phenomenon of homogenization among individuals in the population, resulting in decreased diversity. For this issue, we introduces a multi-population migration mechanism aimed at bolstering population diversity and mitigating local optima.

In order to achieve population migration, it is crucial to implement crossover between individuals from two distinct populations. The purpose of crossover is to alleviate homogenization within the populations. Inspired by prior work [23], we have devised a migration operator that assigns distinct functions to different populations, facilitating genetic exchange across diverse groups. As depicted in Fig. 4, the migration process employs varying crossover and mutation strategies to generate distinct categories of individuals in two populations.

In Fig. 4, lower crossover and mutation rates are executed in Population A to safeguard high-quality individuals, thereby preventing the loss of elite individuals. Consequently, Population A is called as the elite population. Conversely, higher crossover and mutation rates are executed in Population B which is termed as the common population. Figure 5 illustrates the migration process, certain elite individuals from Population A are retained, while another portion individuals transition to Population B. This migration operation provides opportunities for ordinary individuals to engage in crossover with elite individuals, thereby enhancing population diversity and ensuring that both superior and ordinary individuals have the potential to evolve into better individuals through crossover and mutation. Essentially, a mechanism of communication exists between Population A and Population B, bridging the gap between the two populations. In summary, the migration operator accelerates convergence speed, enhances population diversity, breaks local optima, addresses homogenization among individuals in the population, and improves the optimal solution. The following explains the reasons why population migration mechanisms can reduce homogeneity. We employs Jaccard similarity to define the similarity between two paths, as shown in Eq. (13), where sets S_A and S_B represent the sets of points in paths A and B, respectively.

$$J(A, B) = \frac{|S_A \cap S_B|}{|S_A \cup S_B|} \tag{13}$$

Suppose the number of populations is N, then the average similarity J of populations can be defined as in Eq. (14). Before the migration among populations, the J value of two populations is relatively high. After migration, the number of individuals in the population N remains the same, but the J value of the population decreases. The reason is as follows: after migration, there are two types of paths in the two populations, one type belongs to the original population, and the other type is the migrated path. For the path i in the original population, the similarity with other paths is $\sum_{\substack{j=1 \\ j \neq i}}^{N-M} J(i,j)$, M is the number of migrating individuals. The similarity with the migrated paths in the N is relatively low, so $\sum_{\substack{j=1 \\ j \neq i}}^{N} J(i,j)$ will decrease. For the migrated paths m, the similarity with paths in the original population is also relatively low, so $\sum_{\substack{n=1 \\ n \neq m}}^{M} J(n,m)$ will decrease. The average similarity will decrease as a result, which will decrease the homogenization of the population and increase diversity. From the above analysis, it can also be seen that the number of migrated individuals should not be too large, otherwise, it may not significantly change the total similarity.

$$J = \frac{1}{C_N^2} \sum_{i=1}^{N-1} \sum_{j=i+1}^{N} J(i,j) \tag{14}$$

4   Experiments

4.1   Experimental Setup

This article adopts variable-length integer encoding, combining the roulette wheel selection method with the elitist preservation strategy. This section analyzes the performance of the improved genetic algorithm in an artificial simulation environment, including whether the optimal solution is obtained, convergence generations, and the length of the shortest path. The experimental environment is depicted in Table 1.

The robot can move in eight directions: up, down, left, right, upper-left, lower-left, upper-right, and lower-right. In this scenario, several tasks need to be addressed simultaneously: (I) determining the optimal path for the robot based on the positions of static obstacles, and (II) examining the determined path for potential collisions during turns. In this study, the robot's speed during operation is considered constant, and the turning angles are defined as $0°$, $45°$, $90°$, and $135°$.

We records the length of the optimal path and its corresponding fitness function value at each iteration in the experiment. The objective of the experiment

Table 1. Experimental environment

hardware	AMD Radeon Pro 5500M 4 GB
	16 GB 2667 MHz DDR4
	2.3 GHz Intel Core i9
software	MacOS monterey
	PyCharm 2022.2.3 (Community Edition)
	python 3.7

is to find the optimal path with the highest fitness in the grid map, and the termination condition of the experiment is reaching the maximum number of iterations.

The main distinction of this paper from other studies lies in its approach to treating discontinuous ways, specifically focusing on lethal chromosomes. The experimental section is organized as follows: The first segment comprises a comparison experiment with other studies, also serving as an ablation experiment due to the utilization of the deletion operator. This comparison contrasts cases with and without the deletion operator. Following this, the second part conducts a hyperparameter comparison experiment, aimed at illustrating the impact of hyperparameters on experimental outcomes. Lastly, the third part entails the multi-population experiment, which incorporates both the deletion operator and the use of the multi-population strategies, constituting the entirety of the experimental framework in this paper.

4.2 Comparison Experiment On Deletion Operator

This subsection primarily conducts comparative experiments on deletion operators. The experiments ensure that each initial path is continuous. Genetic algorithms tend to retain lethal chromosomes when deletion operators are not uti-

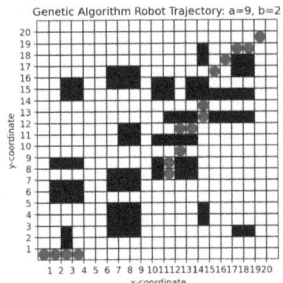

(a) Without deletion operator

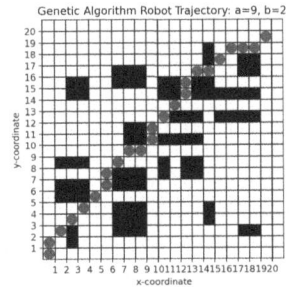
(b) With deletion operator

Fig. 6. Comparison of deletion operator experimental results

lized. Figure 6(a) illustrates the result of the optimal path obtained in a certain experiment without employing deletion operations, ultimately leading to discontinuous experimental outcomes. Figure 6(b) shows the experimental result after the adoption of deletion operations, presenting a continuous path.

4.3   Comparison Experiment on Hyperparameters

This section conducted comparative experiments on hyperparameters and found that setting the hyperparameters to the values in parameter group A as shown in Table 2(a) yielded better results. Deletion operations were utilized to achieve optimal results in this section. Table 2 presents the comparison of two sets of experimental parameters.

Table 2. Parameter Comparison

(a) Parameter Group A		(b) Parameter Group B	
Parameter	Value	Parameter	Value
Population Size	1000	Population Size	1000
Iterations	30	Iterations	30
Crossover Probability	0.9	Crossover Probability	0.7
Mutation Probability	0.3	Mutation Probability	0.05
a	7	a	9
b	3	b	2

In this section, 50 repeated experiments were conducted. Figure 7 presents one of the experimental results using parameter group A, while Fig. 8 displays one of the experimental results using parameter group B. The results from group A are notably superior to those from group B, as evidenced by two evaluation criteria: (1) the average optimal path length for group A is 30.712, compared to 30.836 for group B; (2) the average convergence generation for group A is 11.12, whereas for group B, it is 17. It can be observed that the hyperparameter settings have a significant impact on the experimental outcomes.

4.4   Comparison Experiment on Multiple Population

To validate the effectiveness of the multi-population migration mechanism, this section conducted comparative experiments on the population migration mechanism based on the deletion operator. The experiments aimed to compare results obtained with the same initial population and different initial populations, as well as to investigate the impact of using the migration mechanism on experimental outcomes. In each iteration, six individuals were migrated.

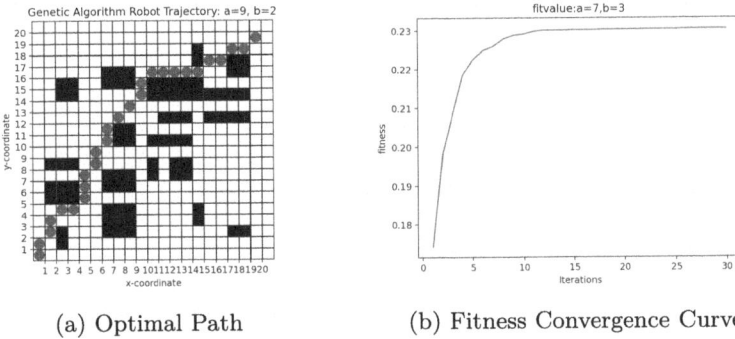

(a) Optimal Path   (b) Fitness Convergence Curve

Fig. 7. Experimental Results for Group A

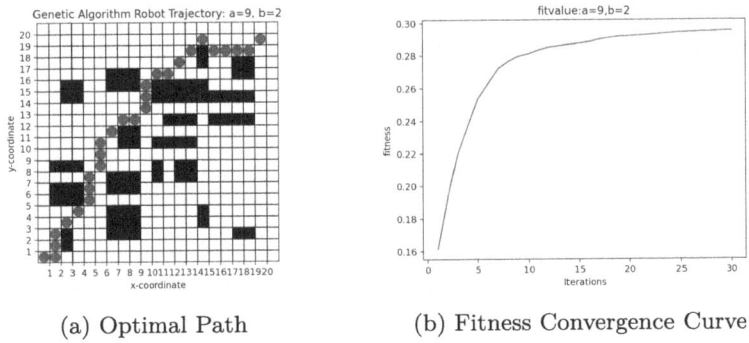

(a) Optimal Path   (b) Fitness Convergence Curve

Fig. 8. Experimental Results for Group B

Comparison Experiment on Initial Population. This experiment aims to evaluate the differences in experimental results obtained under the same initial population conditions and different initial population conditions. The objective is to determine the influence of initial population selection on experimental outcomes, thereby providing a basis for selecting appropriate initialization methods for subsequent experiments. The evaluation metric for this comparative experiment is the shortest path length. The experiment parameters are listed in Table 3. This section conducted 50 repeated trials, and the experimental results are shown in Fig. 9. It can be observed that the same population outperformed different populations in terms of stability and optimal path length. Therefore, subsequent experiments will adopt the same initial population.

Comparison Experiment on Multiple Population. The experimental parameters are shown in Table 4. This section conducted 50 repeated trials to verify the impact of the migration mechanism on the algorithm.

Table 3. Modified Experimental Parameter Settings Table

Parameter	Population A	Population B
Population Size	1000	1000
Iterations	30	30
Crossover Probability	0.9	0.7
Mutation Probability	0.3	0.05
a	7	9
b	3	2

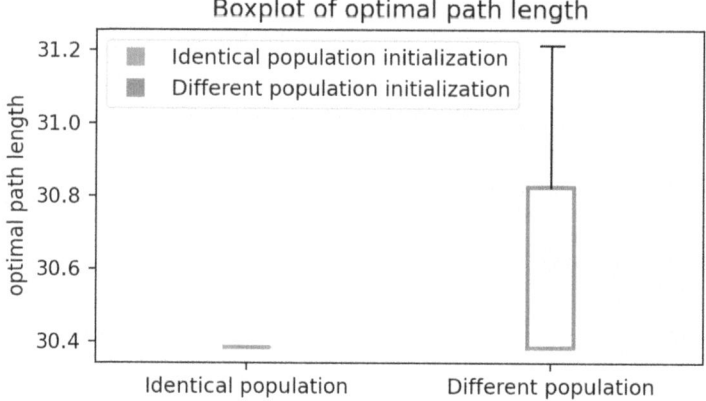

Fig. 9. Comparison of Optimal Path Lengths under Same and Different Initial Populations

The convergence generation experimental results are shown in Fig. 10. It can be observed that without multi-population, the convergence generation is significantly higher than that with the multi-population.

Figure 11 displays the results of the optimal path length under different experimental conditions. It is observed that with using the multi-population, the optimal path length is significantly greater than when the multi-population is not employed. Additionally, the experimental results demonstrate a more significant improvement in stability.

This finding strongly demonstrates the significance of the multi-population in improving algorithm convergence. Specifically, without using the multi-population, the algorithm may be influenced by local optimal solutions, resulting in relatively larger values for the optimal path length and convergence generation. However, with the introduction of the multi-population, the algorithm can better share information between multiple populations, helping to avoid local optimal solutions and convergence generation, thus significantly improving the performance of the optimal path length.

Table 4. Modified Experimental Parameter Settings Table

Parameter	Population A	Population B
Population Size	1000	1000
Iterations	30	30
Crossover Probability	0.9	0.7
Mutation Probability	0.3	0.05
a	7	9
b	3	2

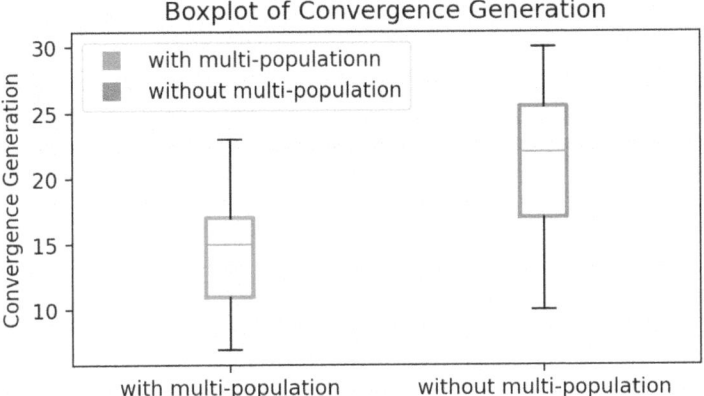

Fig. 10. Multi-population Convergence Generation Comparison Experiment Results

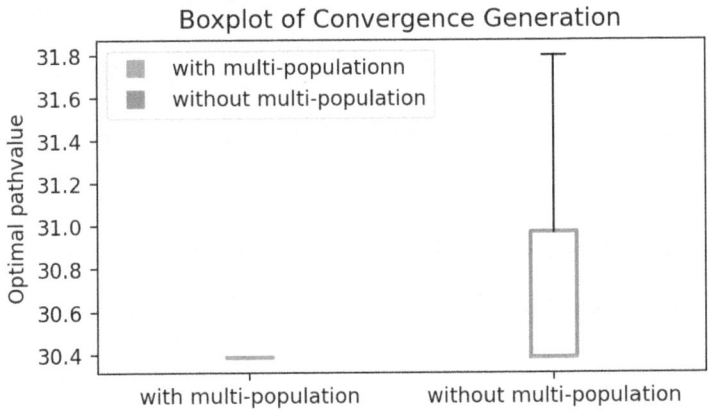

Fig. 11. Multi-population Optimal Path Length Comparison Experiment Results

This experimental result not only achieves significant performance improvement but also provides strong support for the importance of the multi-population in enhancing algorithm robustness and convergence speed.

5   Conclusion

This paper introduces a multi-population genetic algorithm based on lethal chromosomes, which improves the framework and operators of the standard genetic algorithm, enhancing its robustness. Overall, by eliminating lethal individuals from the population to improve convergence speed and further enhancing performance through multi-population, the algorithm achieves better solutions.

The proposed algorithm can find the optimal path within a finite time, demonstrating efficient time performance and the ability to find the optimal path in complex environments, thereby providing strong support for practical applications.

However, there are still areas for further improvement in this paper. Firstly, the influence of hyperparameters has not been eliminated; instead, dependence on hyperparameters has been reduced. Secondly, the assumption that obstacles are static doesn't reflect real-world scenarios, where obstacles are often dynamic. Therefore, one future research direction is to introduce dynamic obstacles into the map to better address challenges in real environments. Additionally, the paper adopts a linear combination of multiple objective functions as the multi-objective function. In the future, we plan to use more advanced multi-objective optimization frameworks to comprehensively optimize path planning problems.

Acknowledgments. The authors acknowledge the support from the National Natural Science Foundation of China under Grants 62277030, 62077029, the Society Development Foundation of Xuzhou under Grant KC23239, and the Postgraduate Research & Practice Innovation Program of Jiangsu Province 2024XKT2598.

References

1. Hasan, K.M., Reza, K.J., et al.: Path planning algorithm development for autonomous vacuum cleaner robots. In: 2014 International Conference on Informatics, Electronics and Vision (ICIEV), pp. 1–6. IEEE (2014)
2. Anqi, X., Viriyasuthee, C., Rekleitis, I.: Efficient complete coverage of a known arbitrary environment with applications to aerial operations. Auton. Robot. **36**, 365–381 (2014)
3. Hameed, I.A.: Intelligent coverage path planning for agricultural robots and autonomous machines on three-dimensional terrain. J. Intell. Robot. Syst. **74**(3–4), 965–983 (2014)
4. Sarkar, R., Barman, D., Chowdhury, N.: Domain knowledge based genetic algorithms for mobile robot path planning having single and multiple targets. J. King Saud Univ.-Comput. Inf. Sci. **34**(7), 4269–4283 (2022)
5. Korayem, M.H., Hoshiar, A.K., Nazarahari, M.: A hybrid co-evolutionary genetic algorithm for multiple nanoparticle assembly task path planning. Int. J. Adv. Manuf. Technol. **87**, 3527–3543 (2016)

6. Mac, T.T., Copot, C., Tran, D.T., De Keyser, R.: A hierarchical global path planning approach for mobile robots based on multi-objective particle swarm optimization. Appl. Soft Comput. **59**, 68–76 (2017)
7. Chen, Y., Jinfeng, W., He, C., Zhang, S.: Intelligent warehouse robot path planning based on improved ant colony algorithm. IEEE Access **11**, 12360–12367 (2023)
8. Chen, Z., Xiong, G., Liu, S., Shen, Z., Li, Y.: Path planning of mobile robot based on an improved genetic algorithm. In: 2022 IEEE 2nd International Conference on Digital Twins and Parallel Intelligence (DTPI), pp. 1–6 (2022)
9. Wang, H.: Continuum robot path planning based on improved genetic algorithm. In: 2022 2nd International Conference on Algorithms, High Performance Computing and Artificial Intelligence (AHPCAI), pp. 23–29. IEEE (2022)
10. Liang, X., Jiang, P., Zhu, H.: Path planning for unmanned surface vehicle with dubins curve based on GA. In 2020 Chinese Automation Congress (CAC), pp. 5149–5154. IEEE (2020)
11. Xiao, Y., Zhao, M.: An improved adaptive genetic algorithm for robot path planning. In: 2022 2nd International Conference on Electrical Engineering and Control Science (IC2ECS), pp. 951–954. IEEE (2022)
12. Orozco-Rosas, U., Montiel, O., Sepulveda, R.: Mobile robot path planning using membrane evolutionary artificial potential field. Appl. Soft Comput. **77**, 236–251 (2019)
13. Ntakolia, C., Platanitis, K.S., Kladis, G.P., Skliros, C., Zagorianos, A.D.: A genetic algorithm enhanced with fuzzy-logic for multi-objective unmanned aircraft vehicle path planning missions. In: 2022 International Conference on Unmanned Aircraft Systems (ICUAS), pp. 114–123 (2022)
14. Li, Y., Dong, D., Guo, X.: Mobile robot path planning based on improved genetic algorithm with a-star heuristic method. In: 2020 IEEE 9th Joint International Information Technology and Artificial Intelligence Conference(ITAIC), vol. 9, pp. 1306–1311 (2020)
15. Lakshmanan, A.K., et al.: Complete coverage path planning using reinforcement learning for tetromino based cleaning and maintenance robot. Autom. Constr. **112**, 103078 (2020)
16. Tharwat, A., Elhoseny, M., Hassanien, A.E., Gabel, T., Kumar, A.: Intelligent bézier curve-based path planning model using chaotic particle swarm optimization algorithm. Cluster Comput. **22**, 4745–4766 (2019)
17. Cheng, K.P., Mohan, R.E., Nhan, N.H.K., Le, A.V.: Multi-objective genetic algorithm-based autonomous path planning for hinged-tetro reconfigurable tiling robot. IEEE Access **8**, 121267–121284 (2020)
18. Ortiz, S., Yu, W., Li, X.: Autonomous navigation using robust slam and genetic algorithm. In: 2021 IEEE 17th International Conference on Automation Science and Engineering (CASE), pp. 1346–1351. IEEE (2021)
19. Lu, A., Lu, Z., Li, R., Tian, G.: Adaptive LQR path tracking control for 4ws electric vehicles based on genetic algorithm. In: 2022 6th CAA International Conference on Vehicular Control and Intelligence (CVCI), pp. 1–6. IEEE (2022)
20. Li, K., Qianqian, H., Liu, J.: Path planning of mobile robot based on improved multiobjective genetic algorithm. Wirel. Commun. Mob. Comput. **1–12**, 2021 (2021)
21. Ma, T., Wang, T., Yan, D., Hu, J.: Improved genetic algorithm based on k-means to solve path planning problem. In: 2020 International Conference on Information Science, Parallel and Distributed Systems (ISPDS), pp. 283–286. IEEE (2020)
22. Xiaojing, W., Zhaohong, D.: Application of improved genetic algorithm based on lethal chromosome in fast path planning of aircraft. In: 2020 5th International

Conference on Intelligent Informatics and Biomedical Sciences (ICIIBMS), pp. 216–220 (2020)

23. Hao, K., Zhao, J., Kaicheng, Yu., Li, C., Wang, C.: Path planning of mobile robots based on a multi-population migration genetic algorithm. Sensors **20**(20), 5873 (2020)

Few-Shot Knowledge Graph Completion Based on Selective Attention and the Transformer

Peng Wang, Chaoxiong Jia, Yongfeng Dong[(⊠)], and Yahui Wang

Hebei University of Technology, Tianjin, China
dongyf@hebut.edu.cn

Abstract. To solve the problem of insufficient entity representation, this paper proposes a few-shot knowledge graph completion algorithm based on a selective attention mechanism and a transformer. First, in the selective encoder stage, selective attention is introduced to help the algorithm distinguish between important neighbors and noisy neighbors. Second, in the relation aggregator stage, the combined structure based on the transformer and LSTM neural network is adopted to encode and output the triplet relationship. Finally, in the matching processor phase, the representation of the reference set is aggregated and compared with the query set for similarity. The proposed model's effectiveness and viability are validated through extensive experiments conducted on publicly available datasets. The results demonstrate the model's ability to enhance knowledge graph completion in few-shot scenarios.

Keywords: Knowledge graph · Few-shot learning · Knowledge graph completion · Attention mechanism · Neighbor coding

1 Introduction

The knowledge graph [1] plays a crucial role in the application of natural language processing [2], which greatly promotes research fields such as search technology, machine reading [3] and entity relationship recognition. It depicts a complex network of relationships by using triples (h, r, t) of entities and the relationships between them.

The knowledge graph is known for its inherent incompleteness, so researchers have developed numerous algorithms to enhance and extend its content, such as the TransE [4], TransR [5], and TransH models. These models rely on large amounts of training data to learn and predict relationships between entities. To solve the problem of more training data, researchers have proposed a method of knowledge graph completion under the condition of few samples.

Gmatching [6] was the first model to address the issue of knowledge graph completion with limited samples. This model uses a neighbor encoder that assigns uniform weights to entity neighbors and employs an LSTM network [7] for multistep matching. FSRL [8] incorporates an attention mechanism [9] to adjust the weight distribution among different neighbors, enhancing the neighbor encoder's performance. The FAAN [10] model, an extension of FSRL, introduces dynamic weight allocation to further

H. Yu et al. (Eds.): CCF NCCA 2024, CCIS 2274, pp. 71–86, 2024.
https://doi.org/10.1007/978-981-97-9671-7_5

increase model performance. MetaR [11] was the first to integrate meta-learning [12] into few-shot knowledge graph completion, leveraging specific meta-information for completion tasks. RSCL [13] expands on this by incorporating second-order neighbors, differentiating relation types into local and global categories, and applying distinct relation representation methods for each type.

Although these models have a certain effect in addressing the problem of knowledge graph completion with few samples, they also have many problems. First, the design of the previous model neighborhood encoder failed to distinguish important neighbors from noisy neighbors effectively, resulting in excessive weights for noisy neighbors that are not conducive to prediction, which affects the accuracy of the results. Second, although the existing models focus mainly on achieving similarity matching and constructing entity representations via using meta-learning mechanisms, they do not fully consider the impact of contextual semantic information on entity representations, and do not fully study few-shot relationships at a more detailed semantic level, thus limiting the model's ability to understand and reason about complex relationships.

To solve these problems, this paper designs a few-shot knowledge graph completion method based on selective attention (FCSA) algorithm. The main contributions of this paper include the following aspects:

(1) By applying the selective attention mechanism, different weights are assigned to important neighbors and noisy neighbors. Because important neighbors and noisy neighbors have different effects on entity embedding, the model uses the selective attention mechanism, and discusses how to assign different weights to different neighbors through the selective attention mechanism, to reduce the adverse effects of noisy neighbors on entity representation.
(2) The model uses the combination structure of the transformer and LSTM to capture the entity context semantic information. Since the context semantics of entities have a certain influence on the representation of entities, the LSTM is added to the model, and the LSTM processes the output of the transformer encoder through the residual connection [14], to build a richer context semantic representation.

2   Related Work

2.1   Few-Shot Learning

Few-shot learning is a technique that enables learning from a limited number of supervised samples. It has applications in areas such as machine vision, product recommendation [15], text classification, and visual question answering [16]. Recent research methods can be categorized into two main types on the basis of their implementation approaches: metric-based methods and meta-optimization-based methods. Metric-based approaches involve classifying by learning a "distance measure" between data points. The fundamental principle of this method is to compare test samples with labeled samples from the training set to assess their similarity and classification on the basis of their distance in the embedded space. If the "distance metric" is effectively learned during the training task, it can be applied to the target task without requiring fine-tuning. For example, the matching network [17] predicts outcomes by comparing input samples with a small set of labeled samples in a support set, functioning as a weighted neighbor classifier in the embedded space. Conversely, the prototypical network [18] uses class representations rather than sample representations, demonstrating the Euclidean distance is more effective than cosine similarity as a measure of similarity. On the other hand, few-shot learning methods rooted in meta-optimization focus on efficiently training models and enhancing generalization with small datasets. Among these, MAML [19] learns a general initialization weight, allowing the model to swiftly adapt to new tasks and achieve better performance with just a few gradient update steps. Additionally, ATAML [20], a meta-learning model that leverages attention mechanisms, aims to address few-shot learning challenges by capturing task correlations and adaptively adjusting the model parameters for new tasks.

2.2 Knowledge Graph Completion Based on Representation Learning

Knowledge graph completion involves predicting or inferring missing relationships, attributes, or entities from existing knowledge graphs via techniques such as machine learning and natural language processing. This process can be categorized into three main types on the basis of the completion methods used: translational distance models, semantic matching models, and neural network models. Translational distance models primarily use distance-based scoring functions to evaluate relationships. These models interpret the rationality of a fact as the distance between the head and tail entities in a vector space after accounting for the relationship. A notable example of this model is TransE, introduced by Antoine Bordes and colleagues. In TransE, the relationship in the knowledge graph is viewed as a translation vector between entities. For each fact triplet (h, r, t), the TransE model represents the entities and the relation within the same vector space and considers the relation vector r as the translation between the head entity vector $\mathbf{h}$ and the tail entity vector $\mathbf{t}$, so $\mathbf{h} + \mathbf{r} \approx \mathbf{t}$. Regarding semantic matching models, RESCAL [21] pioneered semantic matching by using vectors h and t to represent head and tail entities, respectively, while the relation $\mathbf{r}$ is represented by a matrix. This approach models the paired interactions between potential factors in the knowledge graph. ComplEx [22] learns vector representations of entities and relations in complex space, effectively addressing symmetric and antisymmetric relations. In the neural network model [23], the relation in the NTN is represented as a third-order tensor, which makes the semantic types corresponding to each slice different, and is conducive to describing various semantic connections between different entities associated with the same relation. The multilayer convolutional network model ConvE [24] completes the knowledge graph by means of convolution, which can be applied to more complex knowledge graph. Compared with traditional models such as trans-series models, ConvE can capture more complex information. G-GCN [25] models the relationship structure on the basis of a graph neural network.

3 FCSA Model

To solve the problem of knowledge graph completion under the condition of few samples, this paper develops a few-shot knowledge graph completion algorithm (FCSA) based on selective attention. The FCSA algorithm is divided into a selective encoder, transformer aggregator and a matching processor. The selective encoder is represented by selection of attention aggregation entities, and the transformer aggregator uses the combination of transformer and LSTM to enhance contextual semantics. The matching processor assigns a score to predict the relationship. The architecture of the model FCSA is shown in Fig. 1.

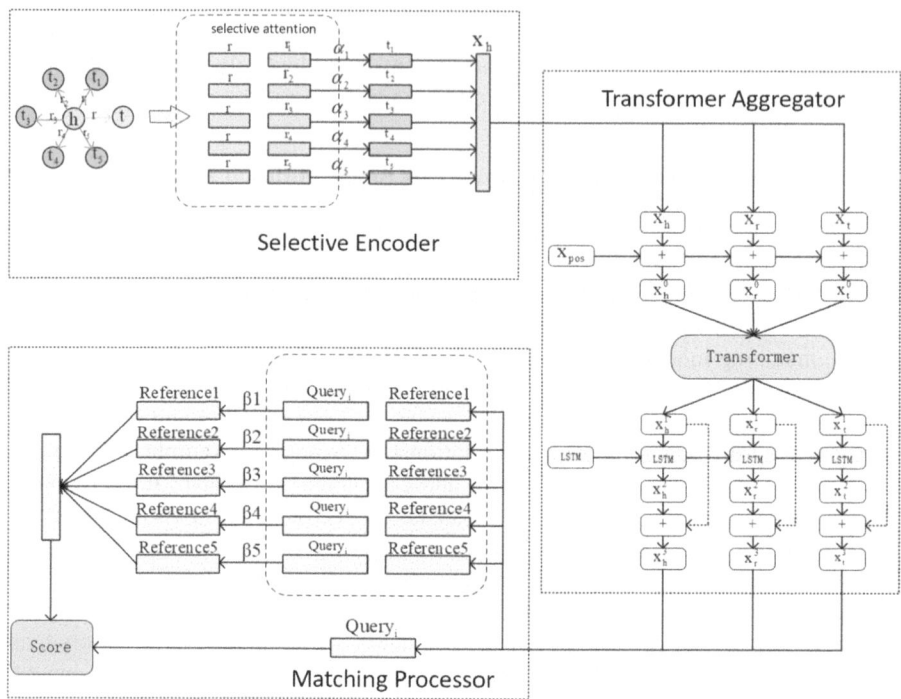

Fig. 1. Diagram of the FCSA.

3.1 Selective Encoder

In a knowledge graph, different types of neighboring entities play different roles in the representation of central entities. In the FAAN, on the basis of a specific relational vector **r**, different neighboring entities have different effects on the central entity, which indicates the importance of adaptive aggregation of neighbor information. Giving different neighbor entities different weights for the representation of central entities can effectively improve the representation quality of central entities. However, when the central entity is represented by integrated neighborhood information, this approach does not fully consider which neighborhood entities have a positive effect on the representation of the central entity and which may be noisy entities. Previous studies [26] have shown that in many real-world knowledge graphs, many neighboring entities have little or no connection with the central entity, which means that removing noisy neighbors can avoid the interference of irrelevant information on entity information aggregation when neighbor information is aggregated. In addition, distinguishing the influence of important neighbors and noisy neighbors on the central entity can better filter out neighbors that are conducive to entity coding, which can improve the accuracy of entity coding and avoid interference from a large number of noisy neighbors, thus enhancing the performance of the encoder [27]. Therefore, to optimize the representation effect of entities, it is necessary not only to flexibly adjust the aggregation strategy of neighbor information on the basis of a specific relation r, but also to consider the importance and relevance of each neighbor, to reduce the influence of irrelevant information and improve the accuracy of entity representation by filtering out the neighbors that make positive contributions to the central entity representation and removing or reducing the weight of noisy neighbors.

In this paper, a selective encoder based on selective attention is designed, which further utilizes aggregation shrinkage on the "adaptive neighbor encoder" in the FAAN model. Aggregation shrinkage means that in the process of aggregation entity representation, the weight values assigned to important neighbors and noisy neighbors are different through certain algorithms, so that the entity representation is more "shrinking" on its closely related neighbor entities. It reduces the effect of noisy neighbor entity coding by increasing the weight of neighbors that are closely associated with a particular task **r**.

Given a specific task triad (h,r,t), the model uses the selective encoder to update the representation of the target entity **h**. First, the first-level neighborhood set of entity **h** is defined as N_h, which can be expressed as $\{(r_i, t_i)|(h, r_i, t_i) \in G'\}$, where G' represents a background knowledge graph containing various entities and relationships. r_i represents the relationship to entity **h**, and t_i represents the tail entity to entity **h**. To measure the degree of correlation between task relation **r** and the adjacency relation r_i of entity **h**, an evaluation function $\varphi()$ is used in this paper. The calculation process is shown in Eq. (1):

$$\varphi(r, r_i) = r^T W r_i + b \tag{1}$$

where r^T represents the initial eigenvector of the task relation **r**; r_i represents a pretrained embedding vector of the relation r_i associated with entity **h**; and $W \in R^{d \times d}$ and $b \in R$ represent a weight matrix and an offset vector that can be learned, respectively. The selective encoder usually assigns a higher similarity score $\varphi(\cdot)$ to adjacent entities t_i, thus giving them a greater attention weight f_i. The calculation process is shown in Eq. (2):

$$f_i = \frac{\exp(\varphi(r, r_i))}{\sum_{r_j \in N_h} \exp(\varphi(r, r_j))} \tag{2}$$

where f_i refers to the magnitude of the influence of entity t_i on the task being processed in the adjacency relation- r_i. A higher score indicates that the neighbor pair (r_i, t_i) plays a larger role in the information aggregation of the target entity embedding.

To reduce the negative effect of noise neighbors on the aggregation of physical information, the selective encoder uses aggregation shrinkage when calculating the weights of neighbors, so that the focus of the selection encoder is more concentrated on those neighbors that are closely related to the task. First, the Euclidean distance $d(r, r_i)$ of task relation **r** and neighbor relation r_i is calculated to indicate the irrelevance of the two, and then the attention mechanism is applied to Euclidean distance $d(r, r_i)$, to obtain the attention weight α_i corresponding to the last neighbor entity. The calculation process is shown in Eqs. (3) and (4):

$$\alpha_i = \frac{f_i \cdot \exp(-\beta d(r, r_i))}{\sum_{r_j \in N_h} f_i \cdot \exp(-\beta d(r, r_j))} \tag{3}$$

$$d(r, r_i) = \sqrt{\sum_{k=1}^{m} (r_k - r_{i_k})^2} \tag{4}$$

where $d(r, r_i)$ refers to the dissimilarity between two different relations, namely, the Euclidean distance between task relation $\mathbf{r}$ and neighbor relation r_i in m-dimensional space. β is a hyperparameter in the interval (0,1), whose function is to adjust the size of the influence of $d(r, r_i)$, that is, to control the effect of differences between different relationships on the calculation of attention weights.

According to formula (3), if the dissimilarity $d(r, r_i)$ between two relationships increases, then the attention weight score of the neighboring entity decreases. By adopting aggregation shrinkage, the selective encoder can further focus on the neighbor entity with the highest task correlation, so that the attention score of the neighbor entity unrelated to the task gradually approaches zero. On the basis of this principle, the neighborhood code c_h of the target entity $\mathbf{h}$ can be obtained, and the calculation equation is shown in (5):

$$c_h = \sum_{t_i \in N_h} \alpha_i \cdot t_i \tag{5}$$

where t_i represents the embedding vector of the neighboring entity.

Finally, the selective encoder adaptively updates the entity representation with its output and the initial characteristics of the target entity, as shown in Eq. (6):

$$x_h = \sigma(w_1 \cdot c_h + w_2 \cdot h) \tag{6}$$

where, $\sigma(\cdot)$ represents the activation function, h represents the initial feature of the target entity, $w_1, w_2 \in R^{d \times d}$ is a parameter that can be learned, and $x_h \in R^d$ represents the entity embedding vector after updating the target entity $\mathbf{h}$.

3.2 Selective Encoder Transformer Aggregator

After the augmented entity embedding vectors are obtained by the selective encoder, the model then generates the embedding of the relation representation for the entity pairs. In the FAAN proposed in reference, each entity pair uses an encoder based on a transformer to generate relational embeddings separately, but this practice does not consider the different effects of contextual semantic information on different entity pairs. In this paper, we improve the transformer encoder, which is influenced by the R-TLM [28] (Recurrent Transformer Language Model). The transformer aggregator incorporates LSTM to better capture contextual semantic information and capture implicit semantic laws between entity pairs together with the transformer encoder. The output of each transformer encoder is connected to an LSTM cell, and the final output of the module is obtained by combining the output of the LSTM cell with the initial input through the residual connection. Moreover, the aggregator concatenates the internal LSTM units of each entity pair and the LSTM units of different entity pairs to build a unified LSTM network, which can effectively capture the semantic information of the entity pair in its context. When learning the relationship representation information related to entities, the LSTM network not only enhances the model's learning of the internal relationship of each entity pair, but also promotes the learning of the interaction between entity pairs.

This paper uses the aforementioned selective encoder module to encode the relation representation of each pair of entities. Let $X = (x_h, x_r, x_t)$ represent the initial input of the triple (h_k, r, t_k) in the transformer aggregator, where x_h represents the neighborhood encoding output of the head entity, x_r is the embedding representation of the relation, and x_t refers to the neighborhood encoding of the tail entity. For tractability, the transformer aggregator combines each input x_i with the corresponding position embedding vector as an input to this module. The calculation process is shown in Eq. (7):

$$x_i^0 = x_i + x_i^{pos} \tag{7}$$

where x_i^{pos} denotes the position embedding vector. When the representations of all the input vectors are constructed, they are fed into the transformer aggregator to encode the entity pairs. The specific calculation process is shown in Eqs. (8), (9) and (10):

$$x_i^1 = Transformer\left(x_i^0\right) \tag{8}$$

$$x_i^2 = LSTM\left(x_i^1, h_{i-1}\right) \tag{9}$$

$$x_i^3 = x_i^1 + x_i^2 \tag{10}$$

In the above formulation, $Transformer(\cdot)$ represents the workflow of the transformer encoder. It includes a multihead self-attention layer, a feedforward neural network layer, residual connections and hierarchical normalization layer normalization. The long short-term memory (LSTM) network takes the current hidden layer state output of the transformer encoder as its input, and then produces an updated LSTM cell state. This newly generated state is merged with the state of the previous LSTM cell to form the final output of the module. Each pair of entities fed into this module will eventually produce a sequence of three vectors $X^3 = (x_h^3, x_r^3, x_t^3)$. The transformer aggregator effectively combines the advantages of transformer in capturing long-distance dependencies and the advantages of long-short-term memory networks in processing time series data.

In the task of few-shot knowledge graph completion, the model pays more attention to the representation of relationships, because the relationship representation can enhance the model's understanding of the relationships between entities. Therefore, the module usually selects the intermediate elements in the sequence as the encoding vector of the relation **r**, which can describe the semantic role relationship between each pair of entities, and provide reference values for the fine semantic relationship between different entity pairs in the few-shot knowledge graph. This encoding mechanism is essential for accurately capturing the relationships in the knowledge graph.

3.3 Matching Processor

To predict the relation r in the query set, the FCSA designs a matching processor to process multiple semantic information associated with the task relation r by comparing the query set with the reference set.

The matching processor module compares the query set with the K-shot reference set and obtains a function that can represent the given reference set S_r. To capture the rich semantics of task relation **r**, the module adopts a metric function $\delta(q_r, s_{rk})$, which measures the semantic similarity between query q_r and reference triple s_{rk}. The calculation equation is shown in (11).

$$\delta(q_r, s_{rk}) = \mathbf{q}_r \cdot \mathbf{s}_{rk} \tag{11}$$

The attention mechanism is subsequently used to obtain a function representation $g(S_r)$ that is adjusted for the query. As shown in Eqs. (12) and (13):

$$g(S_r) = \sum_{s_{rk} \in S_r} \beta_k s_{rk} \tag{12}$$

$$\beta_k = \frac{\exp(\delta(q_r, s_{rk}))}{\sum_{s_{rj} \in S_r} \exp(\delta(q_r, s_{rj}))} \tag{13}$$

where β_k represents the attention weight score of the reference set; $s_{rk} \in S_r$ denotes the k-th reference in the task relation **r**; and s_{rk} is its embedding. Both s_{rk} and q_r can be obtained via Eq. (10). Equation (12) shows that the reference set with similar meaning to the query set will have more reference value; that is, the reference set S_r can adaptively generate entity pair representations according to different query pairs.

To predict the relation **r**, the model defines a metric function $\varphi(q_r, S_r)$, which measures the semantic similarity between the query q_r and the reference set S_r. The goal of this step is to make more accurate predictions by evaluating the degree of semantic similarity between the two. In other words, this process can be understood as a way to predict the result by determining the similarity between the query set and the reference set, as shown in Eq. (14):

$$\varphi(q_r, S_r) = q_r \cdot g(S_r) \tag{14}$$

3.4 Model Training

First, the meta-training dataset T_{mtr} is created to train the FCSA. The construction of the meta-training dataset T_{mtr} is as follows: for each task relation **r**, the model randomly selects K entity pairs from all positive entity pairs that include the relation as a reference set $S_r = \{(h_k, r, r_k)\}$. All remaining positive entities constitute the forward query entity set $Q_r = \{(h_m, t_m) | (h_m, r, t_m) \in T \backslash (h_k, r, t_k)\}$. Moreover, through the entity concentrated random $t_m^- \in \varepsilon \backslash \{t_m\}$, the way to build negative entities to set $Q_{r^-} = \{(h_m, t_{m^-})\}$ can be determined. The model then designs a loss function as follows:

$$L = \sum_r \sum_{q_r \in Q_r} \sum_{q_r^- \in Q_r^-} [\gamma + \varnothing(q_r^-, S_r) - \varnothing(q_r, S_r)] \tag{15}$$

where $[x]_+ = \max(0, x)$ is a standard hinge loss function, γ is expected to query the score difference between the negative to the query and, by minimizing L, make the positive reference set and query semantic similarity scores as large as possible, negative to the query and reference sets of semantic similarity scores as small as possible, and both γ keep a reasonable distance.

4 Experiment

4.1 DataSet

Nell-One Dataset: The NELL dataset was developed by a research team at Carnegie Mellon University to continuously learn and refine knowledge from the internet. The ontology structure of the dataset is initially set, and as a starting point, the system continuously extracts new information from the web data through the self-supervised learning method, to enrich and expand its knowledge graph.

Wiki-One Dataset: Wikidata [29] is a large, free and open knowledge base suitable for both human reading and machine processing. The goal of Wikidata is not only to store data, but also to improve data interoperability and accessibility by linking various sources of information. The information about the dataset is shown in Table 1.

Table 1. The information of the dataset.

Dataset	triple	entity	relation	Train	Validation	Test
NELL-One	181109	68545	358	51	5	11
Wiki-One	58592440	4838244	822	133	16	34

4.2 Evaluation Metrics and Baseline Models

During the validation and testing phases of the model, a crucial step involves scoring the confidence of the generated triples and ranking the alternative tail entities accordingly. To assess the model's performance, two evaluation metrics are employed: the mean reciprocal rank (MRR) and Hits@N. Both metrics are used to measure the accuracy of the model in predicting the correct tail entity.

The mean reciprocal rank is a standard measure of the performance of information retrieval systems, recommender systems, or any other type of ranking system. The mean reciprocal rank is an indicator used to evaluate the prediction ability of a model. The overall prediction performance of the model is measured by calculating the reciprocal position of the correct answer predicted by the model in the ranking, and taking the average of these reciprocal positions, as shown in Eq. (16). This metric reflects the position of the model prediction results among all the prediction results.

$$MRR = \frac{1}{Q}\sum_{i=1}^{Q} \frac{1}{rank_i} \tag{16}$$

Here, Q represents the number of candidate entities and $rank_i$ represents the position of the i-th candidate entity in the entity set after the triple matching score is ranked from high to low.

Hits@N is a widely used metric for assessing model performance in information retrieval, recommendation systems, and other ranking tasks. It quantifies the proportion of relevant items among the top N highest-ranked items. Hits@N is utilized to evaluate a model's ability to identify at least one correct answer within its top N predictions. The calculation method involves determining the ratio of the number of predicted results within the top N positions of the sequence to the total number of predicted results, expressed mathematically in Eq. (17).

$$Hits@N = \frac{1}{s}\sum_{i=1}^{S} \prod (rank_i < N) \tag{17}$$

In the above equations, S denotes the total number of entities in the set of candidate entities, whereas $rank_i$ represents the position of the i-th candidate entity in the ranked set of entities, where the triple matching scores are sorted from highest to lowest. From the provided equations, we can infer that higher values of MRR and Hits@N indicate that the model performs better in addressing the task of few-shot knowledge graph completion.

To assess the performance of the model, two types of baseline models are selected for comparative analysis: the traditional knowledge graph completion model and the few-shot knowledge graph completion (FKGC) model. Four classical knowledge graph embedding methods are chosen as baseline models, namely TransE, DistMult, ComplEx, and SimplE. The traditional knowledge graph embedding model and the few-shot knowledge graph completion model differ in their training approaches: the few-shot knowledge graph completion model relies solely on the background knowledge graph G' and the triples in the training task R_{train} as training data, whereas the triples used during the model's verification and testing phases are not included in the training data. Furthermore, the few-shot knowledge graph completion models discussed in this paper include GMatching, FSRL, FAAN, MetaR, and RSCL.

4.3 Comparative Experiment

The model was evaluated on the NELL-One and Wiki-One datasets. The prediction results of the few-shot knowledge graph completion for all the models on the NELL-One dataset and the Wiki-One dataset are presented in Table 2 and Table 3, respectively.

Table 2 and Table 3 show the following:

(1) The FCSA outperforms the traditional baseline model in both the 1-shot and the 5-shot scenario. This finding demonstrates that selective attention enhances the embedding representation of entity pairs, confirming the model's effectiveness in few-shot scenarios. Compared with earlier few-shot knowledge graph completion models, the FCSA also achieves superior experimental results. The FCSA improves triple prediction accuracy by capturing the contextual semantics of entities through a combination of transformer and LSTM.

In the 1-shot scenario of the NELL-One dataset, the FCSA is 7.25%, 3.24%, 4.67%, and 3.76% higher than the best result of the previous model on MRR, Hit@10, Hit@5,

Table 2. Experimental results on the NELL-One dataset.

Model	MRR		Hit@10		Hit@5		Hit@1	
	1-shot	5-shot	1-shot	5-shot	1-shot	5-shot	1-shot	5-shot
TransE [4]	0.062	0.174	0.169	0.313	0.072	0.204	0.004	0.101
DisMult [30]	0.102	0.200	0.177	0.311	0.126	0.251	0.066	0.137
ComplEx [22]	0.131	0.213	0.233	0.332	0.086	0.225	0.086	0.145
SimplE [31]	0.082	0.158	0.171	0.285	0.081	0.226	0.059	0.097
GMatching [6]	0.185	–	0.313	–	0.260	–	0.119	–
MetaR [11]	0.164	0.209	0.277	0.355	0.226	0.276	0.091	0.141
FSRL [8]	–	0.184	–	0.272	–	0.234	–	0.136
FAAN [10]	–	0.265	–	0.416	–	0.347	–	0.187
RSCL [13]	0.262	0.317	0.401	0.442	0.342	0.386	0.186	0.243
FCSA	**0.281**	**0.327**	**0.414**	**0.474**	**0.358**	**0.397**	**0.193**	**0.251**

Table 3. Experimental results on the Wiki-One dataset.

Model	MRR		Hit@10		Hit@5		Hit@1	
	1-shot	5-shot	1-shot	5-shot	1-shot	5-shot	1-shot	5-shot
TransE [4]	0.052	0.083	0.079	0.117	0.044	0.075	0.021	0.052
DisMult [30]	0.066	0.097	0.108	0.151	0.061	0.119	0.037	0.044
ComplEx [22]	0.075	0.100	0.125	0.161	0.059	0.122	0.032	0.052
SimplE [31]	0.058	0.093	0.096	0.140	0.055	0.101	0.028	0.039
GMatching [6]	0.176	–	0.294	–	0.233	–	0.113	–
MetaR [11]	0.159	0.321	0.277	0.439	0.226	**0.397**	0.091	0.262
FSRL [8]	–	0.137	–	0.269	–	0.181	–	0.091
FAAN [10]	–	0.314	–	0.436	–	0.384	–	0.258
RSCL [13]	0.307	0.348	0.414	0.454	0.365	0.395	0.241	0.281
FCSA	**0.321**	**0.354**	**0.434**	**0.460**	**0.378**	0.389	**0.256**	**0.289**

and Hit@1 respectively. In the 5-shot scenario of the NELL-One dataset, the MRR, Hit@10, Hit@5 and Hit@1 of the proposed algorithm increase by 3.15%, 7.23%, 2.84% and 3.29%, respectively. In the 1-shot scenario of the Wiki-One dataset, the algorithm improves MRR by 4.56%, Hit@10 by 4.83%, Hit@5 by 3.56% and Hit@1 by 6.22%. In the 5-shot scenario of the Wiki-One dataset, the proposed algorithm improves MRR by 1.72%, Hit@10 by 1.32%, and Hit@1 by 2.89%. The experimental results of Hit@5 on the Wiki-One dataset are not as good as those of the baseline model MetaR, which may be because gradient descent meta-learning is adopted in MetaR, and meta-learning is more suitable for datasets with many reference triples.

(3) The prediction accuracy of the few-shot knowledge graph completion model on the Wiki-One dataset is greater than that on the NELL-One dataset in most cases, because there are more reference triples provided by the Wiki-One dataset than those provided by the NELL-One dataset, and the prediction results are more accurate.

(4) The experimental results of MetaR on the Wiki-one dataset are compared with the experimental results of MetaR on the NELL-one model. This is because the number of reference triples on the Wiki-One dataset is larger, which is more suitable for meta-learning.

(5) The GMatching in the table is only for the case of one-shot, and the FSRL and FAAN are only for the case of few shots, so they do not need experimental data in special cases.

4.4 Ablation Experiment

To evaluate the impact of the selective encoder and transformer aggregator on the model, this section employs the control variable method to conduct ablation experiments on the NELL-One dataset in a 5-shot scenario. The experimental results are presented in Table 4.

(1) To assess the efficacy of the selective encoder, we make the following modifications: A1 involves replacing the selective attention mechanism with a normal attention mechanism.

(2) To evaluate the effectiveness of the transformer aggregator module, we make the following modifications: A2_a refers to the model without the transformer encoder, whereas A2_b refers to the model without the LSTM neural network.

Table 4. Ablation experiments on the NELL-One dataset.

Ablation model	MRR	Hit@10	Hit@5	Hit@1
A1	0.266	0.410	0.351	0.201
A2_a	0.283	0.454	0.359	0.223
A2_b	0.264	0.448	0.356	0.214
FCSA	**0.327**	**0.474**	**0.397**	**0.251**

According to Table 4, the FCSA model, which employs a few-shot knowledge graph completion algorithm with selective attention, demonstrates superior performance compared with its various variant models. A comparison between the A1 model and the FCSA model shows that the selective encoder utilizing the selective attention mechanism effectively mitigates the impact of noisy neighbors, thereby enhancing the representation of entity embeddings. Selective attention assigns higher weights to important neighbors, making the aggregated entity information more representative. Additionally, when the A2_a model, A2_b model, and FCSA model are compared, it is evident that the architecture combining LSTM networks and transformer encoders offers a significant performance advantage. This combination surpasses the individual LSTM and transformer applications, with the LSTM network improving the detailed contextual semantics capture, and the fusion with the transformer module further enhances the prediction accuracy.

4.5   The Impact of Sample Size K on the Model

This section describes experiments that examine the influence of the number of small samples K on the accuracy of the model. Figures 2 and 3 showcase the model's performance on the NELL-One dataset for various K values. A comparison is made between the FCSA model and the few-shot knowledge graph completion models MetaR and FAAN. The subsequent figures present the MRR and Hit@1 values for each model as K changes.

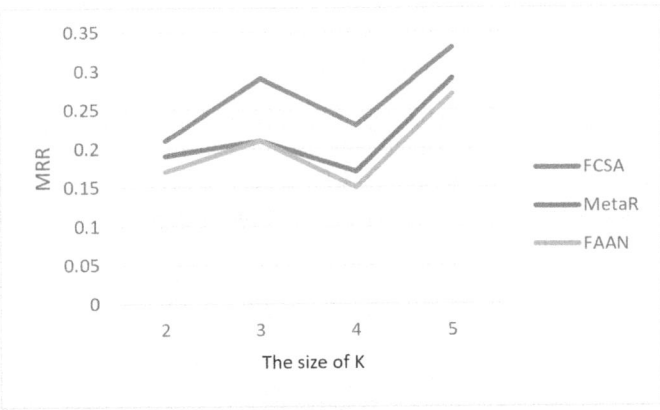

Fig. 2. MRR values for different models with different numbers of few-shot images.

The results from Fig. 2 and Fig. 3 demonstrate that irrespective of the variation in the K values, the proposed few-shot knowledge graph completion algorithm (FCSA) utilizing selective attention consistently outperforms the other models across all the K values. This outcome highlights the effectiveness of the FCSA model in addressing the few-shot relation prediction task and reinforces the ability of the FCSA method to handle few-shot problems effectively.

Figures 2 and 3 also show that as the number of minority samples (K) increases, the model's performance does not consistently improve but instead displays some variability. Although having more samples can introduce additional semantic information, the inclusion of edge or conflicting information within the selected samples may adversely affect the model, causing fluctuations in performance. However, the FCSA model consistently outperforms the other models across different K values, highlighting the effectiveness of the proposed FCSA method in enhancing model performance under various K conditions.

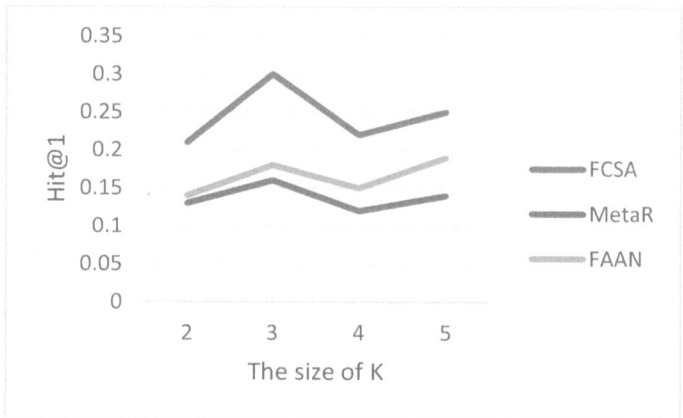

Fig. 3. Hit@1 values of different models with different numbers of few shots.

5 Conclusion

This paper introduces a few-shot knowledge graph completion algorithm that addresses the challenge of accurately differentiating vital neighbors from noise neighbors and comprehensively incorporating contextual semantic information. The proposed algorithm leverages an attention mechanism and an LSTM neural network. Within the neighbor encoder module, a selective attention mechanism assigns varying weights to crucial neighbors and noise neighbors, mitigating the impact of noise neighbors on entity pair representations. Furthermore, the integration of LSTM units in the transformer and LSTM module facilitates the capture of contextual semantic information.

Additionally, the full model was trained on two datasets, with the experimental results confirming the effectiveness of the FCSA model. Extensive testing on benchmark datasets for few-shot knowledge graph completion shows that the FCSA model outperforms existing mainstream algorithms in this area.

References

1. Ji, S., Pan, S., Cambria, E., et al.: A survey on knowledge graphs: representation, acquisition, and applications. In: The IEEE Transactions on Neural Networks and Learning Systems, pp. 1199–1208 (2022)
2. Wu, Z., Pan, S., Chen, F., et al.: A comprehensive survey on graph neural networks. In: The IEEE Transactions on Neural Networks and Learning Systems, pp. 1126–1135 (2021)
3. Baradaran, R., Ghiasi, R., Amirkhani, H., et al.: A survey on machine reading comprehension systems. In: The Natural Language Engineering, pp. 3630–3638 (2022)
4. Bordes, A., Usunier, N., Alberto, G.D., et al.: Translating embeddings for modeling multi-relational data. In: The Conference on Neural Information Processing Systems, pp. 4136–4145 (2013)
5. Lin, Y., Liu, Z., Sun, M., et al.: Learning entity and relation embeddings for knowledge graph completion. In: The AAAI Conference on Artificial Intelligence, pp. 2127–2136 (2015)
6. Xiong, W., Yu, M., Chang, S., et al.: One-shot relational learning for knowledge graphs. In: The Computing Research Repository, pp. 8460–8469 (2018)

7. Cheng, J., Dong, L., Lapata, M., et al.: Long short term memory networks for machine reading. In: Proceeding of the Computing Research Repository, pp. 1125–1134 (2016)
8. Zhang, C., Yao, H., Huang, C., et al.: Few-shot knowledge graph completion. In: The AAAI Conference Artificial Intelligence, pp. 3808–3818 (2020)
9. Aswani, A., Shazeer, N., Parmar, N., et al.: Attention is all you need. In: The 31st Conference on Neural Information Processing Systems, pp. 5998–6008 (2017)
10. Sheng, J., Guo, S., Chen, Z., et al.: Adaptive attentional network for few-shot knowledge graph completion. In: The Computing Research Repository, pp. 1199–1208 (2020)
11. Chen, M., Zhang, W., Zhang, W., et al.: Meta relational learning for few-shot link prediction in knowledge graphs. In: The Conference on Empirical Methods in Natural Language Processing, pp. 3630–3638 (2019)
12. Ma, Y., Zhao, S., Wang, W., et al.: Multimodality in meta-learning: a comprehensive survey. In: Proceeding of the Knowledge-Based Systems, pp. 1126–1135 (2022)
13. Li, Y., Yu, K., Zhang, Y., et al.: Learning relation specific representations for few-shot knowledge graph completion. arXiv Computation and Language 2205.02249 (2022)
14. Luo, J.-H., Wu, J.: Neural network pruning with residual connections and limited data. In: The IEEE Computer Society Conference on Computer Vision and Pattern Recognition, pp. 4136–4145 (2020)
15. Sahraoui, D., Huansheng, Ning., Nyothiri, A., et al.: Personality aware product recommendation system based on user interests mining and metapath discovery. In: The IEEE Transactions on Computational Social Systems, pp. 5998–6008 (2021)
16. Jiang, H., Misra, I., Rohrbach, M., et al.: In defense of grid features for visual question answering. In: The Computer Vision and Pattern Recognition, pp. 2127–2136 (2020)
17. Vinyals, O., Blundell, C., et al.: Matching networks for one shot learning. In: The Advances in Neural Information Processing Systems, pp. 1126–1135 (2016)
18. Snell, J., Swersky, K., Richard, S., et al.: Prototypical networks for few-shot learning. In: The Advances in Neural Information Processing Systems, pp. 8460–8469 (2017)
19. Chelsea, F., Pieter, A., Sergey, L., et al.: Model agnostic meta-learning for fast adaptation of deep networks. In: The Computing Research Repository, pp. 1199–1208 (2017)
20. Zhao, Z., Liang, X., Liu, J., et al.: Adaptive task-mining meta-learning for fast and robust model adaptation. In: The Pattern Recognition and Artificial Intelligence, pp. 1125–1134 (2021)
21. Huang, M., Chen, Q., Sun, A.: RESCAL: a high-order embedding model for semantic knowledge representation. In: The Transactions of the Association for Computational Linguistics, pp. 2127–2136 (2011)
22. Theo, T., Johannes, W., Sebastian, R., et al.: Complex embeddings for simple link prediction. In: The 33rd International Conference on International Conference on Machine Learning, pp. 3808–3818 (2016)
23. Wu, Z., Pan, S., Chen, F., et al.: A comprehensive survey on graph neural networks. In: Proceeding of the IEEE Transactions on Neural Networks and Learning Systems, pp. 3630–3638 (2021)
24. Tim, D., Pasquale, M., Pontus, S., et al.: Convolutional 2D knowledge graph embeddings. In: The Thirty-Second AAAI Conference on Artificial Intelligence and Thirtieth Innovative Applications of Artificial Intelligence Conference and Eighth AAAI Symposium on Educational Advances in Artificial Intelligence, pp. 1811–1818 (2018)
25. Chami, I., Ying, R., Ré, C., et al.: Hyperbolic graph convolutional neural networks. In: Advances in Neural Information Processing Systems, pp. 5998–6008 (2019)
26. He, T., Zhou, H., Ong, Y., et al.: Not all neighbors are worth attending to: graph selective attention networks for semi-supervised learning. arXiv preprint arXiv, 2212.03928 (2022)

27. Zhang, M., Wang, X., Zhu, M., et al.: Robust heterogeneous graph neural networks against adversarial attacks. In: Proceedings of the AAAI Conference on Artificial Intelligence, pp. 4363–4370 (2022)
28. Sun, G., Zhang, C., Woodland, P.C.: Transformer language models with LSTM-based cross-utterance information representation. In: The 2021 IEEE International Conference on Acoustics, Speech and Signal Processing, pp. 7363–7367 (2021)
29. Stephan, G., Jens, B., Julian, M., et al.: Wikidata: a free collaborative knowledge base. In: The Linked Data on the Web, pp. 2127–2136 (2012)
30. Yang, B., Yih, W., He, X., et al.: Embedding entities and relations for learning and inference in knowledge bases. In: The International Conference on Learning Representations, pp. 4136–4145 (2014)
31. Kazemi, S.M., Poole, D., et al.: SimplE embedding for link prediction in knowledge graphs. In: The Advances in Neural Information Processing Systems, pp. 5998–6005 (2018)

Node Embedding of the Abstract Syntax Tree for Source Code Representation

Chang-Feng Chen[1,2](✉), Azlan Mohd Zain[2], and Kai-Qing Zhou[3]

[1] College of Computer Science and Engineering, Jishou University, Jishou 416000, China
Ccf_cise@jsu.edu.cn
[2] Faculty of Computing, Universiti Teknologi Malaysia, 80310 Skudai, Johor, Malaysia
[3] School of Communication and Electronic Engineering, Jishou University, Jishou 416000, China

Abstract. Source code representation has garnered significant attention owing to its critical role in solving software engineering problems. There are various methodologies for representing code, and the abstract syntax tree (AST) is one of the most widely used techniques. However, current AST-based approaches need to pay more attention to the impact of AST generation, resulting in high duplication of the tree structure. Additionally, most AST-based methods struggle to balance homophily and structural equivalence in sequence sampling while failing to express semantic information in node feature learning. These limitations lead to poor source code representation and reduced performance in solving software engineering problems. To address these challenges, we propose a novel model that combines a deduplication algorithm, a hybrid sampling strategy, and an optimized skip-gram feature learning approach to represent source code. To evaluate the efficacy of our proposed model, we conduct experiments on OJ datasets for program classification. Our results demonstrate that the proposed model reduces the duplication of the tree structure and improves the representation performance of source code compared with other methods.

Keywords: Source code representation · abstract syntax tree · skip-gram · code classification

1 Introduction

Representation learning, as defined by Bengio et al. [1], involves transforming raw data into low-dimensional vectors, also known as embeddings, that maintain the properties of the original data and can be used in downstream deep learning models. Source code representation is a crucial step in solving software engineering problems, as it involves transforming the textual program into a standard input format that can be utilized by the corresponding model [2]. To improve the performance of source code representation, researchers have proposed various techniques, given their importance in tasks such as code clone detection [3], code classification [4], code search, and code translation. Previous research on source code representation can be classified into various categories, such as textual-based [5], token-based [6], tree-based [7], graph-based [8], and other approaches [9].

© The Author(s), under exclusive license to Springer Nature Singapore Pte Ltd. 2024
H. Yu et al. (Eds.): CCF NCCA 2024, CCIS 2274, pp. 87–111, 2024.
https://doi.org/10.1007/978-981-97-9671-7_6

Among the mentioned methodologies, tree-based approaches are most widely utilized because of their robust performance in programming language representation. It transfers source code into an abstract syntax tree or parses a tree on the basis of syntax analysis. Since the first development of an abstract syntax tree (AST) by Baxter for code clone detection in 1998 [10], it has received widespread attention. It has been applied in various software engineering domains [11–13]. While promising, two significant drawbacks limit the uptake of ASTs for source code representation: the high duplication of the AST structure and the insufficient model capacity in learning program semantic and structural information in AST node embedding.

AST-based approaches tend to be more complex than other representations because they involve representing the code in a highly detailed, hierarchical manner. The AST structure contains many nodes, each of which may represent a different element of the code, such as variables, functions, or control flow statements. As the code becomes more complex, the number of nodes in the AST can grow exponentially, making it harder to understand and analyze. Phan et al. [7] presented a heuristic technique to prune redundant branches, which can reduce the duplication of ASTs. Moreover, they introduced a minor procedure subtrees pruning algorithm to cut all the AST branches of other procedures and retain only the main function. The verification experiment revealed that pruning redundant branches can efficiently eliminate noisy information without the loss of useful information. Pruning minor procedures results in the cutting down of meaningful information. Zhang et al. split each significant AST into a sequence of small statement trees, which can decrease the duplication of the AST to some extent. Liu et al. reduced the AST structure by removing the syntax tree leaf nodes, which affects judgment [14]. Yang et al. proposed a variant of AST to achieve more abstract code representations using defined node types to replace the original node representations [15]. To better represent a program with an AST, Michal Duracik et al. divided nodes into relevant and irrelevant nodes, which can help transform such a tree into a linear structure or a collection of structures [16]. Shi et al. hierarchically split a large AST into a set of subtrees and encoded them with the RNN model [17]. These methods can mitigate the influence of the high duplication of ASTs. However, the substantial redundant and vital related feature nodes in the AST still block the further processing of source code.

Insufficient learning of program semantics also hinders code representation learning for the model. This is because some AST-based approaches tend to achieve initial feature learning via one-hot encoding, which represents AST nodes with a binary number sequence. For example, Gan et al. proposed an AST-based software vulnerability detection method in which one-hot is employed to encode the nodes and extract the feature information of source code. Moreover, the concatenation of different nodes relies on the simple addition of the generated vector from one-hot encoding. This causes low feature learning performance, as it can capture only the lexical and partial semantic information of the source code. To improve the semantic representation learning ability of source code in ASTs, numerous unsupervised learning techniques have been developed to use the abundance of unlabeled data that are easily accessible. As a result, highly valuable pretrained token embeddings such as Mikolov et al. [18], Pennington et al. [19], and Bojanowski et al. [20] have emerged. However, the current methods for feature learning in ASTs need to be improved in expressing the wide range of connectivity nodes in the AST structure.

To address the above challenges, we propose the novel node embedding of AST for source code representation (MAST-SG). Specifically, MAST-SG is designed as follows: (i) A tree-based deduplication algorithm is developed to remove the redundant node information in the AST and integrate the strongly related nodes. (ii) Inspired by node2vec [21], which uses the hybrid BFS-DFS strategy for learning continuous feature representations for nodes in networks, we design a sequence sampling strategy with a combination of Walker and Visitor to construct the node sequence of the AST. (iii) To further improve the feature learning ability of nodes in the AST, a negative sampling-based skip-gram method is employed to identify different types of node information in the AST.

To evaluate the effectiveness of the proposed node embedding approach, we conducted a series of experiments on the public datasets [22] shared by Mou et al. We compared them with a series of baseline models, such as BOW-based approaches, several other embedding layer-based models, and several classical source code representation algorithms. The experimental results indicate that MAST outperforms BOW-based methods and Embedding-based methods in terms of the metrics of Micro-F1 and Macro-F1. In addition, the deduplication algorithm can decrease the redundant nodes of the AST.

The main contributions of this paper are as follows.

We propose a novel AST node deduplication algorithm based on the Trie tree and node relations, which can eliminate redundant nodes, integrate strongly related nodes and improve the representation effectiveness of the AST structure.

Considering the imbalance of sequence sampling in traditional node embedding, we propose a hybrid sampling strategy based on Walker and Visitor. With the combination of Walker and Visitor, semantic and structural information in nodes can be embedded in further feature learning.

For further simultaneous embedding of the semantic and structural information of nodes, a negative sampling-based skip-gram is combined with a hybrid sampling strategy to achieve feature learning of different nodes.

The evaluation results on the statistics of the node numbers and program comprehension tasks show that our proposed approach can achieve state-of-the-art performance compared with popular and classical strategies.

The rest of this paper is organized as follows. The second section discusses related works on source code representation. The details of the proposed approaches are described and analyzed in Sect. 3. In Sect. 4, verification experiments are designed, and the results are discussed. Section 5 concludes this paper and briefly discusses future research work.

2   Background and Related Works

The transformation of source code into a numerical representation involves a comprehensive analysis based on the compiling process. Tree-based approaches typically convert the source code into an abstract syntax tree (AST), which focuses on the syntax structure of the code. To establish standard mathematical models for plagiarism detection,

Zhang et al. [23] utilized ANTLR as the processing model to transform code into the corresponding AST. In this AST-based model, extensions and modifications to grammar files were made to simplify the formulation process.

To address the issue of "garbage" in programs, Son et al. [24] proposed a novel program processing model based on parse tree kernels. Tuo et al. [25] developed a modified AST processing technique by rearranging the nodes of the AST into a linear structure. Resmi et al. [26] introduced an improved grammar AST method for processing C, C++, and Java programming languages, reducing the parse tree size by modifying each program's grammar. White [27] developed an RNN-based software language model, whereas Mou et al. [22] proposed a CNN-based programming language processing method based on tree structures, which was effective in program classification. Phan et al. [7] combined a tree-based CNN and a support vector machine to create a TBCNN+SVM model. Yang et al. [15] combined different variants of the AST and proposed a novel function-based processing approach in which nodes in the AST are classified according to other types and functions are measured as code fragments.

In conclusion, on the basis of the reviewed literature, tree-based approaches have been shown to provide good representations of source code with suitable computational complexity. However, some approaches still face the challenge of the high duplication and redundant nodes of the AST structure and limited expression of semantic information.

3 The Proposed Model

In this section, to address the problems mentioned above, a novel node embedding of the AST model for source code representation is developed. Figure 1 shows the structure of the proposed embedding model in detail. Unlike regular tree-based code representation techniques, we propose a novel framework that combines a node deduplication algorithm, a hybrid walker and visitor sampling strategy, and a negative sampling-based skip-gram to overcome the shortcomings discussed. Specifically, it mainly includes three phases.

Phase 1 - AST deduplication: In this phase, we design a key node extraction algorithm to integrate strongly related nodes and eliminate the in-related nodes from the generated AST text. Depending on the AST deduplication process in this phase, the redundant nodes are removed, and the high-dimensional data of the generated AST decrease.

Phase 2- Sequence sampling and encoding: In this phase, on the basis of the obtained AST, we introduce a hybrid sampling strategy with a combination of Walker and Visitor. With the help of a hybrid sampling strategy, it can balance homophily and structural equivalence in sequence sampling and fully leverage the structural information and naturalness of statements. Moreover, the obtained node sequence is encoded with one hot encoding.

Phase 3- Feature learning: In this phase, after we obtain the initial encoding of each node in the AST, the optimized skip-gram is used to learn unsupervised vectors of the node, and the trained embeddings of a node are generated as the final representation of the AST. We did this for two reasons. First, compared with CBOW, which counts only the symbols of source codes, skip-gram is a good feature learning technique with many strengths that can map words (even phrases) to vectors of real numbers. Second, an optimized skip-gram with negative sampling is used to specify one set of multinomial distributions for each type of node.

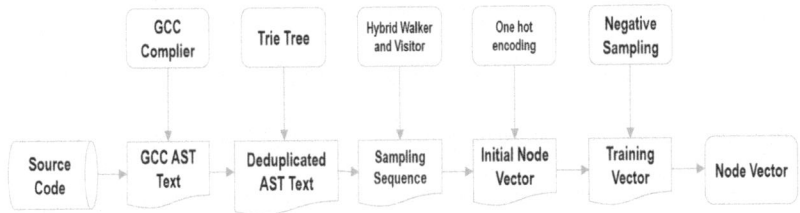

Fig. 1. Framework of the proposed node embedding approach

3.1 AST Deduplication

On the basis of the discussion in Sect. 2, we choose the AST as the middle representation of the source code in this paper. To further improve the expressive ability of ASTs by decreasing their duplication, we propose the key node extraction algorithm to reduce the duplication of ASTs. The process of AST deduplication can be described as follows in Fig. 2.

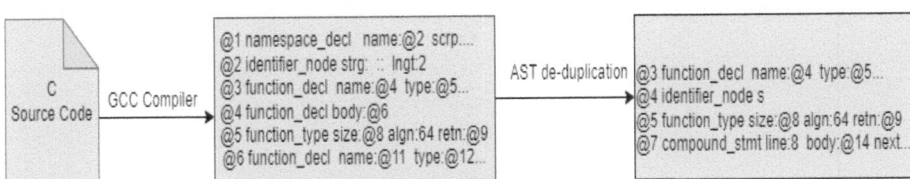

Fig. 2. Process of AST deduplication

3.1.1 AST Text Generation

GCC-AST text is the syntax tree file produced by C source code through the GCC compiler. Each input source code is compiled on the basis of the compiling command "gcc-dump-translation-unit C" or "gcc-fdump-tree-all" and produces an AST file with the name "source code name xxxt.tu". Table 1 in the appendix lists the partial nodes of the generated AST file from the source code.

GCC-AST saves the data with a unit of the node. There are eight main types of nodes in the GCC-AST text, which are listed in Table 1. Each node includes node order, node identification, and subnode information. The node order is unique compared with other nodes. Moreover, GCC-AST text files are stored in ascending order according to node numbers.

Table 1. The main types of nodes in GCC-AST text

Node Type	Suffix	Example
Expression node	*_expr	addr_expr
Type node	*_type	integer_type
constant node	*_cst	integer_cst
declaration node	*_decl	parm_decl
Identifier node	*_identifier	Ht_identifier
List node	*_list	tree_list
Statement node	*_stmt	Outer_stmt

On the basis of the AST text file, we can construct an AST sample, as shown in Fig. 3, which is an example of a subtree of an AST from the source code.

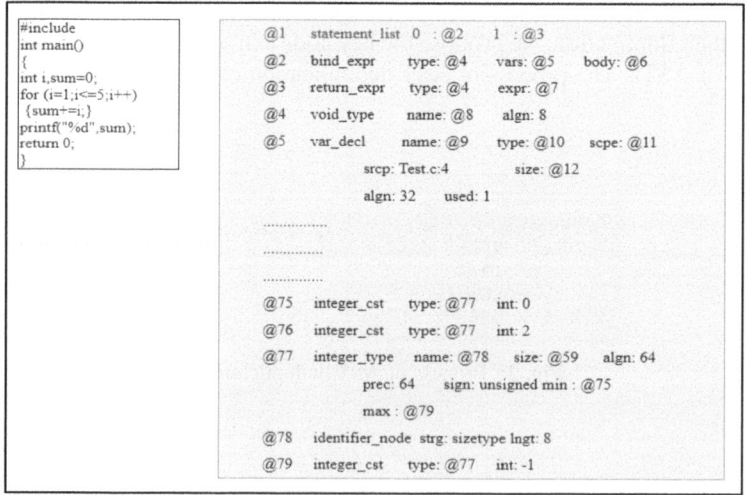

Fig. 3. Partial node information of the original abstract syntax tree text

3.2   AST Text Duplication

As discussed before, an AST-based representative of source code has the disadvantage of redundant nodes in the generated AST structure. Thus, in the proposed model, a key node extraction algorithm is proposed to filter the related nodes and integrate strongly related nodes, effectively decreasing the duplication of the AST structure and avoiding the loss of useful nodes.

Definition 1 (Related node): those nodes are strongly related to the program analysis data flow and control flow. All node types are included in a simplified syntax parse tree, which are nodes extracted from a complete parse tree with the exclusion of in-related nodes. The mathematical definition of the source code can be described as follows:

$K = \{k_1, k_2..., k_i..., k_n\}(1 <= i <= n)$, where n is the number of related node types, k_i represents the i-th related node, $k_i = \{ki_1, ki_2..., k_{ij}...k_{im}\}$, k_{ij} represents the j-th subnode of related node k_i, and m represents the number of subnodes [28].

The related node represents features in the source code with critical values that must be emphasized in the research. It is usually processed to produce a feature vector or matrix.

Definition 2 (In-related node): These nodes do not influence the program analysis data flow or control flow.

Definition 3 (Unknown node): These nodes cannot decide whether they are related nodes or unrelated nodes.

By judging whether node information is helpful for static analysis, the related nodes can be recognized from the related node information. It can achieve the initial exclusion of inherent and system-redundant information. On the basis of the obtained related nodes, we can build the Trie tree for key node extraction via Algorithm 1. In Algorithm 1, related nodes are input, and different types of nodes are set with separate tags. To increase the query efficiency, node datasets are built on the basis of the ordering sequence of the first characters of related nodes. When nodes are indexed with the same prefix, they are stored in the same node group, which can effectively decrease the number of saved nodes.

Algorithm 1   Key Node Extraction

Input: Set of keyword fields K with node information contained in GCC
    abstract syntax tree text

Output: GCC AST Related Trie tree

 1: initialize:

 2:     $i = 1, j = 1, q = 1$, q records the serial number of the child node

 3:     C_i = ROOT, C_i is the node being detected

 4: **while** $i \leq n$ **do**

 5:    acquire the first letter of the keyword of the type node S

 6:    enter the S child node tree to query

 7:    **if** $K_{ij} == child_q$ **then**

 8:       $j + +$

 9:       **if** $j < m$, **then**

10:          $S = child$;

11:          $q = 1$;

12:          **Goto 6;**

13:       **else**

14:          $i + +$;

15:       **end if**

16:    **else**

17:       **if** $child \rightarrow next\ node == null$ **then**

18:          create a new node $child_{q+1}$;

19:          $child_{q+1} = K_{ij}$;

20:          $j + +$;

21:          **if** $j < m$, **then**

22:             create a child node;

23:             assign $child_{q+1}$;

24:             $j + +$;

25:          **else**

26:             initialize: $i + +$;

27:          **end if**

28:       **end if**

29:       **if** $child \rightarrow next\ sibling\ node \neq null$, **then**

30:          $q + +$;

31:          **continue;**

32:       **end if**

33:    **end if**

34:    **if** $i \geq n$, **then**

35:       **return;**

36:    **else**

37:       **continue;**

38:    **end if**

39: **end while**

In this phase, the produced GCC-AST text file is input, and the Trie tree works as the critical node extraction algorithm to eliminate redundant information. The GCC-AST text data flow is queried with the fundamental rule when several related nodes are included in the token. The pseudocode of the process can be described as Algorithm 2.

Algorithm 2   AST deduplication based on Trie trees

Input: GCC abstract syntax tree text data stream AST_{test}, GCC AST
    Related node Trie tree

Output: Optimized redundant GCC abstract syntax tree text AST_{test}

1: initialize: $i = 1$, $q = 1$, q used to record the character sequence number
    entering the branch;

2: input ast_i

3: **if** $i < n$ **then**

4:    $q = i, j = 1$;

5:    acquire S, the first letter of the node type of ast_i;

6:    query child node *child* whose node type starts with S in the keyword Trie tree;

7:    **if** $ast_i == child$ **then**

8:        output @*number*, the unique identifier line number of ast_i, and
    node information to AST_{new};

9:        output the child node information in the output ast_i description
        field to AST_{new};

10:    **else**

11:        **Goto 14**;

12:    **end if**

13: **else**

14:    **if** *the next node $\neq$ Null* **then**

15:        **Goto 6**;

16:    **else**

17:        $i = q + 1$;

18:        **if** $i < n$, **then**

19:            **Goto 2**;

20:        **else**

21:            **return** ;

22:        **end if**

23:    **end if**

24: **end if**

On the basis of Algorithms 1 and 2, the deduplicated GCC-AST text file, which is part of the text file corresponding to the deduplicated GCC-AST text file, is produced as shown in Fig. 4. Moreover, on the basis of the deduplicated GCC AST text, the user node is extracted, and the simplified AST is generated. Figure 5 shows a sample of sub-ASTs from the deduplicated GCC-AST text.

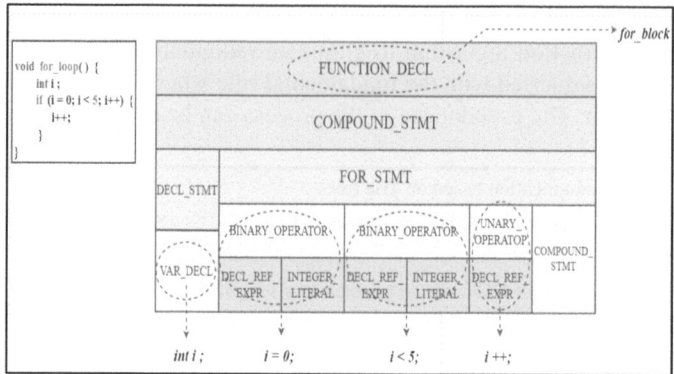

Fig. 4. Simplified AST based on SAMPLE source code

3.3 Sequence Sampling and Encoding

After the generation of the AST, we can use the traversal algorithm to obtain the sequence of nodes in the AST. The sequence sampling of the previous AST-based code embedding approach faces the problem of in-balance between homophily and structural equivalence, as most previous researchers have used one of two extreme sampling strategies for generating sample sequences: Walker and Visitor.

3.3.1 Walker

Walker is one of the most commonly used AST traversal algorithms and is based on the depth-first search strategy. With the Walker algorithm, only the generated response function should be processed. For example, Fig. 5 shows a traversal process of Walker. The figure shows that when the parser reaches the decl_stmt node, it starts executing the enterdecl() operation automatically. Hence, only the addition of start logic is needed. Moreover, when the traversal of the decl_stmt node is finished, the exit deal() operation is executed automatically.

Sequence sampling under Walker tends to closely embed nodes that are highly inter-connected and belong to similar clusters, as it emphasizes the connectivity of nodes between each other.

3.3.2 Visitor

Walker is another often-used AST traversal algorithm that is based on the breadth-first search strategy. Visitors provide a dynamic traversal method that treats corresponding nodes as parameters and transfers them into the visitor() function. Finally, traversal nodes information with the visitor() function. For example, Fig. 6 shows a typical traversal process of the visitor. The figure clearly shows that the implementation of the visitmt() method should call the visit() method and pass all child nodes to it as parameters to continue the traversal process. Alternatively, the visitX() method can explicitly call the visitdecl() method.

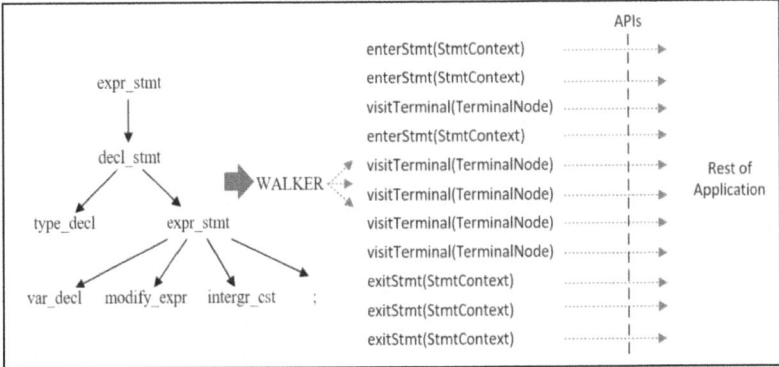

Fig. 5. AST traversal process of Walker

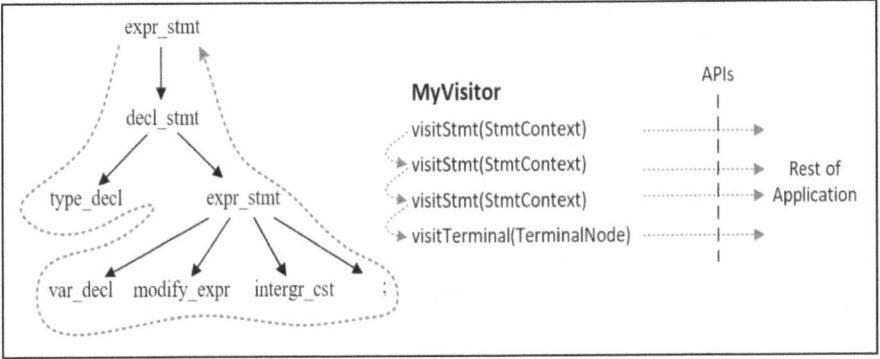

Fig. 6. AST traversal process of visitors

In a visitor-based sampling strategy, nodes that have similar structures are embedded closely together. Unlike the Walker strategy, the visitor strategy highlights the importance of structure equivalence.

3.3.3 Hybrid Walker and Visitor Sampling Strategy

The Walker and Visitor strategies play a vital role in producing representations that can reflect either homophily or structural equivalence, as the emphasis on homophily equivalence will stick in a microscopic view of the neighbor of each node. In contrast, a focus on structure equivalence reduces the variance in characterizing the distribution of nodes. However, in the embedding of an AST, both connectivity and structural similarity should be considered to achieve the best representation of learning. Therefore, a strategy should be proposed to balance homophily and structural equivalence in the node sampling of the AST, in which we design a hybrid sampling strategy to produce the node sequence. In the proposed sampling strategy, a random node (u) is chosen as the root node; the walk length is fixed. Let wt denote the t-th node in the walk, starting with root node w_0

$= u$. Node w_t is generated with Eq. (1).

$$P(w_t = m | w_{t-1} = v) = \begin{cases} 1, & \text{if } (m, v) \in \text{leaf nodes} \\ \pi_{mv}, & \text{otherwise} \end{cases} \qquad (1)$$

where π_{mv} represents the unnormalized transition probability from node m to n.

Considering the structural information and homogeneity representation ability of Visitor and Walker, respectively, a biased search method based on combining Walker and Visitor is utilized in the sampling strategy.

We define a second-order random walk from a subsample of the simplified abstract tree, which traverses from the key node and reaches another key node. Then, the strategy must decide which node is the next sampling node and evaluate the transition probability. The transition process can be demonstrated via Eq. (2).

$$\pi_{mv} = \alpha_{kj}(m, x) = \begin{cases} \frac{i}{k}, & d_{mx} = l \\ \frac{1}{j}, & d_{mx} = 2l \end{cases} \qquad (2)$$

The distance between two neighboring nodes is fixed as l, and d_{mx} represents the straight distance between nodes m and x. In addition, parameters k and j guide the walk change between Walker and Visitor. The pseudocode of the hybrid Walker and Visitor sampling strategy is shown in Algorithm 3.

Algorithm 3   Hybrid walk and visitor sampling strategy

Input: AST $T=(V, B, \pi)$, start node u, walk length l.

Output: *walk*

1: initialize *walk* =[u]

2: **for** *walk_iter* = 1; *walk_iter* < l − 1; *walk_iter* + + **do**

3:      *newnum*=*walk*[-1]

4:      *V_newnum*=GetNeighbors(*newnum*, *T*) according to Eq 2.

5:      *s*=AliasSample(*V_newnum*, *T*)

6: Append *s* to *walk*

7: **Return** *walk*

3.3.4   Initial Node Encoding

After the generation of a sampling sequence, the work of feature encoding can start [29]. Feature encoding is an important operation in transforming ASTs into numerical features. Here, one-hot encoding is selected as the encoding method. The process of encoding is described in Fig. 7.

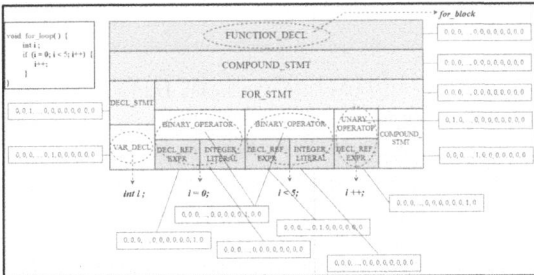

Fig. 7. One-hot encoding of nodes in the AST

3.4 Feature Learning

In this section, a framework based on an optimized skip-gram is designed to achieve simultaneous modeling of semantic and structural information in the feature learning of an AST. The developed framework uses a negative sampling-based skip-gram to learn the source code's effective AST node representations. In the framework, the AST is represented as $T = (U, n, l, d, k)$, where U represents the set of nodes, n represents the number of walks per node, l represents the walk length, k represents the neighbor size and d represents the embedding dimensions.

With the guidance of hybrid sampling in the framework, it facilitates the transformation of AST structures into skip grams. However, it ignores the node type information in Softmax. To further improve the code representation learning capacity of an AST with skip-grams, negative sampling is utilized to optimize the feature learning process. In the proposed framework, the maximum probability of having the following node u_t on the basis of the given node u is described by Eq. (3).

$$arg \max_{\theta} \sum_{u \in U} \sum_{t \in T_U} \sum_{u_t \in N_{t(u)}} \log p(u_t | u; \theta) \tag{3}$$

where $Nt(u)$ denotes *the* neighborhood of u with the *t-th* type of node and where $p(u_t | u; \theta)$ in Eq. (4) serves as a SoftMax function [30]:

$$p(u_t | u; \theta) = \frac{e^{X_{u_t} \cdot X_u}}{\sum_{v \in U} e^{X_v \cdot X_u}} \tag{4}$$

where X_{ut} represents the embedding vector for node u [31]. On the basis of the unique structure of an AST, a negative sampling method is applied to optimize skip-gram, which can achieve a more efficient and accurate representation of code. With the optimization of negative sampling, Eq. (5) is updated as follows:

$$F(X) = \log \delta(X_{u_t} \cdot X_u) + \sum_{n=1}^{N} \mathbb{E}_{v_t^n \sim P_t(v_t)}[\log \delta(-X_{v_t^n} \cdot X_u)] \tag{5}$$

where δ is the sigmoid function and $P_t(u_t)$ represents the predefined distribution from which a negative node u_t^n is drawn N times [32] (S. Kim et al. 2021). The gradients are then derived via Eq. (6).

$$\frac{\partial F(X)}{\partial X_{v_t^n}} = \left(\sigma\left(X_{v_t^n} \cdot X_u - \mathbb{I}_{u_t}[v_t^n]\right)\right)X_u$$

$$\frac{\partial F(X)}{\partial X_u} = \sum_{n=0}^{N} (\delta(X_{v_t^n} \cdot X_u - \mathbb{I}_{u_t}[v_t^n]))X_{v_t^n} \tag{6}$$

The pseudocode of the proposed framework is shown in Algorithm 4.

Algorithm 4 Negative Sampling-based Skip-Gram

Input: The abstract syntax tree $T=(U, n, l, d, k)$, a path chooses scheme C, walks per node n, walk length l, embedding dimension d, neighborhood size k

Output: The latent node embedding $X \in R^{|U| \times d}$

1: initialize X;
2: **for** $i = 1$; $i < w$; $i + +$ **do**
3:     **for** $u \in U$ **do**
4:         $walk = Walker\text{-}VisitorRandowWalk(T, C, u, l)$;
5:         $X = NegativeSamplingSkipGram(X, k, walk)$;
6:     **end for**
7: **end for**
8: **return** X;
9: **function** $NegativeSamplingSkipGram(X, k, walk)$
10:     **for** $i = 1$; $i < l$; $i + +$ **do**
11:         $u = walk[i]$;
12:         **for** $j = max(0, i - k)$; $j < min(i + k, 1)$; $j + +$ **and** $j \neq i$ **do**
13:             $w_r = MP[j]$;
14:             $X^{new} = X^{old} - \eta \cdot \partial O(X)/\partial X (Eq.6)$
15:         **end for**
16:     **end for**
17: **end function**

The hybrid sampling strategy combined with skip-gram ensures that the semantic relationships between different types of nodes can be properly incorporated into the model.

4 Experimental Setup

This study used Python 3.8 for the experiment, and all programs were coded in PyCharm 2020.3.5 (win64) and executed with Intel(R) Xeon(R) CPU E5-2680 v4 @ 2.40 GHz, NVIDIA GeForce RTX 3070 under the Ubuntu 18.04.5 operating system.

For the proposed MAST-SG model, we first conduct a deduplication verification to test the effectiveness of the AST deduplication algorithm. Then, we implement a parameter sensitivity test to examine how the different choices of parameters affect the performance of the proposed model on a frequently used dataset. Finally, we design a comparison experiment to implement source code classification. We use several current presented source code representation approaches as benchmark datasets. Additionally, several classical approaches were compared to further verify the effectiveness of the proposed model.

4.1 Dataset

The datasets originated from a pedagogical programming open judge (OJ) system by Mou [3]. There are many programs based on 104 programming questions, each of which includes 500 programs. Programs with the same programming questions are put on the same target label.

4.2 Experimental Design

The main goal of this research is to propose a novel node embedding model based on an AST for source code representation and compare it with other methods: four tokenizers with BOW, four tokenizers with embeddings, and three classical methods.

4.2.1 Verification of Deduplication

To test the effectiveness of the applied AST deduplication algorithm, the nodes before and after AST deduplication were counted and compared with those of two baseline methods, KMP1 [33] and KMP2 [34]. Table 2 provides a simple description of the chosen datasets for verification of AST deduplication in this research. The table shows that the number of AST nodes in the original datasets is more than 3000 [7]. It can reflect the truth that some ASTs of programs are too large, which is not suitable for further processing. Thus, it is necessary to conduct preprocessing to reduce the complexity and noise of the AST.

4.2.2 Parameter Sensitivity

In the developed code representation learning model, several standard parameters exist. We conduct a sensitivity analysis of MAST-SG for these parameters. Moreover, in the experiments, we vary the size of the training set from 10% to 80% and the remaining for the testing set, which can better find the best parameter values.

4.2.3 Verification of the Code Representation

To verify the effectiveness of the proposed model, we employ source code classification experiments on the classical OJ datasets with the proposed approach and various source code representation approaches. The details of those approaches are defined as follows.

First, four tokenizers are constructed and fitted with datasets using words, token words, category tokens, and AST nodes separately [35]. These four tokenizers are combined with the BOW model to achieve vectorization of the source code. Hence, four representation methods are produced: BOW with words, BOW with word tokens, BOW with category tokens, and BOW with AST nodes. Second, a simple model is developed using an embedding layer [36]. Furthermore, the embedding code model is combined with four tokenizers to produce another four representation methods: Embedding & Words, Embedding & Wokrd Tokens, and Embedding & Category Tokens. Finally, three classical code representation approaches are also surveyed and compared with the proposed methods, which are based on a support vector classifier (SVC), to verify the effectiveness of the proposed approaches. Those approaches are LDA [37], TF-IDF [38], and N-gram [39].

4.3 Evaluation Metrics

4.3.1 Evaluation of AST Deduplication

The statistics figures indicate the number of nodes of the generated AST. To verify the effectiveness of the proposed AST deduplication algorithm, the statistics of the number of nodes are listed and analyzed, and the results are compared with the original number of AST nodes.

4.3.2 Evaluation of Parameter Sensitivity

Evaluation of accuracy: The accuracy is evaluated statistically as the number of true samples out of the total number of samples. The prediction accuracy is expressed in Eq. (7).

$$Accuracy(\%) = \frac{TruePositive + TrueNegative}{Total\ Samples} \tag{7}$$

4.3.3 Evaluation of Code Representation Ability

The confusion matrix is the metric most commonly used to understand the performance of a classifier in machine learning. The details of the confusion matrix can be described as follows. It is a specific matrix used to visualize the mixtures between actual and predicted conditions, which include true negative (TN), true positive (TP), false negative (FN), and false negative (FP). TN represents the actual valid case and is predicted as unfavorable. TP represents the real solid case and is indicated as positive. FP represents the actual false case that is expected to be positive. FN represents the actual wrong case that is predicted as unfavorable. On the basis of the confusion matrix, seven metrics were generated to evaluate the performance of the proposed approach.

 i. Evaluation of accuracy
ii. Evaluation of precision

The precision (Eq. (8)) is used to evaluate the number of true positive samples among the true samples.

$$Precision(\%) = \frac{TruePositive}{TruePositive + FalsePositive} \tag{8}$$

iii. Evaluation of sensitivity

The sensitivity (Eq. (9)) is used to evaluate the number of true positive samples in the sum of the true positive and false negative models.

$$\text{Recall}(\%) = \frac{TruePositive}{TruePositive + FalsePositive} \tag{9}$$

By analyzing the importance of precision and recall, Micro-F_1 and Macro-F_1 are chosen to measure the effectiveness of the proposed strategy.

iv. F-$_1$ score

By analyzing the importance of precision and recall, the F-1 score in Eq. (10) is chosen to measure the effectiveness of the proposed strategy.

$$F_{1-score} = \frac{2 \times Precision \times \text{Re}call}{Precision + \text{Re}call} \tag{10}$$

v. Micro-F_1

The Micro-F_1 (Eq. (11)) computes the global average F-$_1$ score by counting the sums of the TP, FN, and FP.

$$Micro - F_1 = \frac{2 \times \sum_{i=1}^{n} Precision_i \times \sum_{i=1}^{n} \text{Re}call_i}{\sum_{i=1}^{n} (Precision_i + \text{Re}call_i)} \tag{11}$$

vi. Macro-F_1

Macro-F_1 (Eq. (12)) is employed to compute the macro average F-$_1$ score via the arithmetic mean of all the per-class F-1 scores.

$$Macro - F_1 = \frac{\sum_{i=1}^{n} F_i}{n} \tag{12}$$

5 Experiment Results and Discussion

5.1 Deduplication Results

In the experiments of deduplication verification, the statistical results of the original ASTs and deduplicated AST node numbers of the five samples are described in Table 2. The table shows that the number of nodes in the original AST sample varies from 3216–6221. After tree deduplication, the number of nodes ranges from 218–396. Compared with the original ASTs, the optimization rate of the deduplication algorithm is greater than 92%. The results indicate that the proposed deduplication algorithm can optimize the AST structure and eliminate redundant nodes. Compared with those of KMP1 and KMP2, the deduplication ability of KMP2 is not superior. This is because KMP1 tends to tag the subnodes of related nodes as the unknown nodes in the first traversal. KMP2 has the shortcoming of eliminating valuable nodes in the previous term of traversal. However, the proposed deduplication-based algorithm avoids the loss of useful nodes, such as some constants and variables in the simplification process, which can ensure data integrity.

Table 2. Statistics of the number of AST nodes with different deduplication algorithms

No. Program	Original_AST node	KMP1	KMP2	AST_Trie node	Optimization rate of AST_Trie
Program 1, Label 1, File 13. c	4367	259	271	282	93.5%
Program 2, Label 19, File 267. c	5179	325	344	365	93.0%
Program 3, Label 65, File 82. c	6221	341	359	396	93.6%
Program 4, Label 85, File 320. c	4385	284	297	319	92.7%
Program 5, Label 104, File 499. c	3216	193	205	218	93.2%

5.2 Parameter Sensitivity Results

Several standard parameters exist in skip-gram-based models. To better represent nodes in the AST, we conduct a sensitivity analysis of MAST-SG to confirm better parameter values. Figure 8 shows the classification results, in which the studied parameter is tested with other parameters fixed.

From Fig. 8 (a), we can see that the embedding dimensions d can achieve better classification performance when the values are set to 128 and 192. Although the performance is close between both values, we assign d to 128, as high embedding dimensions increase the computational cost. It is also clear from Fig. 8 (b) that the neighbor size k is negative for the classification results. The performance reaches its top when the values are set to 2, which also meets the characteristics of source codes, as shown in Fig. 8 (c). In addition, when the number of walks per node n and the number of walks of length l are set to 15 and 100, respectively, the classification performance can reach better results.

On the basis of the above analysis, the settings of the proposed embedding method are described below:

(1) Embedding dimensions d: 128;
(2) Neighborhood size k: 2;
(3) The number of walks per node w: 15;
(4) The walk length l: 100;
(5) The size of the negative samples: 5.

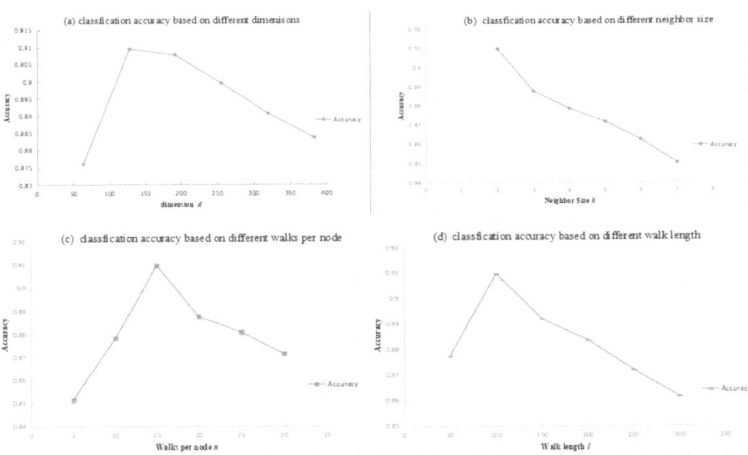

Fig. 8. Parameter sensitivity in code classification (training set: test set = 8:2)

5.3   Source Code Classification Results

Then, program classification is performed with the given parameters above. The macro-F1 and micro-F1 classification results achieved by different algorithms are listed in Table 4, and better results are marked with bold font. For a clear comparison, the results are illustrated in Fig. 9 and Fig. 10.

Figure 9 shows the program classification macro-F1 values obtained via those approaches. It is evident from Fig. 9 that the macro F-1 values experience considerable development with increasing training set from 10% to 30%. However, the macro F-1 values still need to be higher, and between these phases, the MAST-SG does not show any superiority over the other approaches. This is because the common training set contributes to overfitting of the model, which blocks the representation ability of the source code. Compared with the other approaches, the proposed MAST-SG model has greater representation ability, with the macro F-1 value reaching almost 92. However, all the remaining six methods experienced a slow development phase and exhibited poor representation ability compared with MAST-SG. From 70% to 90%, the MAST-SG decreases slightly, as the high training set led to underfitting of the mode.

Figure 10 shows the Micro F-1 values of the program classification results with different training rates. The figure shows that MAST-NG performed poorly when the training set was set between 10% and 30%. The Micro F-1 value is the same as that of Embedding & Word Tokens and Categories. When the training set rate continues to increase, it outperforms the other approaches and reaches the top 70%.

As shown in Table 3 and Figs. 9, 10, the proposed novel node embedding of the AST representation model can achieve better results in Micro-F1 and Macro-F1 than the other algorithms can achieve. Specifically, embedding-based methods tend to perform better than BOW-based methods do, as the embedding layer allows the model to learn better patterns and representations of the code. Moreover, treating code as text can achieve better classification results with the BOW-based approaches than with the other three methods. One reason can explain this result. The transformation of code into tokens and

Table 3. Program classification results of different representations

Metric	Macro-F1								Micro-F1							
Method	Text with BOW	BOW & Category Tokens	BOW & Word Tokens	BOW & AST Nodes	Embeddings & Words	Embeddings & Category Tokens	Embeddings & Word Tokens	MAST-SG	Text with BOW	BOW & Category Tokens	BOW & Word Tokens	BOW & AST Nodes	Embeddings & Words	Embeddings & Category Tokens	Embeddings & Word Tokens	MAST-SG
10%	0.5213	0.4871	0.5815	0.5614	0.5306	0.5?41	**0.6592**	0.5373	0.5472	0.5054	0.6334	0.6015	0.5772	0.6136	**0.6719**	0.5412
20%	0.6225	0.5907	0.6933	0.6739	0.6523	0.6?13	**0.7017**	0.6832	0.6514	0.6327	0.7116	0.6943	0.5714	0.6821	**0.7427**	0.6965
30%	0.7264	0.6991	0.6946	0.6875	0.7675	0.7?07	0.7686	**0.7856**	0.7483	0.7386	0.7319	0.7161	0.7839	**0.7912**	0.7849	0.7944
40%	0.7291	0.7039	0.6934	0.6918	0.7704	0.7?14	0.7693	**0.8011**	0.7497	0.7142	0.7473	0.7195	0.7967	0.8066	0.8105	**0.8163**
50%	0.7303	0.7054	0.6947	0.6992	0.7761	0.7?65	0.7718	**0.8315**	0.7592	0.7259	0.7602	0.7312	0.3104	0.8192	0.8331	**0.8386**
60%	0.7316	0.7088	0.6992	0.6999	0.7812	0.7?39	0.7672	**0.8701**	0.7671	0.7422	0.7229	0.7467	0.3356	0.8449	0.8109	**0.8729**
70%	0.7388	0.7127	0.7048	0.6975	0.7866	0.7?63	0.7712	**0.9144**	0.8119	0.7351	0.7386	0.7557	0.8449	0.8634	0.8302	**0.9187**
80%	0.7296	0.7153	0.7121	0.7105	0.7829	0.7?21	0.7759	**0.9068**	0.7964	0.7512	0.741	0.7213	0.8315	0.8704	0.8416	**0.8914**
90%	0.7287	0.7127	0.7093	0.7142	0.7713	0.7?97	0.7806	**0.8936**	0.7621	0.7307	0.7308	0.7347	0.8002	0.8515	0.8557	**0.8903**

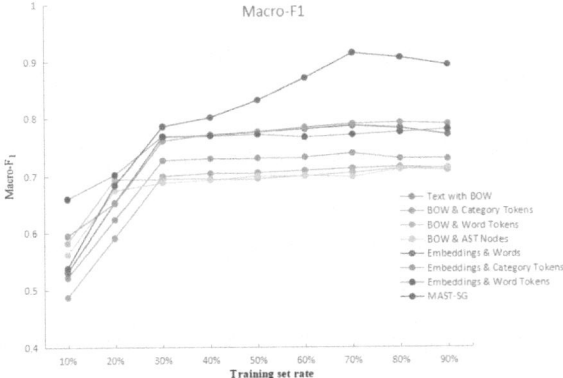

Fig. 9. Macro F-1 values of the program classification results with different training rates.

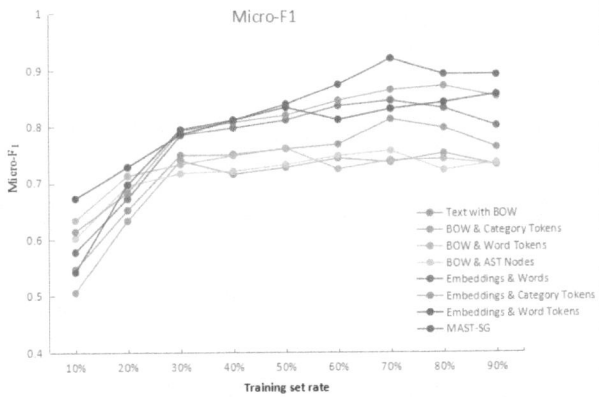

Fig. 10. Micro F-1 values of the program classification results with different training rates.

ASTs results in the loss of some characteristic information of the code, which includes syntax and semantic information. However, among the embedding-based approaches, the extraction of AST nodes can achieve better results than the other three methods do, as the embedding layer can help the model capture the syntax and semantic information hidden in the nodes of the AST. Finally, on the basis of the results, it is clear that the suitable training and test ratio is 7:3.

Furthermore, on the basis of the basis of the best training rate, the accuracy of the proposed MAST-SG was compared with that of LDA, TF-IDF, and N-gram. The results are listed in Table 4 and described in Fig. 11, which shows that the MAST-SG far exceeds the other three approaches, reaching 91%. This is because regardless of the LDA, TF-IDF, or N-gram, they do not consider the structure and semantic information in the AST of the source code. These two results verify the effectiveness of the proposed novel embedding model.

Table 4. Program classification of different approaches with a suitable training rate

Method	Accuracy	Recall
SVC+LDA	46.57	45.26
SVC+TF-IDF	78.32	76.97
SVC+N-gram	85.69	85.21
SVC+MAST-SG	**91.94**	**91.15**

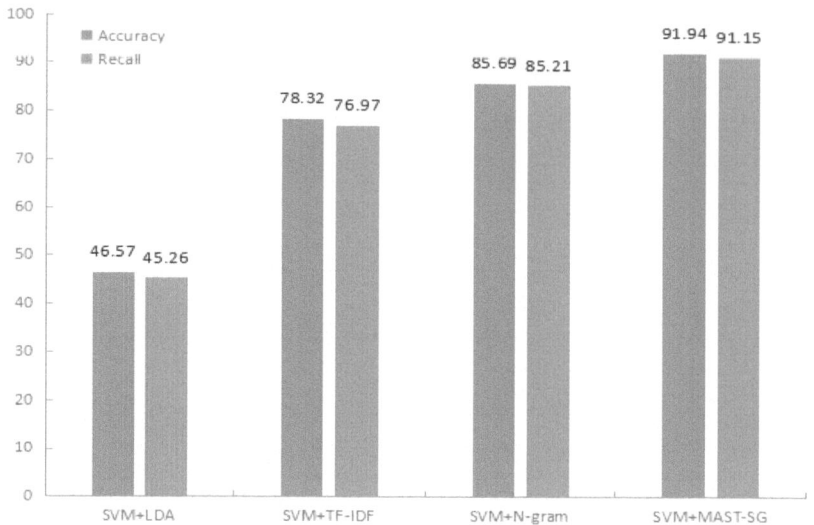

Fig. 11. Classification accuracy and recall values among different representation approaches

6　Conclusion and Future Work

In this work, we propose a novel model based on the Trie tree and optimized skip-gram to solve the challenges in source code representation. Facing the problem of duplication in traditional AST representation, the Trie tree-based deduplication algorithm was utilized to decrease duplication by combining the same prefix node and removing redundant nodes. Moreover, the Hybrid Walker and Visitor sequence sampling strategies are designed to balance homophily and structural equivalence in sampling. Finally, a negative sampling-based skip-gram was applied to capture the structure and semantic information of the source code. To verify the effectiveness of the proposed approach, comparison experiments are designed for program classification. The experimental results show that the proposed model performs better than previous methods do. Future work can be conducted in the following areas. First, a novel embedding approach can be developed to capture the influence of relationships between different node types. Second, other embedding techniques, such as code2vec and edge2vec, can be used to find the best one. Third, the embedding technique presented can be used to explore other program representations (e.g., trees and graphs). Finally, a processing model based on multiple languages can be analyzed to achieve the processing of different programming languages.

Acknowledgments. This work was supported by the National Natural Science Foundation of China under Grants 62066016 and 62266019, the Natural Science Foundation of Hunan Province of China, under Grant 2024JJ7412, the Research Foundation of Education Bureau of Hunan, China, under Grant 21C0363, and the Fundamental Research Grant Scheme by the Ministry of Higher Education, Malaysia, under grants (FRGS/1/2022/ICT02/01/1).

Disclosure of Interests. The authors have no competing interests to declare that are relevant to the content of this article.

References

1. Bengio, Y., Courville, A., Vincent, P.: Representation learning: a review and new perspectives. IEEE Trans. Pattern Anal. Mach. Intell. **35**, 1798–1828 (2013). https://doi.org/10.1109/TPAMI.2013.50
2. Samoaa, H.P., Bayram, F., Salza, P., Leitner, P.: A systematic mapping study of source code representation for deep learning in software engineering. IET Softw. **16**(4), 351–385 (2022). https://doi.org/10.1049/sfw2.12064
3. Li, L., Feng, H., Zhuang, W., Meng, N., Ryder, B.: CCLearner: a deep learning-based clone detection approach. In: Proceedings of the 2017 IEEE International Conference on Software Maintenance and Evolution, ICSME 2017, pp. 249–260 (2017). https://doi.org/10.1109/ICSME.2017.46
4. Reyes, J., Ram, D., Paciello, J.: Automatic classification of source code archives by programming language: a deep learning approach (2016). https://doi.org/10.1109/CSCI.2016.102
5. Kodhai, E., Kanmani, S., Kamatchi, A., Radhika, R., Vijaya Saranya, B.: Detection of type-1 and type-2 code clones using textual analysis and metrics. In: ITC 2010 - 2010 International Conference on Recent Trends in Information, Telecommunication and Computing, pp. 241–243 (2010). https://doi.org/10.1109/ITC.2010.55
6. Wang, P., Svajlenko, J., Wu, Y., Xu, Y., Roy, C.K.: CCAligner: a token based large-gap clone detector. In: Proceedings of the International Conference on Software Engineering, pp. 1066–1077 (2018). https://doi.org/10.1145/3180155.3180179
7. Phan, A.V., Chau, P.N., Le Nguyen, M., Bui, L.T.: Automatically classifying source code using tree-based approaches. Data Knowl. Eng. **114**, 12–25 (2018). https://doi.org/10.1016/j.datak.2017.07.003
8. Wang, W., Li, G., Ma, B., Xia, X., Jin, Z.: Detecting code clones with graph neural network and flow-augmented abstract syntax tree. In: SANER 2020 – Proceeding of the 2020 IEEE 27th International Conference on Software Anal Evolution Reengineering, pp. 261–271 (2020). https://doi.org/10.1109/SANER48275.2020.9054857
9. Mi, Q., Hao, Y., Ou, L., Ma, W.: Toward using visual, semantic and structural features to improve code readability classification. J. Syst. Softw. **193**, 111454 (2022)
10. Koschke, R., Falke, R., Frenzel, P.: Clone detection using abstract syntax suffix trees. In: Proceeding of the Working Conference on Reverse Engineering, WCRE, pp. 253–262 (1998). https://doi.org/10.1109/WCRE.2006.18
11. Rahman, M.M., Watanobe, Y., Nakamura, K.: Source code assessment and classification based on estimated error probability using attentive LSTM language model and its application in programming education. Appl. Sci. **10**(8) (2020). https://doi.org/10.3390/APP10082973

12. Pradel, M., Darmstadt, T.U.: DeepBugs: a learning approach to name-based bug detection **2** (2018). https://doi.org/10.1145/3468264.3477221
13. Okutan, A.: Use of source code similarity metrics in software defect prediction, pp. 1–14 (2018). http://arxiv.org/abs/1808.10033
14. Nan, L., Li-fang, H., Kun-feng, X., Tong, Q.: An improved algorithm based on abstract syntax tree for source code plagiarism detection (2014). https://doi.org/10.3969/j.issn.1671-1122.2014.01.009
15. Yang, Y., Ren, Z., Chen, X., Jiang, H.: Structural function based code clone detection using a new hybrid technique. In: Proceeding of the International Computational Software Applied Conference, vol. 1, pp. 286–291 (2018). https://doi.org/10.1109/COMPSAC.2018.00045
16. Duracik, M., Hrkut, P., Krsak, E., Toth, S.: Abstract syntax tree based source code antiplagiarism system for large projects set. IEEE Access **8**, 175347–175359 (2020). https://doi.org/10.1109/ACCESS.2020.3026422
17. Shi, E., et al.: CAST: enhancing code summarization with hierarchical splitting and reconstruction of abstract syntax trees (2021)
18. Mikolov, T., Chen, K., Corrado, G., Dean, J.: Efficient estimation of word representations in vector space. CoRR abs/1301.3781. http://arxiv.org/abs/1301.3781 (2013)
19. Pennington, J., Socher, R., Manning, C.D.: GloVe: global vectors for word representation. in empirical methods in natural language processing (EMNLP), pp. 1532–1543 (2014). http://www.aclweb.org/anthology/D14-1162
20. Bojanowski, P., Grave, E., Joulin, A., Mikolov, T.: Enriching word vectors with subword information. Trans. Assoc. Comput. Linguist. **5**, 135–146 (2017)
21. Grover, A., Leskovec, J.: node2vec: scalable feature learning for networks. In: Proceedings of the 22nd ACM SIGKDD International Conference on Knowledge Discovery and Data Mining, pp. 855–864, August 2016
22. Mou, L., Li, G., Zhang, L., Wang, T., Jin, Z.: Convolutional neural networks over tree structures for programming language processing. In: 30th AAAI Conference on Artificial Intelligence, AAAI 2016, pp. 1287–1293 (2016)
23. Zhang, L., Liu, D., Li, Y., Zhong, M.: AST-based plagiarism detection method. In: Wang, Y., Zhang, X. (eds.) Internet of Things. Communications in Computer and Information Science, vol. 312, pp. 611–618. Springer, Heidelberg (2012). https://doi.org/10.1007/978-3-642-32427-7_87
24. Son, J.W., Noh, T.G., Song, H.J., Park, S.B.: An application for plagiarized source code detection based on a parse tree kernel. Eng. Appl. Artif. Intell. **26**(8), 1911–1918 (2013) https://doi.org/10.1016/j.engappai.2013.06.007
25. Tao, G., Guowei, D., Hu, Q., Baojiang, C.: Improved plagiarism detection algorithm based on abstract syntax tree. In: Proceeding of the 4th International Conference onEmerging Intelligent Data and Web Technologies, EIDWT 2013, pp. 714–719 (2013). https://doi.org/10.1109/EIDWT.2013.129
26. Resmi, N.G., Soman, K.P.: Abstract syntax tree generation using modified grammar for source code plagiarism detection. IJCAT Int. J. Comput. Technol. **1**(6), 319–326 (2014). www.IJCAT.org
27. White, M., Tufano, M., Vendome, C., Poshyvanyk, D.: Deep learning code fragments for code clone detection. In: Proceeding of the 31st IEEE/ACM International Conference on Automated Software Engineering, ASE 2016 , pp. 87–98 (2016). https://doi.org/10.1145/2970276.2970326
28. Han, L., Hu, J.: GCC abstract syntax tree redundancy elimination algorithm based on keyword trie tree. Comput. Sci. **47**(09), 47–51 (2020)
29. Li, W., Guo, C., Ma, X., Pan, Y.: A strictly predefined-time convergent and noise-tolerant neural model for solving linear equations with robotic applications. IEEE Trans. Ind. Electron. **71**(1), 798–809 (2024)

30. Xie, Z., Jin, L.: Hybrid control of orientation and position for redundant manipulators using neural network. IEEE Trans. Syst. Man Cybern. Syst. **53**(5), 2737–2747 (2023). https://doi.org/10.1109/TSMC.2022.3218788
31. Peng, H., et al.: Dynamic network embedding via incremental skip-gram with negative sampling **63**, 1–19 (2020)
32. Xiao, L., et al.: Design and analysis of a novel distributed gradient neural network for solving consensus problems in a predefined time. IEEE Trans. Neural Netw. Learn. Syst. **35**(3), 3478–3487 (2024). https://doi.org/10.1109/TNNLS.2022.3193429
33. Li, X., Wang, T.T., Su, X.H., Ma, P.J.: Research on algorithm for eliminating redundant information in GCC abstract syntax tree text. Comput. Sci. **10**, 170–172 (2008)
34. Tian, B.Y., Sun, K., Chao, H.Q.: A new algorithm for simplifying GCC abstract syntax tree. Comput. Sci. **42**(S1), 516–530 (2015)
35. Azcona, D., Hsiao, I.H., Arora, P., Smeaton, A.: User2Code2vec: embeddings for profiling students based on distributional representations of source code. In: ACM International Conference Proceeding Series, pp. 86–95 (2019). https://doi.org/10.1145/3303772.3303813
36. Xingli, G., Hongliang, G., Zhan, L.: A new multi-agent reinforcement learning method based on evolving dynamic correlation matrix. IEEE Access **7**, 162127–162138 (2019)
37. Li, H.: Program code plagiarism detection system based on TF-IDF. Digit. Technol. Appl. **38**(09), 136–138 (2020). https://doi.org/10.19695/j.cnki.cn12-1369.2020.09.52
38. Li, M., Gao, Q., Ma, S., Zhang, S.K., Hu, W.H., Zhang, X.M.: Enhanced simhash algorithm for code similarity detection. Ruan Jian Xue Bao/J. Softw. **32**(7), 2242–2259 (2021). https://doi.org/10.13328/j.cnki.jos.006271
39. Zhao, W., Joshi, T., Nair, V.N., Sudjianto, A.: SHAP values for explaining CNN-based text classification models, pp. 1–17. http://arxiv.org/abs/2008.11825 (2020)

End-to-End Deep Reinforcement Learning for Inclined Ladder Steps Grasping in Humanoid Robots

Peng Lin⬤, Guodong Zhao⁽✉⁾ ⬤, Haoyu Zhang⬤, Jianhua Dong⬤,
Shuaiqi Zhang⬤, Mingshuo Liu⬤, and Xuan Liu⬤

College of Computer Science and Technology, Harbin Engineering University, No. 145 Nantong Street, Liaoyuan Street, Nangang District, Harbin, Heilongjiang, China
zhaoguodong@hrbeu.edu.cn

Abstract. When it comes to grasping inclined ladder, the conventional approach involves using pre-determined fixed actions to control the humanoid robot. However, this method necessitates manual design and imposes strict initial position requirements on the robot. To overcome this challenge, we propose an autonomous grasping method for an humanoid robot using a deep reinforcement learning algorithm called Deep Q-Network (DQN). Our approach involves developing an end-to-end network model that takes camera images and servo angles as inputs and generates optimal action policies for the humanoid robot. By utilizing this strategy, the humanoid robot achieves a high success rate in grasping the inclined ladder. To verify the effectiveness of our method, we conducted performance tests on the model in various scenarios and compared it with the fixed action control method.

Keywords: Ladder Grasping · Deep Reinforcement Learning · Humanoid Robot · End-to-end Networking

1 Introduction

1.1 A Subsection Sample

Natural disasters like tsunamis, earthquakes, and floods have inflicted significant losses on human society, making post-disaster search and rescue operations highly challenging. In such circumstances, employing humanoid robots to replace humans for hazardous tasks would be significant. Traditional manual coding methods are often utilized to design actions for humanoid robots, but they possess limitations in decision-making specificity and adaptability to dynamic scenarios. However, recent years have witnessed remarkable advancements in deep reinforcement learning techniques, empowering robots with powerful learning capabilities to enhance the intelligence of their decision-making processes.

Deep reinforcement learning [1–3] empowers robots to continuously learn and enhance their behavior strategies through trial and error and feedback. Compared to traditional coding methods that rely on manual design and adjustment of robot actions

H. Yu et al. (Eds.): CCF NCCA 2024, CCIS 2274, pp. 112–122, 2024.
https://doi.org/10.1007/978-981-97-9671-7_7

and decision rules, deep reinforcement learning enables robots to autonomously adjust and improve their behavior strategies through interactions and feedback from the environment. This autonomy enables robots to adapt to diverse tasks and scenarios, even in previously unencountered situations. Consequently, robots become better equipped to adapt to complex and dynamic environments and respond flexibly to various task requirements.

This study explores a method for action decision-making based on deep reinforcement learning for the task of humanoid robot climbing inclined ladder. Specifically, we created an environment in Webots for the Robotis-Op2 robot to perform the task of grasping inclined ladder steps, and equipped the robot's hands with grippers. The RGB images captured by the robot's head camera and the current servo angles recorded by the servo sensors were used as inputs, which were fed into a neural network. The neural network selected and executed actions from eight predefined action groups. We trained the neural network using the DQN algorithm to enable the robot to make reasonable decisions and accurately place the grippers on the target stair step.

2 Related Work

There have been some research efforts in the field of robot climbing stairs. These studies primarily utilize technologies such as point clouds [4, 5] and laser sensors to obtain geometric information about obstacles in the environment, aiding the robot in perceiving and understanding the structure and position of the stairs. By reconstructing a three-dimensional model of the stairs, the robot can generate the required motion trajectory for climbing using a predetermined decision algorithm model.

For instance, Prashanta Gyawali et al. [6] utilized point cloud data to locate stairs and estimate the centroids of the steps. They used this centroid information to control the PR2 robot for grasping the stairs. X. Sun et al. [7–9] focused on the WAREC-1 quadruped robot and estimated the relative coordinates of the stairs. They employed laser sensors for grasp correction to optimize the grasping performance. However, these methods require collecting estimated coordinates [10, 11] of the target steps and involve a substantial amount of manual programming for planning the hand motion trajectory. Therefore, these methods heavily rely on accurately estimating coordinate information and perform inadequately when faced with significant variations in the task environment.

To address these issues, this paper adopts deep reinforcement learning techniques and, for the first time, uses images and servo motors as inputs to solve the problem of human-like robots grasping inclined ladder steps. By constructing an end-to-end control decision model, our aim is to enable human-like robots to autonomously learn the ability to perform climbing tasks. By reducing reliance on manual programming and model design through the construction of an end-to-end control decision model, we provide a new approach for human-like robots to grasp inclined ladder steps by allowing them to master the skill through autonomous learning.

3   Preliminaries

3.1   Arkov Decision Process (MDP)

The mathematical foundation of reinforcement learning is Markov Decision Processes (MDPs). In an MDP, an intelligent agent interacts with an environment E and continuously receives rewards r. An MDP is typically represented by a tuple (S, A, π, R, γ), where S represents the state space, A represents the action space, π: S × A represents the policy function, R represents the reward function, and γ represents the discount factor. As shown in Fig. 1, reinforcement learning is a sequential decision-making process that aims to find a policy π that maximizes the cumulative reward, i.e., achieves maximum value.

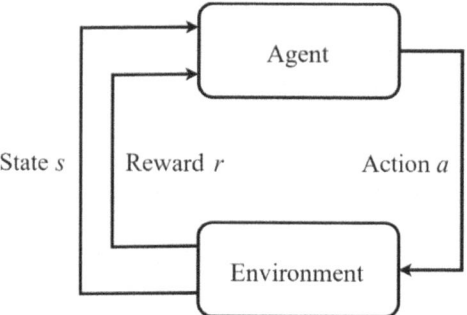

Fig. 1.   Agent-environment interaction diagram.

3.2   Deep Q-Network (DQN)

DQN, which stands for Deep Q-Network, is a pioneering work that combines deep learning with reinforcement learning. It introduced the application of deep learning in the field of reinforcement learning. DQN utilizes an experience replay mechanism, where the training data is stored in a replay buffer for later random sampling during training. The benefits of using experience replay are: 1. High data utilization, as the stored experiences can be reused multiple times for training; 2. Reduced correlation between consecutive samples, which helps to decrease variance in the training process.

As shown in Fig. 2, the DQN algorithm utilizes two neural networks: the evaluate network and the target network. These two networks have identical structures. The evaluate network is responsible for calculating Q-values for policy selection and performing Q-value iteration updates. Gradient descent and backpropagation are also performed on the evaluate network. On the other hand, the target network is used to calculate the Q-values for the next state in the TD target. The parameters of the target network are updated by copying the parameters from the evaluate network.

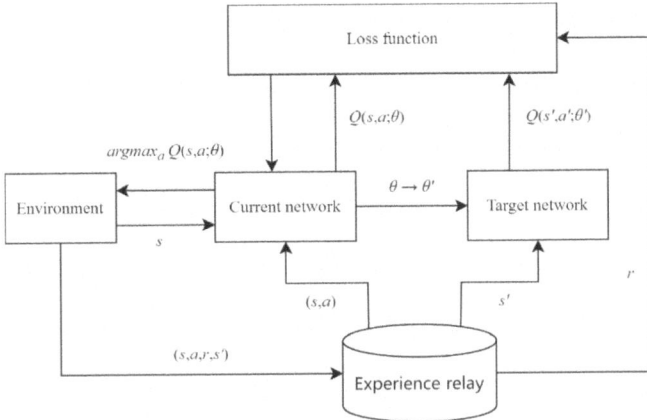

Fig. 2. DQN Algorithm Flowchart.

4   Method

4.1   Construction of Experimental Environment

As shown in Fig. 3, the problem explored in this study is how to use only the robot's vision to guide the decision-making of its arm servos. We used the Webots simulation platform, a professional software package for simulating mobile robots, to build the simulation environment. In this experiment, we constructed an inclined climbing ladder with a vertical height of 1 m, an inclination angle of 45°, a step spacing of 10 cm, a step length of 50 cm, and a step cross-section of 3 cm × 3 cm.

Fig. 3. Agent-environment interaction diagram.

The study utilized the Robotis-op2 humanoid robot, renowned for its sophisticated computing capabilities, diverse sensor array, substantial payload capacity, and dynamic motion capabilities. Figure 4 and Fig. 5 all illustrates the robot model, with 4 depicting the original Robotis-op2 robot and 5 showcasing the modified version. In the modified model, the arm shell was eliminated, and a 1-degree-of-freedom gripper was incorporated, enabling the Robotis-op2 robot to adeptly grasp the inclined ladder steps.

In addition to the gripper, we also added pressure sensors to the newly added gripper on the Robotis-op2 robot. This ensures that collision information is collected when the gripper comes into contact with the ladder steps. We utilize the collected collision information for tasks such as setting reward functions and early termination criteria, providing assistance to vision-based control.

Fig. 4. Comparison chart of Robotis-op2 robot before modification.

Fig. 5. Comparison chart of Robotis-op2 robot after modification.

We utilize the gym environment architecture as the foundation for our experiment. The state space in our experiment is divided into two parts. The first part consists of the external image obtained from the Robotis-op2 robot's head camera. This image serves as the primary input to the neural network for feature extraction. The second part includes the servo angle information of the Robotis-op2 robot. The robot has 20 servos, and each servo has a specific angle when the robot is in motion. We store the angle information of the servos as secondary inputs to the neural network, which collectively contribute to the decision-making of the action group.

The action space in this experiment consists of 2 discrete action groups, specifically designed for the Shoulder and ArmLower servo groups of the Robotis-op2 robot. Each servo group operates in a mirrored fashion, with one servo on the left side and one on the right side of the robot. Each servo group has 2 action options: increase the angle by

0.1 radians, or keep the angle unchanged. By combining these options in different ways, we obtain a total of 4 discrete action groups. By removing the set of actions where all parameters are set to 0 and 0.1, we obtain the final set of 2 action groups.

By utilizing these 8 discrete action groups, the Robotis-op2 robot will be able to select the most optimal action at each step, enabling it to accomplish the current goal of a ladder grasping task. The robot can evaluate the environment and its current state, and based on the learned policy, it can choose the action group that maximizes the chances of successfully completing the grasping task. Through reinforcement learning, the robot can iteratively improve its decision-making process and achieve better performance in accomplishing the goal of the grasping task.

4.2 Neural Network Architecture

In general, the training of a neural network is influenced by various factors, especially the network architecture, including the number of layers, the structure of each layer, and the choice of activation functions. As the number of layers increases, the neural network has a better chance of capturing complex features, which is particularly helpful for addressing complex problems. Therefore, we propose the neural network architecture used in this study, as shown in Fig. 6. We train the decision model for the robot to grasp the target rung using this network architecture.

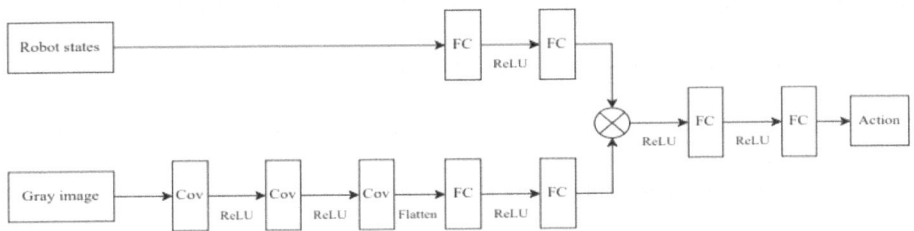

Fig. 6. Neural network architecture diagram used in this experiment.

We adopt an end-to-end neural network architecture, where we use the image information captured by the Robotis-op2 robot's head camera as the input to the network. The images captured by the Robotis-op2 robot's head camera are color images with dimensions of 160px height and 120px width. We convert these images into grayscale images with a single channel for network training.

After feature extraction through convolutional layers, the initial image features are extracted. At this point, we flatten the features into a one-dimensional tensor and aim to combine them with the robot's full-body servo angle parameters. However, it is not reasonable to directly concatenate two data with such a significant difference in magnitude and use them as the state information for training in the fully connected layers. The feature information extracted from the images can have magnitudes in the thousands, while the robot's full-body servo angle states typically have magnitudes in the tens. Therefore, before performing the concatenation operation, it is necessary to normalize the feature data to ensure a balanced representation of both types of information.

Therefore, we feed the image features and the robot's arm servo angles into two separate sets of fully connected layers. After passing through these fully connected layers, the dimensions of the two sets of data will be adjusted to the same size. Then, we concatenate the two sets of feature data and feed them into the fully connected network to calculate the expected reward Q-value. Finally, the decision model selects the action with the highest expected reward Q-value for execution.

4.3 Reward Function

The reward function plays a crucial role in guiding the Robotis-op2 robot's decision-making process during the cascaded grasping task, with the objective of training it to achieve an optimal action strategy. The experiment aims to enable the robot's gripper to successfully grasp specific steps in a cascade. Consequently, the design of the reward function focuses on directing the gripper to reach the desired position of the step accurately.

To devise the reward function, a series of tests was conducted to establish a reference point for calculating the gripper's relative position with respect to the step. The built-in GPS plugin in Webots was utilized to track the coordinates of both the gripper (x, y, z) and the target step (x′, y′, z′), serving as the two reference points. Due to the approximate parallelism between the robot and the steps, the calculation of the relative distance ignores the relative distance along the z-axis. Therefore, Eq. 1 can be used to calculate the relative distance between the gripper and the target step.

$$Distance = \sqrt{(x - x')^2 + (y - y')^2} \tag{1}$$

We have designed a segmented reward function based on the relative distance between the robotic gripper and the target rung during the grasping process. The specific formulation is presented in Eq. 2. When the relative distance exceeds 0.06 m, the robot will not receive any reward. However, when the relative distance is below 0.06 m, the robot will receive different magnitudes of rewards based on different distance stages. The purpose of this reward function is to encourage the robot to get as close as possible to the target rung, enabling efficient execution of the grasping task. By setting up segmented rewards, we can provide more positive reinforcement to encourage the robot to maintain contact with the target at closer distances.

$$Reward = \begin{cases} 100 & success \\ 2 & distance \leq 0.03\,\text{m} \\ 0.05 & 0.03\,\text{m} < distance < 0.06\,\text{m} \\ 0 & distance \geq 0.06\,\text{m} \end{cases} \tag{2}$$

During the motion process of the Robotis-OP2 robot, the range of motion for the servo motors is limited. If the desired action exceeds the maximum range of the joint servo, the robot will be unable to perform those actions and may potentially damage the servo motors. Therefore, when executing actions, the robot needs to take into account the range of motion of the servos and ensure that the intended actions fall within this range. Additionally, the joint servos of the robot may sometimes fail to fully execute the desired actions due to jamming or other reasons, which can have detrimental effects on model training. If the joint servos of the robot cannot fully execute the target actions, the model will struggle to calculate the current reward, potentially leading to less accurate or unstable trained models.

When the pressure sensor on the robot gripper is triggered or the total time steps for movement are exhausted, the robot gripper will attempt to grasp the target ladder by closing. When the gripper successfully grasps the target ladder, the robot will receive a reward of 100; otherwise, the robot will not receive any reward. The objective of the decision model is to train the model parameters by acquiring high rewards for successfully completing the specified task, thereby guiding the model towards obtaining the maximum reward.

In each training episode, we will set 20 timesteps, during which the Robotis-op2 robot will execute a selected set of actions. After each set of actions is completed, we will calculate the reward value to define the value of the current action. By assigning different reward values, the gripper controller of the robot will be guided to a state where it can accurately grasp the target ladder.

5 Simulated Experiments and Results

5.1 Training Parameter Settings

When training decision-making models using deep reinforcement learning, the choice of parameters has a significant impact on the experimental results. In our experiments, we used a host equipped with an RTX 3050 GPU and an AMD Ryzen 5 5600H CPU for training. Since the action space is discrete, we chose the DQN algorithm as the underlying algorithm for training the model. We set the learning rate to 0.0001, which is a commonly used initial learning rate. To stabilize the training process, we also introduced a target network that is updated every 100 episodes.

In reinforcement learning, the discount factor γ is used to measure the importance of future rewards. We set the discount factor γ to 0.99, which means the agent program places more importance on future rewards. By choosing an appropriate discount factor, we can balance the importance of short-term and long-term rewards.

By adjusting and optimizing these parameters, we can improve the performance of the decision-making model and make it better suited to specific tasks and environments.

5.2 Experimental Results

We conducted 4000 episodes of experiments in a simulated environment using the specified parameter settings and recorded the cumulative rewards obtained in each episode. To validate the effectiveness of the method, we calculated the average cumulative rewards over 5 repeated experiments and plotted them on a graph, as shown in Fig. 7.

By observation, it is evident that during the initial 1500 episodes of the training process, the model's cumulative rewards show significant fluctuations. This behavior is expected as the model explores the action space and refines its grasping strategy. However, between episodes 1500 and 4000, the cumulative rewards of the model stabilize,

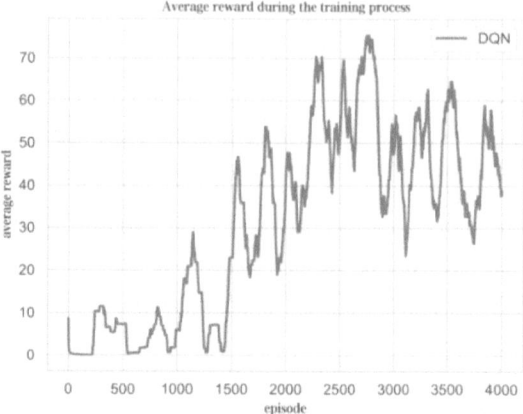

Fig. 7. The average cumulative reward per episode during the training process.

indicating that the decision model has converged. We select the decision model trained during this stable period for further comparative testing against a fixed set of actions.

As depicted in Fig. 8, the robot's initial position is at the origin O, facing the target inclined ladder along the Y-axis. Throughout the training process, the robot's initial coordinates are dynamic, varying within the range of X [−0.075, 0.075] and Y [−0.02, 0].

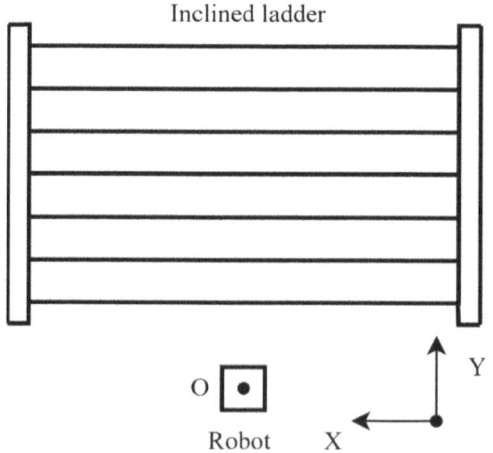

Fig. 8. Diagram of robot and inclined ladder position.

We saved the decision model when it reached a stable state and used it for comparative analysis with a fixed set of actions. The comparison scenarios are categorized based on the dynamic range of the robot's initial position, as presented in Table 1. Scenario A represents the origin point O, Scenario B represents a scenario with small fluctuations, and Scenario C represents a scenario with large fluctuations.

Table 1. The range of variation for the x and y coordinates in each scene.

Scenario	Variation range of X	Variation range of Y
A	0	0
B	[−0.03, 0.03]	[−0.01, 0]
C	[−0.075, 0.075]	[−0.02, 0]

As presented in Table 2, we conducted a total of 1000 tests for each scenario to evaluate the robot's success rate in grasping the target ladder. In Scenario A, where the robot's initial position remained constant, both the decision model and the fixed set of actions achieved a 100% success rate. However, in Scenario B, where the robot's initial position varied, the success rate of the fixed set of actions dropped to 56.4%. In contrast, our proposed decision model demonstrated adaptability to the changing environment and maintained a success rate of 79.2%.

Table 2. Comparison of success rates for DQN and fixed actions in each scene over 1000 grasping attempts.

Model	A	B	C
DQN	100%	79.2%	76.6%
Fixed actions	100%	56.4%	33.9%

6   Conclusion

This paper proposes an end-to-end action decision network based on DQN. By taking the images from the camera of the Robotis-op2 robot and the numerical values of each servo angle as inputs, the network outputs the action sequence to be executed next. We conducted tests on the proposed method using the Webots simulation platform and observed that employing the end-to-end decision-making network model for action selection resulted in successful grasping of the steps of a fixed inclined staircase in a given scenario.

Furthermore, we conducted tests to compare the performance of the trained decision model with a fixed set of actions. The fixed set of actions can only execute pre-determined actions, leading to a notably low success rate in grasping the ladder steps when the robot's initial position undergoes significant variation. In contrast, our proposed decision model demonstrates adaptability to environmental changes and achieves a higher success rate, even in scenarios where the robot's initial position deviates significantly.

Despite the limitations of the current experimental model, such as the potential for improving the success rate of grasping ladder steps in dynamic scenarios, we have successfully employed this method to enable the Robotis-OP2 robot to grasp target ladder steps successfully in the Webots simulation software. In future experiments, our research will continue to focus on the robot climbing process and training the robot with an autonomous climbing decision-making model. This ongoing work aims to enhance the robot's autonomy and exploration capabilities even further.

References

1. Chen, X., Atkeson, C.G., Huang, Q.: 3D model based ladder tracking using vision and laser point cloud data. In: 2015 IEEE International Conference on Robotics and Biomimetics (ROBIO), pp. 1365–1370 (2015)
2. Gyawali, P., McGough, J.: Simulation of detecting and climbing a ladder for a humanoid robot. In: IEEE International Conference on Electro-Information Technology, EIT 2013, pp. 1–6 (2013)
3. Huang, C., Wang, G., Zhou, Z., Zhang, R., Lin, L.: Reward-adaptive reinforcement learning: dynamic policy gradient optimization for bipedal locomotion. IEEE Trans. Pattern Anal. Mach. Intell. (2022)
4. Luo, J., et al.: Robust ladder-climbing with a humanoid robot with application to the DARPA robotics challenge. In: 2014 IEEE International Conference on Robotics and Automation (ICRA), pp. 2792–2798 (2014)
5. Nishikawa, K., et al.: Disaster response robot's autonomous manipulation of valves in disaster sites based on visual analyses of RGBD images. In: 2019 IEEE/RSJ International Conference on Intelligent Robots and Systems (IROS), pp. 4790–4797 (2019)
6. Saputra, A.A., Chin, W.H., Toda, Y., Takesue, N., Kubota, N.: Dynamic density topological structure generation for real-time ladder affordance detection. In: 2019 IEEE/RSJ International Conference on Intelligent Robots and Systems (IROS), pp. 3439–3444 (2019)
7. Shimooka, S., et al.: Development of automatic ladder climbing inspection robot using extension type flexible pneumatic actuators. Int. J. Autom. Mech. Eng. 19(1), 9593–9605 (2022)
8. Sun, X., et al.: Error compensation system with proximity sensors for vertical ladder climbing of the robot "WAREC-1". In: 2018 IEEE-RAS 18th International Conference on Humanoid Robots (Humanoids), pp. 40–46 (2018)
9. Sun, X., Ito, A., Matsuzawa, T., Takanishi, A.: Limb stiffness improvement of the robot WAREC-1R for a faster and stable new ladder climbing gait. J. Bionic Eng. 20(1), 57–68 (2023)
10. Sun, X., Naito, H., Namiki, A., Liu, Y., Matsuzawa, T., Takanishi, A.: Assist system for remote manipulation of electric drills by the robot "WAREC-1R" using deep reinforcement learning. Robotica 40(2), 365–376 (2022)
11. Zhao, J., Sun, J., Cai, Z., Wang, L., Wang, Y.: End-to-end deep reinforcement learning for image-based UAV autonomous control. Appl. Sci. 11(18), 8419 (2021)

Cooperative Coverage Path Planning for Air-Ground Heterogeneous Robots in Aircraft Skin Inspection Tasks

Minnan Piao, Jia Luo, Haifeng Li[✉], Yuhan Zhou, and Longfei Fan

College of Computer Science and Technology, Civil Aviation University of China,
Tianjin 300300, China
hfli@cauc.edu.cn

Abstract. Aircraft skin inspection is crucial for ensuring flight safety. Due to the complexity of aircraft structures, traditional manual operations and the use of a single type of robot such as Unmanned Ground Vehicles (UGVs) or Unmanned Aerial Vehicles (UAVs) have different degrees of limitations. Therefore, this paper proposes a method for Collaborative Coverage Path Planning (CCPP) using aerial and ground heterogeneous robots. Firstly, an obstacle map is constructed for collision detection. Secondly, viewpoints are classified based on the accessibility constraints of different robots. Then, the Digital Differential Analyzer (DDA) algorithm is combined with the Rapidly-exploring Random Tree Star (RRT*) algorithm to generate collision-free paths based on the classified viewpoints. Additionally, a time-cost matrix is built considering the differences in maneuverability among different robots. Finally, a tailored dual-population genetic algorithm is designed to achieve fast solutions for complete coverage paths, and the fitness function incorporates considerations for endurance constraints. The effectiveness of the proposed method is validated through simulation experiments in this paper.

Keywords: Aircraft Skin Inspection · Collaborative Coverage Path Planning · Heterogeneous Robots · Dual-population Genetic Algorithm

1 Introduction

Aircraft skin damage, such as scratches, cracks, lightning strikes, and dents, poses a significant threat to flight safety. Skin damage detection is crucial for ensuring flight safety and performance. Currently, the majority of aircraft skin damage detection tasks still rely on manual visual inspection, supplemented by handheld detection devices. Inspection personnel approach the areas to be inspected through methods such as walking around the aircraft, using work platforms, or climbing ropes. However, manual inspection is often unsafe, difficult to ensure coverage, subjective, and prone to issues such as false positives and false negatives.

The rapid development of robotics technology offers the possibility of effectively addressing the aforementioned issues. UGVs have strong payload and endurance capabilities, allowing them to simultaneously carry various detection devices such as high-definition cameras, active infrared thermal imagers, and laser scanners. They exhibit stable motion, enabling good positioning accuracy and ensuring operational safety. However, they cannot cover higher areas such as the aircraft's back, wings, and tail. In contrast, UAVs can flexibly cover these areas, but they have weaker payload and endurance capabilities. Therefore, there is an urgent need for UGVs and UAVs to collaborate in order to achieve comprehensive coverage detection of aircraft skin.

The CCPP of UAVs and UGVs is one of the core technologies for heterogeneous robot operations. The CCPP comprises several subproblems, including viewpoint set generation, viewpoint allocation, and path planning. Viewpoint set generation involves planning a set of sensor poses (referred to as viewpoints) capable of covering the aircraft skin surface. These viewpoints are then allocated to each robot using a certain strategy to obtain the set of viewpoints that each robot needs to traverse. Finally, the sequence for each robot to traverse the viewpoints is determined to obtain the motion paths. While viewpoint set generation is relatively independent, viewpoint allocation and path planning can be solved separately or jointly.

Existing research on CPP for aircraft skin inspection primarily focuses on single robots, utilizing either a single UGV or a single UAV. Viewpoint generation for UGV-based CPP is addressed in [1], where a mixed sampling method combining dual and primal sampling is proposed to meet photographic requirements and practical constraints. Subsequently, [2] plans a coverage path based on the generated viewpoints from [1], reducing UGV waypoints and path length through a viewpoint merging strategy validated in apron environment experiments. Adaptive primal sampling methods are introduced in [3] and [4], ensuring coverage by resampling regions with finer resolution until sufficient samples are obtained. Due to the limitations of Travelling Salesman Problem (TSP) algorithms with numerous viewpoints, [3] adopts a graph search method. [5] proposes a multi-resolution hierarchical framework to enhance CPP efficiency, planning paths at high and low levels for subspaces and detailed coverage paths, respectively. Considering potential unavailability of airplane CAD models, [6] proposes a two-stage approach. Initially, a UAV-camera system follows a predefined path to quickly generate a coarse model, followed by computation of an optimal scanning path for full coverage using Monte Carlo tree search and max-min ant system strategies.

In other domains, such as large-scale terrain search and rescue, research on robot CCPP has also been conducted. In [7], the region is partitioned into multiple equally sized subregions based on the initial positions of the robots, transforming the collaborative CPP problem into multiple CPP problems. An improved ant colony algorithm is then employed for solving. [8] initially employs a mixed-integer linear programming method to assign each robot to designated regions, followed by solving individual robot CPP problems using an enhanced LKH algorithm. [9] employs cluster deep Q-learning reinforcement learning to select inspection areas for each leader in UAV groups. Subsequently, the inspection areas are further divided into several equally sized subregions, with each follower employing the Spiral-Zigzag pattern for CPP. Although separate solving can reduce problem complexity, it may result in a significantly smaller feasible solution space compared to real-world scenarios.

Due to the differences in workspace, motion characteristics, and endurance between UAVs and UGVs, only by jointly solving viewpoint allocation and path planning can a better solution be obtained, such as shorter detection time. To address this, a joint optimization strategy is employed in this study. Firstly, an environment map is constructed based on the aircraft model to facilitate collision detection and local path planning. Then, considering the different reachable spaces for UAVs, viewpoints are categorized into three types: reachable by UAVs, reachable by UGVs, and reachable by both, along with the corresponding design of a reachability matrix. Next, a time-cost matrix between viewpoints is constructed, and collision-free paths between viewpoints are planned using DDA [10] and RRT* [11] algorithms, with the robot travel time between viewpoints calculated based on path length. Finally, based on the reachability and cost matrices, a dual-chromosome genetic algorithm is designed. One chromosome records the traversal sequence of viewpoints, and the other records the allocation results of viewpoints. Population update operators are tailored to achieve relatively optimal convergence results in a short time frame.

The remainder of this paper is organized as follows. In Sect. 2, the problem is formulated. In Sect. 3, the CCPP for air-ground heterogeneous robots is presented. Simulations are performed in Sect. 4. In Sect. 5, we give the concluding remarks.

2 Problem Description

The objective of this work is to achieve efficient path planning for comprehensive coverage detection of aircraft skin by utilizing a collaborative approach between a UAV and a UGV. Currently, the detection task in this paper is for a static environment, assuming that the geometry of the aircraft is known and represented by a triangular mesh. Therefore, the planning problem is formulated as an off-line planning issue based on the aircraft model, where the UAV and UGV carry cameras respectively to conduct detour inspection around the aircraft. The purpose is to minimize the collaborative inspection time under the constraints of full coverage, robot accessibility and collision avoidance. The research work in this paper is based on the existing viewpoint, which refers to the sensor pose for aircraft skin detection, as obtained from our previous work [1]. The following is the mathematical representation of the CCPP problem:

$$\min_{X} \max_{k} \left(\sum_{i=1}^{n_{X_k}-1} \frac{D(X(k,i), X(k,i+1))}{V_k} \right), \ k \in \{0, 1\} \tag{1}$$

which is subject to the following constraints:

$$X(0,:) \cup X(1,:) = S_{\min} \tag{2}$$

$$X(0,:) \cap X(1,:) = \varnothing \tag{3}$$

$$T(X(k,:)) \leq T_k \tag{4}$$

$$h_k^{\min} \leq h_{X(k,i)} \leq h_k^{\max}, \ i = 1, 2, \cdots, n_{X_k} \tag{5}$$

where k represents the robot, $k = 0$ denotes a UAV, and $k = 1$ denotes a UGV; D is the collision-free distance matrix, V_k represents the linear velocity of the robot k, n_{X_k} represents the number of viewpoints allocated to the robot k, and $X(k, i)$ represents the ith viewpoint for the robot k; $X(0, :)$ and $X(1, :)$ represent all the assigned viewpoints for the UAV and the UGV respectively, while S_{min} represents the minimum set of viewpoints; For robot k, $T(X(k, :))$ represents the total time required to traverse the viewpoints, and T_k represents the maximum endurance time; $h_{X(k,i)}$ represents the height of the ith viewpoint for robot k. Because the UAV cannot fly in narrow spaces beneath aircraft such as the fuselage, and the sensor height on the UGV is also limited, the upper and lower bounds of $h_{X(k,i)}$, h_k^{min} and h_k^{max}, are introduced.

3 Proposed Method

This paper proposes a novel algorithm framework for cooperative coverage path planning of air-ground heterogeneous robots to solve the task of full coverage detection of aircraft skin. Firstly, an obstacle map is constructed for subsequent viewpoint classification and collision-free path generation. Additionally, considering the accessibility issues of robots, a reachability matrix is designed. Subsequently, collision detection and path planning for collision-free paths are performed using the DDA algorithm [10] and 3D RRT* algorithm [11]. Based on the collision-free paths, a time-cost matrix is constructed. Finally, a targeted genetic algorithm is designed in this paper, utilizing a dual-chromosome approach for viewpoint allocation and designing population update operators to achieve optimal convergence results within a short period of time. The proposed method framework is depicted in Fig. 1.

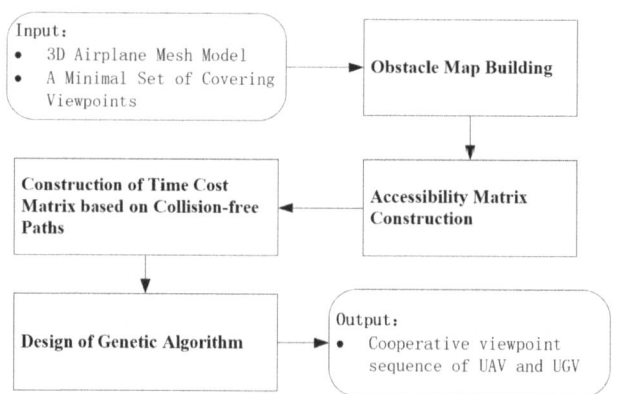

Fig. 1. Framework of CCPP for Air-Ground Heterogeneous Robots

3.1 Obstacle Map Building

In practical applications, the movement path of a robot cannot directly pass through the fuselage of the aircraft. Therefore, it is necessary to create an obstacle map to detect whether the planned robot path collides with the aircraft. If a collision with the obstacle map occurs, it indicates that the path does not meet the practical requirements, and a collision-free path needs to be replanned between the two viewpoints. Additionally, for unmanned ground vehicles, it would be unreasonable if the camera carried by the vehicle would have to pass through the aircraft model to reach the viewpoint. Such viewpoints need to be reassigned. Thus, in order to meet the practical requirements of the two collision detections mentioned above, this study first adopts the method of rasterizing the aircraft model to construct an obstacle map.

The specific implementation of the algorithm begins by initializing a three-dimensional space map slightly larger than the occupied space of the aircraft model. Each point in the aircraft model is divided by the grid size and rounded down, and the resulting value is used as the map index, representing the grid occupied by that point. Each grid has a value of 0 or 1, where 0 indicates that the grid does not contain any points from the aircraft model, and 1 indicates that the grid contains points from the aircraft model. For example, $Grid\_Map(x, y, z) = 0$ indicates that the grid with the map index (x, y, z) is not occupied by the aircraft model, while $Grid\_Map(x, y, z) = 1$ indicates that the grid with the map index (x, y, z) is occupied by the aircraft model (where $Grid\_Map$ represents the obstacle map).

3.2 Accessibility Matrix Construction

Considering the height constraints of the aircraft fuselage and the operational height of UGVs and UAVs, the set of minimum coverage viewpoints is initially divided into three categories based on a height threshold. The first category consists of viewpoints accessible only by UGVs, the second category consists of viewpoints accessible only by UAVs, and the third category consists of viewpoints accessible by both. Subsequently, for the viewpoints in the first and third categories, this study applies a constraint condition to determine UGV accessibility. Specifically, if a viewpoint, when a perpendicular line is drawn downwards, intersects with the aircraft model, that viewpoint is assigned to the category accessible only by UAVs. This is because the camera carried by the UGV would face difficulties in passing through the aircraft body to reach that viewpoint. By combining this method with the distance threshold and UGV accessibility constraint, we obtain three categories of viewpoints that meet the practical requirements. Based on these three categories of viewpoints, an accessibility matrix is constructed.

The purpose of this paper is to allocate the third view reasonably to the UGV or UAV through the designed allocation algorithm, thereby achieving optimal collaborative path planning and minimizing the overall time required for collaborative detection.

The specific construction of the accessibility matrix involves using a two-dimensional matrix with two rows and N columns (where N represents the total number of viewpoints in the viewpoint set). The first row represents the accessibility of unmanned drones, while the second row represents the accessibility of unmanned vehicles. If a viewpoint is accessible, the corresponding element is set to 1; otherwise, it is set to 0. For example, if the Nth viewpoint is accessible by the unmanned vehicle, then $AM[1, N] = 1$ (where AM represents the accessibility matrix).

3.3 Construction of Time Cost Matrix Based on Collision-Free Paths

Considering the direct connection between viewpoints, there is a possibility of collision with the aircraft model. This paper makes the following processing:

a) If there is no collision between the viewpoints connected directly without intersecting with the aircraft model, the Euclidean distance of that path is considered as the distance cost between the two viewpoints.
b) If there is a collision between the viewpoints connected directly and the aircraft model, a collision-free path is planned using the combination of DDA algorithm [10] and 3D RRT* algorithm [11]. The length of this path is considered as the distance cost between the two viewpoints.

Furthermore, considering the different mobility performance of UAV and UGV, simply using Euclidean distance to calculate the distance cost between two viewpoints to ensure the shortest collaborative inspection time is not in line with the actual needs. Therefore, this paper introduces velocity constraints of the robots and converts the distance cost into time cost to make it more practical.

The specific implementation of the algorithm adopts two N-by-N two-dimensional arrays, which are used to store the time cost for the unmanned ground vehicle and unmanned aerial vehicle respectively. Firstly, the Euclidean distance of the collision-free path between the two viewpoints is calculated, and then this result is divided by the linear velocity of the robot to obtain the time cost matrix of the robot.

3.4 Design of Genetic Algorithm

Dual-Chromosome Construction. Since the genetic algorithm in this study aims to solve the multi-robot collaborative path planning problem, it not only needs to obtain a time-optimal path that achieves full coverage but also needs to address the problem of which robot visits which viewpoints. By adding an allocation chromosome to the traditional single-viewpoint chromosome, this problem can be effectively solved. The viewpoint chromosome refers to a list used for the traversal order of viewpoints, and the allocation chromosome corresponds to the viewpoint chromosome through list indices, indicating that a specific viewpoint is allocated to a particular robot. Given the existing accessibility matrix, the first and second category viewpoints can be easily assigned to the corresponding robots. For the third category viewpoints, both types of robots can access them. During initialization, this study assumes that each viewpoint in the third category is randomly assigned to a robot with a probability of 0.5. Thus, an initial allocation chromosome is obtained.

Population Initialization. The original viewpoint chromosome is rearranged, and the corresponding assignment chromosome is constructed after each rearrangement, so as to obtain the initial viewpoint chromosome population and the corresponding assignment chromosome population.

Fitness Function Design. By combining each viewpoint chromosome with its corresponding allocation chromosome, the viewpoints belonging to different robots are separated, resulting in paths for different robots. Then, based on the time cost matrix for each robot, the time cost of the corresponding robot's path is calculated. At the same time, considering the endurance time constraint for each robot, this paper compares the obtained time cost with the corresponding endurance time. If both robot paths exceed their respective endurance time, priority is given to optimizing the drone's path. In this case, the reciprocal of the time cost for the drone is returned as the fitness value. If both robot paths are within their respective endurance time, the reciprocal of the longer time cost between the two robot paths is returned as the fitness value, as the objective of this paper is to minimize the longest time.

Selection of Parents. The parents are generated in two ways. One way is by directly inheriting the elite population, which involves sorting the chromosomes based on the fitness calculation results and selecting the top-ranked chromosomes as elite individuals for direct inheritance. The other way is through roulette wheel selection, where there is a tendency to select chromosomes with higher probabilities as parents. Specifically, the chromosome with the longest shortest time, or the shortest completion time, is chosen as a parent.

Population Update. Considering the randomness and uncertainty of genetic algorithms, in order to achieve faster and better convergence, this study adopts a combination of micro fine-tuning and macro major adjustments, and utilizes the segments of the best chromosomes as much as possible for genetic operations. The specific implementation is as follows:

For the elite population obtained through direct inheritance, each elite individual is subjected to elite-guided mutation to generate N offspring. The mutation positions are randomly generated for each mutation.

For the parent population selected through roulette wheel selection, a parent generates ten offspring [12]. Among these offspring, two are generated by performing a reordering operation on the parent chromosome. The generation of the other four offspring is based on a comparison between the parent and the best solution. If the parent is the same as the best solution, the offspring are generated by performing mutations near the positions of 0%, 25%, 50%, 75%, and 100% of the chromosome. If they are different, the offspring are generated by inserting segments from the best solution into the parent chromosome segments of 0–25%, 25–50%, 50–75%, and 75–100%, respectively. Additionally, the four offspring are generated by reversing the segments of the parent chromosome within the ranges of 0–25%, 25–50%, 50–75%, and 75–100%.

The final population consists of three parts: the elite population, the offspring obtained by fine-tuning and mutation of the elite population, and the offspring obtained through the roulette wheel selection from the parents selected in ten lifecycles.

Finally, the convergence result is obtained by constantly iterating the loop.

4   Result

4.1   Setup

In this paper, the effectiveness of the proposed planning method is verified by simulation experiments. The computer configuration used features an Intel® Core™ i7-8550U CPU with a base clock speed of 1.80 GHz and a maximum turbo frequency of up to 2.00

GHz., and the programming language employed is Python. The target structure is a high-precision Boeing 737–300 aircraft model with a wingspan of 28.8 m, a fuselage length of 33.5 m and a height of 11.0 m. The UAV is set with a linear speed of 1 m/s and a flight endurance of 1800 s, while the UGV is set with a linear speed of 0.6 m/s and a flight endurance of 7200 s. The minimum coverage viewpoint set used in the experiments is derived from our previous work [1], comprising 380 viewpoints. The map grid size is 0.1 m. Among the genetic algorithm parameters, the population size is 300, the number of iterative updates is 2000, and the mutation rate is 0.2.

4.2 Simulation Experiment

Due to the higher operational altitude of the UAV and the lower operational altitude of the UGV, this study defines the airspace above 2 m as the UAV operational space and the airspace below 4 m as the space where the UGV with lifting equipment can operate. Through accessibility calculations, the quantities of the three categories of viewpoints are 160, 194, and 26, respectively. The simulation results are shown in Figs. 2, 3 and 4 (where the coordinate unit dm represents decimeters):

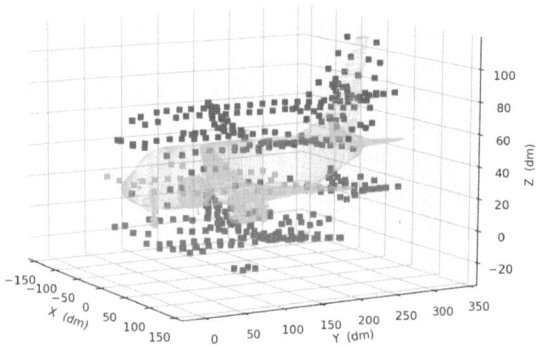

Fig. 2. Viewpoint classification results. The colors red, green, and orange respectively represent the first, second, and third categories of viewpoints.

To evaluate the performance of the proposed algorithm, this study compared it with traditional TSP algorithms that separately solve the viewpoints for UAV and UGV. The evaluation metric used is the collaborative completion time. The results are shown in Table 1. Compared to traditional TSP algorithms that separately solve the problem, although the proposed method in this study has slightly longer program execution time, it achieves shorter collaborative completion time for the planned paths.

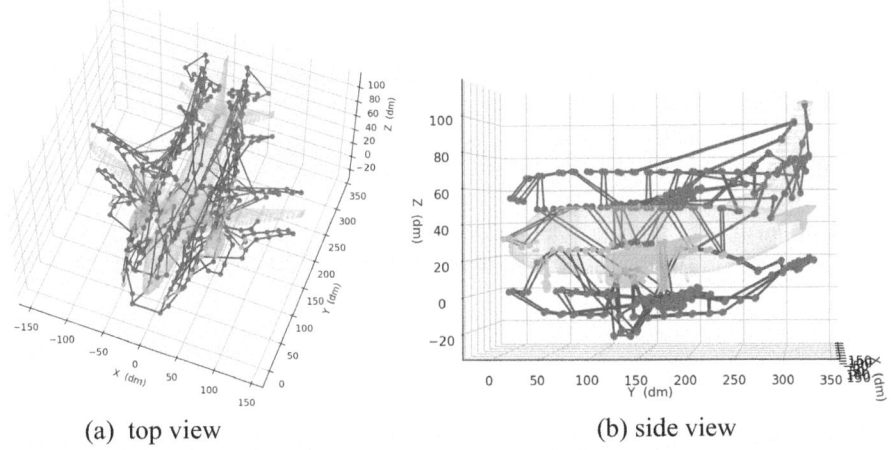

(a) top view   (b) side view

Fig. 3. Collaborative path planning results. The colors blue and green represent the paths of the UGV and UAV respectively. (Color figure online)

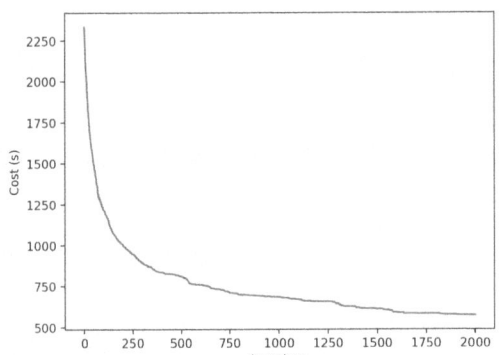

Fig. 4. Changes in the collaborative completion time during the iteration process.

Table 1. Planning result.

Algorithm	Program execution time	Collaborative completion time
Our CCPP	124.6 s	552 s
Traditional TSP	117.5 s	579 s

5   Conclusion

For aircraft skin inspection tasks, this paper proposes a CCPP method using aerial and ground heterogeneous robots. The method takes into account the maneuverability, endurance, and collision constraints between the robots and the aircraft. It consists of four steps: constructing an obstacle map, generating an accessibility matrix, creating a collision-free path time-cost matrix, and designing dual-chromosome genetic algorithm. The ultimate goal is to minimize the collaborative inspection time while ensuring full coverage of the aircraft skin. The effectiveness of the proposed method has been validated through simulations and comparative experiments. Future research on heterogeneous robot CCPP includes collision avoidance between robots, path smoothing, optimization, and handling other uncertainties.

Acknowledgement. The authors acknowledge the financial support from the Fundamental Research Funds for the Central Universities under Grant 3122022QD09.

References

1. Piao, M., Li, H., Zhang, F., Du, X., Yue, T., Huang, Y.: Viewpoint generation for coverage inspection of airplane fuselage surface based on unmanned ground vehicles. In: 2023 42nd Chinese Control Conference (CCC), Tianjin, China, pp. 4413–4418 (2023)
2. Piao, M., Du, X., Li, H., et al.: Design of an unmanned ground vehicle system for image coverage of fuselage surface (accepted for publication). Manufacturing automation (2024, to be published)
3. Almadhoun, R., Taha, T., Gan, D., Dias, J., Zweiri, Y., Seneviratne, L.: Coverage path planning with adaptive viewpoint sampling to construct 3D models of complex structures for the purpose of inspection. In: 2018 IEEE/RSJ International Conference on Intelligent Robots and Systems (IROS), Madrid, Spain, pp. 7047–7054 (2018)
4. Silberberg, P., Leishman, R.C.: Aircraft inspection by multirotor UAV using coverage path planning. In: 2021 International Conference on Unmanned Aircraft Systems (ICUAS), Athens, Greece, pp. 575–581 (2021)
5. Cao, C., Zhang, J., Travers, M., Choset, H.: Hierarchical coverage path planning in complex 3D environments. In: 2020 IEEE International Conference on Robotics and Automation (ICRA), Paris, France, pp. 3206–3212 (2020)
6. Sun, Y., Ma, O.: Automating airplane scanning for inspection or 3D model creation with a UAV and optimal path planning. Drones 6(4), 87 (2022). ARTN
7. Gao, C., Kou, Y., Li, Z., et al.: Optimal multirobot coverage path planning: ideal-shaped spanning tree. Math. Probl. Eng. 23(9), 1024–1032 (2018)
8. Modares, J., Ghanei, F., Mastronarde, N., Dantu, K.: UB-ANC planner: energy efficient coverage path planning with multiple drones. In: 2017 IEEE International Conference on Robotics and Automation (ICRA), Singapore, pp. 6182–6189 (2017)
9. Mou, Z., Zhang, Y., Gao, F., Wang, H., Zhang, T., Han, Z.: Three-dimensional area coverage with UAV swarm based on deep reinforcement learning. In: ICC 2021 - IEEE International Conference on Communications, Montreal, QC, Canada, pp. 1–6 (2021)
10. Grayeli, R., Hatami, K.: Implementation of the finite element method in the three-dimensional discontinuous deformation analysis (3D-DDA). Int. J. Numer. Anal. Methods Geomech. **32**, 1883–1902 (2010)
11. Pharpatara, P., Hérissé, B., Pepy, R., Bestaoui, Y.: Shortest path for aerial vehicles in heterogeneous environment using RRT*. In: 2015 IEEE International Conference on Robotics and Automation (ICRA), Seattle, WA, USA, pp. 6388–6393 (2015)
12. Tappe, M., Dose, D., Alpen, M., Horn, J.: Autonomous surface inspection of airplanes with unmanned aerial systems. In: 2021 7th International Conference on Automation, Robotics and Applications (ICARA), Prague, Czech Republic, pp. 135–139 (2021)

A Redis Cache-Based Approach to High Concurrency Response in Applications of Large Language Models

Peng Liu[1] , Zhaoyang Xu[2(✉)] , and Changjie Wang[2]

[1] Changchun University of Science and Technology, Changchun, China
[2] North China Electric Power University, Beijing, China
xuzhaoyang@ncepu.edu.cn

Abstract. The deployment of large language models (LLMs) in various domain applications has led to an urgent need for efficient response. This study proposes a Redis cache-based conversation matching approach to address the challenge of highly concurrent processing, with the aim of optimizing the power load and response speed of a large language model (LLM). The server- side maintains a cache of high-frequency requests, and responds to user requests by optimally combining the predicted content that reaches the matching threshold through the Bloom filter's filtering algorithm and Jaccard's similarity algorithm. This approach saves costs on API recalls and makes response times much faster. To comprehensively assess the effectiveness of this method, a series of experiments are designed that not only focus on the balance between speed and accuracy, but also clarify the best processing solution by analytically comparing different solutions side-by-side. The experiments begin with the determination of Bloom's initial screening threshold and the Jaccard similarity threshold directly on the target text and dataset, followed by a side-by-side comparison of the accuracy and processing speed between this study's method and other methods. The two datasets selected for the experiments include the public English dataset STS-B, which is derived from the GLUE benchmark for standard text semantic similarity evaluation, and a self-constructed Chinese text dataset covering a variety of industry domains, such as pharmaceuticals, the internet, electrical power safety, and machinery manufacturing, with a total of 3,000 samples, which is used to test the model's ability to generalize to multii-domain text similarity understanding. The experimental results demonstrate that the method significantly improves the processing power and response speed, and effectively addresses the challenge of high concurrent response.

Keywords: Large language model · High concurrency · Redis cache · Bloom filter · Jaccard similarity algorithm

ⓒ The Author(s), under exclusive license to Springer Nature Singapore Pte Ltd. 2024
H. Yu et al. (Eds.): CCF NCCA 2024, CCIS 2274, pp. 133–149, 2024.
https://doi.org/10.1007/978-981-97-9671-7_9

1 Introduction

The potential of large language models (LLMs) has been demonstrated in a wide range of application scenarios, including but not limited to risk prediction, text generation, and sentiment analysis. This is due to their superior ability to understand and generate natural language. As these applications are deployed, achieving fast and efficient processing while maintaining high quality responses has become a critical issue [4]. In the context of large-scale concurrent requests, traditional processing methods frequently encounter difficulties in meeting the dual requirements of performance and efficiency [2]. Servers are required to process a considerable number of requests within a limited timeframe, which places significant pressure on computing resources and may impact the end-user experience [13]. Consequently, the development of an approach that enhances the response speed while reducing the power load is of paramount importance for large-scale language modeling applications [24].

In practice, large language models deployed on different hardware platforms may take seconds or even tens of seconds to process text generation [31]. However, the use of caching techniques can significantly reduce the response time. This study proposes an innovative Redis cache-based conversation matching approach with the objective of optimizing the response speed and power load of large language models in highly concurrent environments. The server-side component rapidly filters the predicted content, akin to the text of user-requested conversations, from the cached database. This is achieved by maintaining a cache of high-frequency request results and optimally combining the Bloom filter filtering algorithm with the Jaccard similarity algorithm. This enables the server to respond rapidly to user demands and reduce the computing pressure on the server [25]. The findings of this research are not only significant in enhancing the efficacy of large-scale language models in practical applications, but also offer novel insights into the development of highly concurrent processing technology [1]. The comprehensive design is illustrated in Fig (see Fig. 1).

2 Related Work

2.1 Accelerating LLM Inference

The reasoning time and resource consumption of large language models (LLMs) represent significant research topics within the current field of artificial intelligence. To address this problem, researchers have proposed a range of optimization methods, including quantization, pruning, compression and reference inference.

Quantization is a method of reducing the size of a model by reducing the number of bits required to represent each parameter. In a 2023 paper, Dettmers and Zettlemoyer noted that quantization can be effective in reducing the memory footprint of a model, but this can create a trade-off between accuracy and the memory footprint [5]. The key to quantification is to find a balance between maintaining model performance and reducing resource consumption.

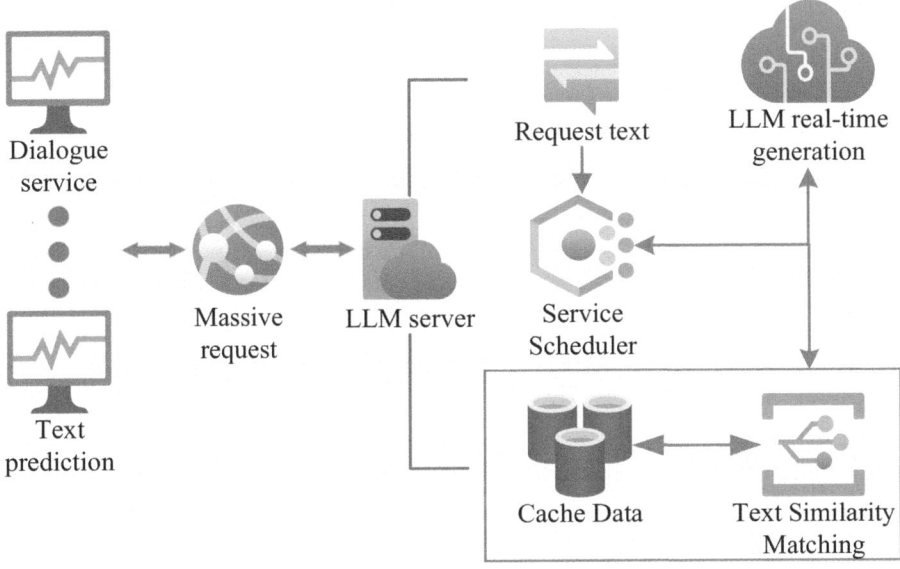

Fig. 1. Overall architecture.

Pruning methods permit the generation of sparse pretrained transformer (GPT) models without the necessity of retraining. Frantar and Alistarh demonstrated SparseGPT in 2023, which is a method to achieve model sparsity without the necessity of retraining. Pruning reduces model complexity by removing some parameters from the model, but requires careful design to maintain model performance [7].

Xu et al. (2020) introduced the concept of model compression, which typically involves modifying the model structure and may affect the model's generalizability [28]. Reference reasoning is employed to accelerate the reasoning process by utilizing existing reasoning results. Yang et al. (2023) introduced this approach, which requires sufficient reference data to support it [30].

Several other optimization methods have been included in recent studies. For example, Chang et al. investigated efficient cueing methods in a 2024 study [3]. These methods can be broadly classified into two categories: computationally efficient cues and design efficient cues. The former encompasses various techniques for compressing cues, whereas the latter employs automatic cue optimization techniques.

Furthermore, Yang et al. (2024) proposed a method utilizing LLM optimizers, designated OPRO [29]. This method describes the optimization task in natural language and generates new solutions at each optimization step, which are then evaluated and added to the cue for the next optimization step.

The selection and application of these methods must be determined on the basis of specific application scenarios and requirements to achieve optimal performance and efficiency balance. In practice, a combination of optimization methods may be employed to reduce the inference time and resource consumption of the model.

2.2 Widespread Application of Caching

A great deal of research has been conducted on the subject of handling highly concurrent requests for web services. One of the key areas of focus has been the application of caching techniques. Memory is widely acknowledged to be an essential resource [14]. The initial findings of this research provide promising evidence that the throughput limits of cache-enabled communication systems are in fact far beyond what can be achieved by existing networks [12]. Currently, Redis cached databases are widely used in web services. Researchers have employed Redis to increase the responsiveness and processing power of the system by developing efficient data storage structures and update strategies.

Additionally, with the rapid growth of mobile data traffic and the limitations of traditional network capacity, intranet caching technology has garnered significant interest [15,17,26]. In addition to reducing network costs, intranet caching improves the efficiency of content delivery and the user experience. It reduces network bandwidth usage, decreases data transmission latency, and lowers energy consumption, primarily by storing popular or frequently accessed content on the internal nodes of the network [22].

Caching techniques have also received significant attention in improving content delivery in wireless networks. The exponential growth of mobile data traffic, coupled with the limitations of traditional networks, has led to the realization that traditional methods, such as increasing the physical rate at the service side or adding base stations, are costly and have limited effectiveness [20]. Consequently, caching techniques are regarded as an effective means to address this gap, and a variety of innovative application scenarios and ideas have been proposed [6]. First, caching can be performed at the evolved packet core (EPC) for deep caching to reduce content delivery delay [27]. This move helps to optimize network performance and improve the efficiency of content delivery. Second, caching at the base station relieves congestion on the backhaul link, which in turn improves the overall throughput of the network [9]. This caching operation at the base station optimizes the data transmission path, reduces the data transmission delay, and improves the user experience. Furthermore, caching on mobile devices utilizes communication between devices to further enhance content delivery efficiency [8].

In conclusion, previous research has proposed a range of techniques and methodologies to address the challenge of high concurrent request processing. This provides a theoretical and practical foundation for this paper, which proposes a similar matching method for dialog text based on Redis.

3   Approach

3.1   System Architecture

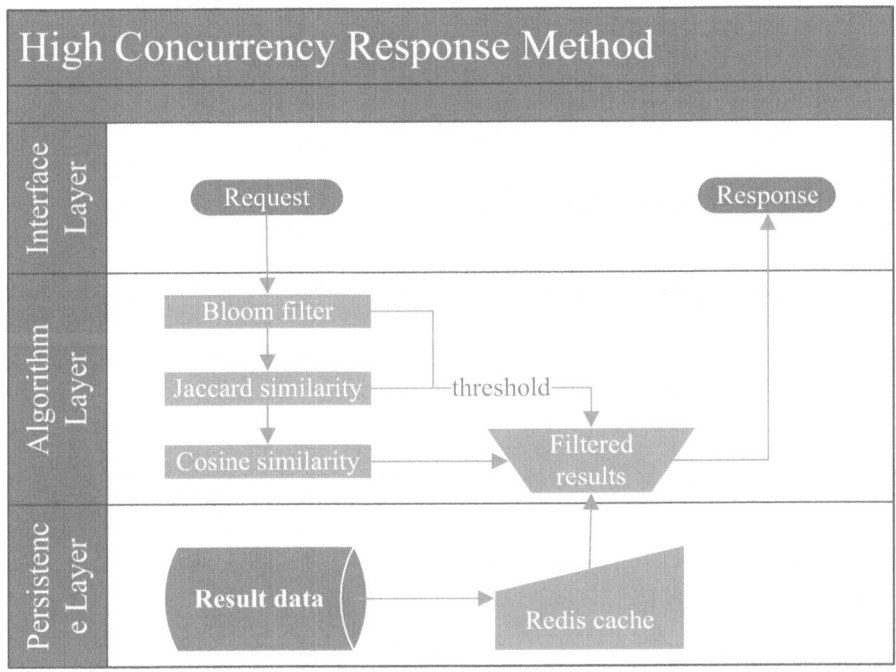

Fig. 2. Key architecture.

The architectural design of this method comprises two key components: the Redis data layer and the similarity matching algorithm layer. Redis data layer, one of the system's principal components, not only receives and responds promptly to data from the interface layer but also effectively manages memory usage, preventing data expiration and overflow by setting reasonable expiration policies and elimination mechanisms. The result cache stores three data points: the request text, the vector of the request text, and the result of the text generation. The vector of the requested text and the generated result are generated by the large language model in the past when processing the text and stored in the result cache. Furthermore, the persistence function of Redis ensures the security and stability of data, and protects the data from being lost even in the case of system failure. The similarity matching algorithm layer is the core processing logic, which includes the Bloom filter matching algorithm and the Jaccard similarity algorithm. The system initially employs the Bloom matching module to prefilter the request prediction content in comparison with the high-frequency request cache. If the matching threshold is reached, the Jaccard similarity algorithm is then utilized for accurate matching. The prediction result that best meets the user's needs is selected as the response object to improve the matching accuracy and efficiency. The key architecture is shown in Fig (see Fig. 2).

3.2 Algorithm Design

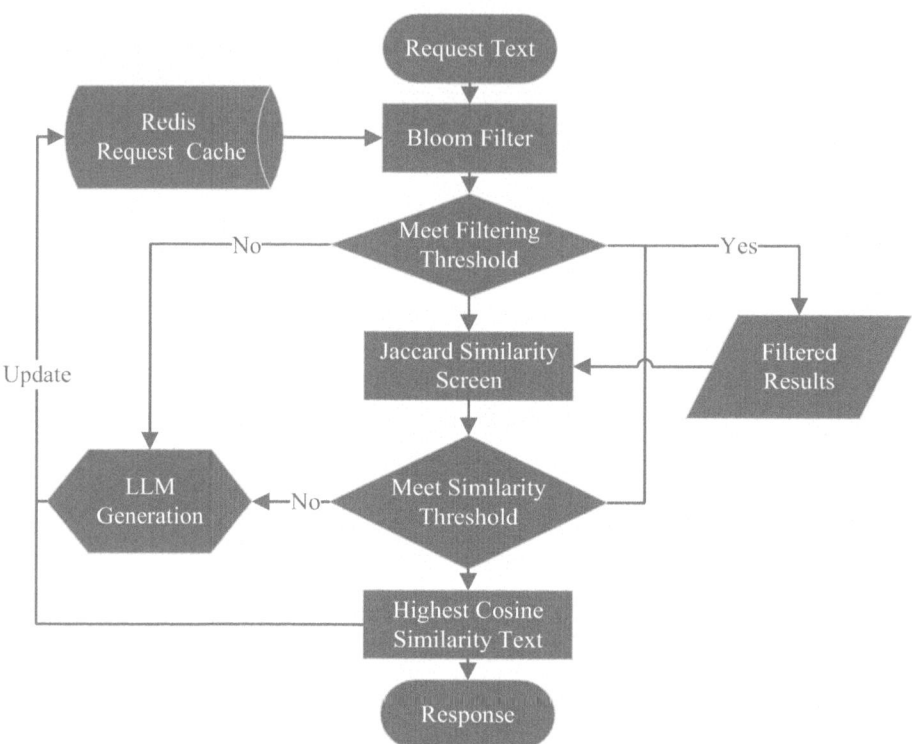

Fig. 3. Algorithmic flowchart

The text is initially preprocessed, including text cleaning and stemming extraction [19]. The filtering algorithm based on the Bloom filter subsequently filters the request content and the historical request items in the HF request cache. If the coverage of the filtering results exceeds the expected threshold, the next step of similarity matching is performed. However, if the coverage of text words in the filtering result is less than the filtering threshold T_b, the similarity matching is abandoned in real time by the AI algorithm. The cached content that meets the screening requirements is then subjected to the Jaccard similarity optimization algorithm, which determines whether the similarity with the requested content reaches the similarity threshold T_j. If this is the case, the result corresponding to the historical request item with the highest similarity is returned. The overall flow of the algorithm is depicted in Fig (see Fig. 3).

A Bloom filter is a space-efficient probabilistic data structure for determining whether an element belongs to a set. It allows for the possibility of false positives, but not misses. When dealing with large-scale datasets, Bloom filters are used to quickly detect the existence of an element by mapping the element to a bit array via multiple hash functions [16]. In this study, we utilize Bloom filters to design text filtering algorithms to determine whether a specified number of words, denoted by N_r, are present in a collection of texts from a history of request items in a cache. First, we map the words of all the texts in the database into a bit array of m bits via k independent hash functions. For the text of the words of the historical request entries, we similarly process them through these hash functions and check whether the corresponding bits have been set to 1. If all the relevant positions are set to 1, we consider that the word may exist in the cache. The number of words covered in this manner, denoted by N_m, can be used to compute the degree of string matching between the request text and the set of cached texts, which is referred to as the Bloom coverage, denoted by P_m.

$$P_m = \frac{N_r}{N_m} \tag{1}$$

Importantly, owing to the false-alarm property of Bloom filters, the final results may require subsequent verification through an exact matching process. Furthermore, the selection of the size of the bit array (m) and the number of hash functions (k) has a direct effect on the false alarm rate (p) [10]. Consequently, these parameters must be meticulously designed in accordance with the specific requirements of the intended application to achieve an optimal balance between the false alarm rate and performance.

$$m = \frac{n \ln p}{(\ln 2)^2} \tag{2}$$

$$k = \frac{m}{n} \ln 2 \tag{3}$$

The Jaccard similarity is computed for cache history result entries that reach the top portion of the Bloom filtering threshold T_b. The Jaccard similarity algorithm is based on set theory and is used to measure the similarity between two sets [18]. Its core principle is to determine the similarity by comparing the ratio of the number of common elements to the total number of elements in two sets. In particular, the Jaccard coefficient is defined as the ratio of the size of the intersection of two sets to the size of the concatenated set. This ratio ranges from 0 to 1, with 0 indicating no similarity and 1 indicating complete similarity. In this study, similarity is determined by comparing the number of common elements between sets. In this study, the Jaccard similarity algorithm is employed to assess the degree of similarity between a long string of text in a sentence and a long string of text stored in a database. First, the intersection and concatenation of the set of words in the query text with the set of words in each text in the database are calculated. The Jaccard similarity coefficient is defined as the ratio of the size of the intersection of two sets to the size of the concatenation, with the following formula:

$$J(A, B) = \frac{|A \cap B|}{|A \cup B|} \tag{4}$$

In Eq. (4), A and B represent the sets of words in the query text and database text, respectively. By calculating the Jaccard similarity coefficients between all database texts and query texts, we can sort them and select the text with the highest similarity as the optimal match. Furthermore, the Jaccard threshold, denoted by T_j, may be set. Only when the similarity coefficient exceeds this threshold will the two texts be considered to be significantly similar. The TF-IDF vectorization of several historical request items and request contents with high similarity is subsequently employed to construct a text vector table [23]. The cosine similarity is then calculated by the vector, with the result of the historical request item with the highest cosine similarity taken as the response content [21]. The formula for calculating cosine similarity is as follows:

$$CS = \frac{\boldsymbol{A} \cap \boldsymbol{B}}{\|\boldsymbol{A}\|\|\boldsymbol{B}\|} \tag{5}$$

In Eq. (5), the symbols $\boldsymbol{A}$ and $\boldsymbol{B}$ represent the TF-IDF vectors of two texts. The symbol $\boldsymbol{A} \cdot \boldsymbol{B}$ denotes the dot product of the two vectors, whereas the symbols $\|\boldsymbol{A}\|$ and $\|\boldsymbol{B}\|$ represent the Euclidean norms of the vectors.

4   Experiments and Results

4.1   Purpose

To assess the efficacy of the analogous matching algorithm in this study, it is essential to consider both the speed and accuracy components. Consequently, it is imperative to identify the optimal method with the highest accuracy within an acceptable processing time range. To this end, it is necessary to conduct screening experiments on both the target text and the dataset text directly, employing Bloom's first screening threshold T_b to ascertain the optimal threshold. Similarly, the same method can be employed to determine the optimal Jaccard screening threshold T_j. Finally, the experiment quantifies the final accuracy and processing time of the similarity matching method.

4.2   Experimental Dataset

To comprehensively test the effectiveness of the proposed Redis cache-based high concurrency response approach in large-scale language modeling applications, two datasets were used for the experiments: a public English dataset, STS-B (Semantic Textual Similarity Benchmark), and a self-constructed Chinese text dataset covering a wide range of industry domains. The Antic Textual Similarity Benchmark (ATSB) was employed in conjunction with a self-constructed dataset comprising 3,000 Chinese text datasets, which spans a diverse range of industry domains, including pharmaceuticals, the internet, electric power safety, machinery manufacturing, and so forth.

The STS-B dataset is derived from the GLUE benchmark, which is a widely recognized standard tool for evaluating the semantic similarity of text. The dataset comprises pairs of English sentences, each pair of which is accompanied by a manually assigned similarity score on a scale of 0–5. This reflects the degree of semantic relatedness between the sentences, with 5 indicating complete semantic agreement and 0 indicating no association. The diversity of STS-B and its specialized scores make it an ideal tool for testing the ability of models to understand text similarity in different contexts.

The self-built Chinese dataset was created with the specific aim of simulating the need for text similarity matching that may arise in real-world multidomain applications. It contains texts from a variety of sources, including summaries of specialized literature from the pharmaceutical industry, posts from internet technical forums, descriptions of electrical safety codes, and technical specifications from the field of machine manufacturing. The texts cover a wide range of industry terminologies and technical languages, and have diverse practical application scenarios, thus providing a rigorous testing environment for the model's generalizability and domain adaptability.

By conducting experiments on these two datasets, we validate the generalisability of the method in a crosslinguistic environment and examine its applicability and accuracy within a specific domain.

4.3   Experimental Indicators

The primary metrics of the experiment are the accuracy rate, hit ratio and processing time. Among these, the precision rate is defined as the ratio of the quantity of screened similar texts to the total quantity of screened texts. The success of matching is determined by comparing the manual labeling score of the text in the dataset and the target text, with a score of 3.5 or above indicating a match. The weighted average of the precision rate is then calculated on the basis of the manual labeling score.

$$Accuracy = \frac{\sum_{i=1}^{n} score_i \cdot I(score_i \geq T)}{\sum_{i=1}^{n} score_i} \tag{6}$$

In Eq. (6), n represents the total number of texts. The manual labeling score for the ith text is denoted by $score_i$. The indicator function, denoted by $I()$, returns a value of 1 when the condition in parentheses is valid and 0 otherwise. T is the score threshold, which is valid when the score is greater than 3.5 and not valid otherwise.

The hit ratio is the proportion of similar texts in the total dataset after screening. In the collection of texts screened by a certain threshold, the number of similar standards reaches the proportion of the total collection of similar standards, but also needs to be considered in the context of the weighted share of the manual labeling score.

$$HitRatio = \frac{\sum_{i=1}^{n} score_i \cdot I(score_i \geq T)}{\sum_{i=1}^{N} score_i \cdot I(score_i \geq T)} \tag{7}$$

In Eq. (7), N is the number of texts in the dataset that meet the similarity criterion, and n is the number of filtered texts that meet the similarity criterion. $score_i$ is the manual labeling score corresponding to the ith text. $I()$ is an indicator function that returns a value of 1 when the condition in the parentheses is valid, and 0 otherwise. T is the score threshold, and the condition is valid when the score is greater than 3.5, and not valid otherwise. The above formula ensures that the degree of text similarity is more fully considered in the calculation by weighting the accuracy of each text with its manual annotation score.

4.4   Experimental Design

In this study, we use a dataset consisting of paired texts as well as manually assigned text similarity scores, typically using scores greater than or equal to 3.5 to define similarity [11]. In this study, the accuracy of the Bloom screening and Jaccard similarity matching results is used to confirm the reasonable interval of the Bloom screening threshold Tb and the Jaccard similarity threshold T_j. A series of four experiments was designed to provide a systematic evaluation and validation of the proposed Redis cache-based response optimization approach for large language models. The following provides an overview of each experiment.

Experiment 1: Bloom Filter Initial Screening Threshold Testing. This experiment examines the initial filtering efficiency of Bloom filters in a highly concurrent environment. By setting different filtering thresholds, the experiment measures the performance of the Bloom filter in rapidly identifying and excluding nonmatching requests, while also recording instances of genuine requests that are omitted to optimize the balance between accuracy and efficiency of filtering.

Experiment 2: Jaccard similarity exact match threshold testing. This experiment examines the application of the Jaccard similarity algorithm to cached content for filtering. The experiment evaluates the accuracy and efficiency of the Jaccard similarity algorithm in determining matching results under different threshold calculations by applying different Jaccard similarity threshold calculations to cached content.

Experiment 3: Comprehensive Performance Test. Experiment 3 integrates the Bloom filter and the Jaccard similarity algorithm to simulate the entire process of handling highly concurrent requests in real applications. The experiment seeks to identify the optimal parameter configuration by adjusting the similarity thresholds of the Bloom filter and the Jaccard algorithm to achieve an optimal balance between the false alarm rate and performance. Moreover, the experiment quantifies the specific improvement in processing time and accuracy of the final method, thereby ensuring the feasibility and effectiveness of the research method in practical deployment.

Experiment 4: Multimodal horizontal comparison test. The final experimental model employs the Chinese and English datasets to cross-sectionally compare caching with only bloom filtering, caching with only Jaccard similarity matching, and caching with the combined approach.

4.5   Results and Analysis

Table 1. Results of Experiment 1.

No.	P_m	accuracy	HitRatio
1	$0.05 < $ P_m ≤ 0.15	0.118	0.001
2	$0.15 < $ P_m ≤ 0.25	0.411	0.033
3	$0.25 < $ P_m ≤ 0.35	0.472	0.088
4	$0.35 < $ P_m ≤ 0.45	0.633	0.164
5	$0.45 < $ P_m ≤ 0.55	0.718	0.221
6	$0.55 < $ P_m ≤ 0.65	0.727	0.214
7	$0.65 < $ P_m ≤ 0.75	0.713	0.202
8	$0.75 < $ P_m ≤ 0.85	0.571	0.071
9	$0.85 < $ P_m ≤ 0.95	0.131	0.060

In Experiment 1, the accuracy rate and hit ratio under different Bloom coverage P_m ranges were calculated for the Bloom-screened text set, and Table 1 was obtained. The reasonable interval of the Bloom screening threshold T_b was subsequently analyzed by plotting line graphs of the accuracy rate and hit ratio.

A line graph is constructed to plot Bloom's accuracy and hit ratio against the results presented in Table 1. Figure 4 shows that the reasonable interval of Bloom's screening threshold T_b is (0.35, 0.75). Furthermore, the final accuracy after item 8 of the experiment decreased. This is because higher screening thresholds lead to the screening of similar texts in the dataset, which in turn leads to a reduction in the final accuracy rate.

In Experiment 2, the precision rate and hit ratio for different Jaccard similarity J ranges are calculated for the Jaccard-matched text set. The results are presented in Table 2.

The reasonable interval of the Jaccard similarity threshold, designated T_j, is subsequently analyzed through the construction of line graphs representing the precision rate and hit ratio. Figure 5 shows that the threshold of the Jaccard similarity J is set within a reasonable interval of T_j at (0.45, 0.65). The final accuracy rate of experimental items 7 to 9 is negatively impacted by the higher threshold of Jaccard similarity, which results in the inability to match similar

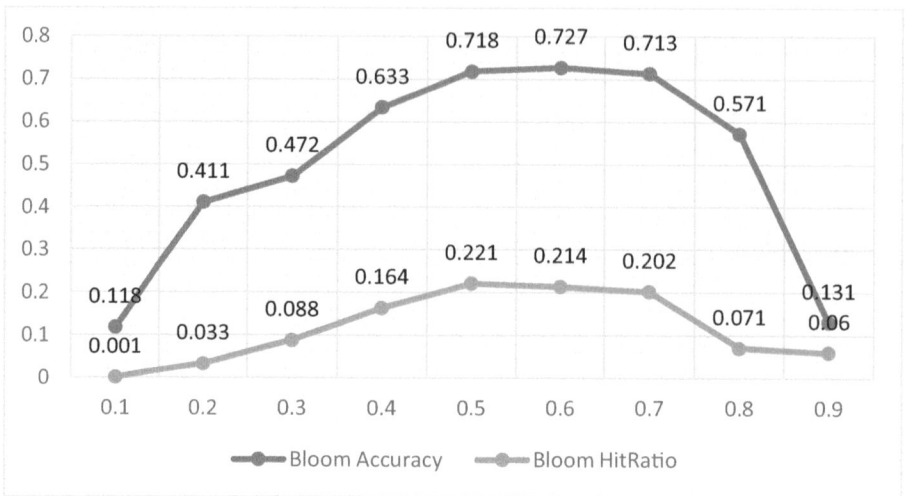

Fig. 4. Bloom Accuracy and HitRatio.

Table 2. Results of Experiment 2.

No.	J	Accuracy	HitRatio
1	$0.05 < J \leq 0.15$	0.057	0.003
2	$0.15 < J \leq 0.25$	0.124	0.042
3	$0.25 < J \leq 0.35$	0.312	0.097
4	$0.35 < J \leq 0.45$	0.685	0.113
5	$0.45 < J \leq 0.55$	0.882	0.236
6	$0.55 < J \leq 0.65$	0.893	0.234
7	$0.65 < J \leq 0.75$	0.815	0.122
8	$0.75 < J \leq 0.85$	0.745	0.102
9	$0.85 < J \leq 0.95$	0.717	0.051

text within the dataset. This, in turn, leads to a reduction in the final accuracy rate.

Experiment 3 was conducted to determine the final accuracy rate of this study's method. Two combinations of different Bloom screening thresholds T_b and different Jaccard similarity thresholds T_j were tested in the reasonable intervals, and the results are plotted in Table 3.

As illustrated in Fig. 6, the accuracy rate of this method reaches 0.921 when the Bloom screening threshold $T_b = 0.4$ and the Jaccard similarity threshold $T_j = 0.6$, and the accuracy rate of this method reaches 0.921. In this experiment, we investigated the impact of varying similarity thresholds on the final accuracy rate. The experimental results indicate that a reasonable threshold interval for

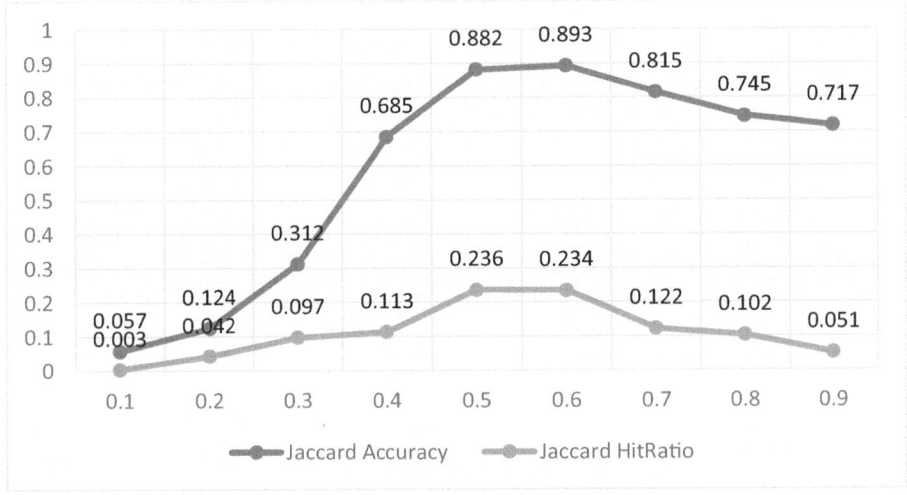

Fig. 5. Jaccard Accuracy and HitRatio.

Table 3. Results of Experiment 3.

No.	Tb	Tj	Final Accuracy	Time
1	0.4	0.5	0.894	2.1 s
2	0.4	0.6	0.921	1.3 s
3	0.5	0.5	0.803	1.5 s
4	0.5	0.6	0.910	2.0 s
5	0.6	0.5	0.868	1.3 s
6	0.6	0.6	0.813	1.1 s
7	0.7	0.5	0.773	1.0 s
8	0.7	0.6	0.756	0.9 s

Bloom filter coverage is (0.35, 0.75). Within this interval, the accuracy remains high, particularly when the coverage is in the range of (0.45, 0.65), where the accuracy reaches more than 70%. The Jaccard similarity is within the range of (0.45, 0.65), during which the assessment of text similarity achieves a high accuracy rate, particularly when the Jaccard similarity is within the range of (0.55, 0.65), with an accuracy rate greater than 80%. However, after both exceed a reasonable threshold, the accuracy begins to decrease. This may be because too high a similarity threshold results in the erroneous exclusion of many text items that are actually similar.

In Experiment 4, to investigate the effectiveness of different caching strategies in improving the high concurrency responsiveness of large-scale language modeling applications, we designed and implemented an experiment that comparatively analyzes the use of Bloom filter caching only, Jaccard similarity caching

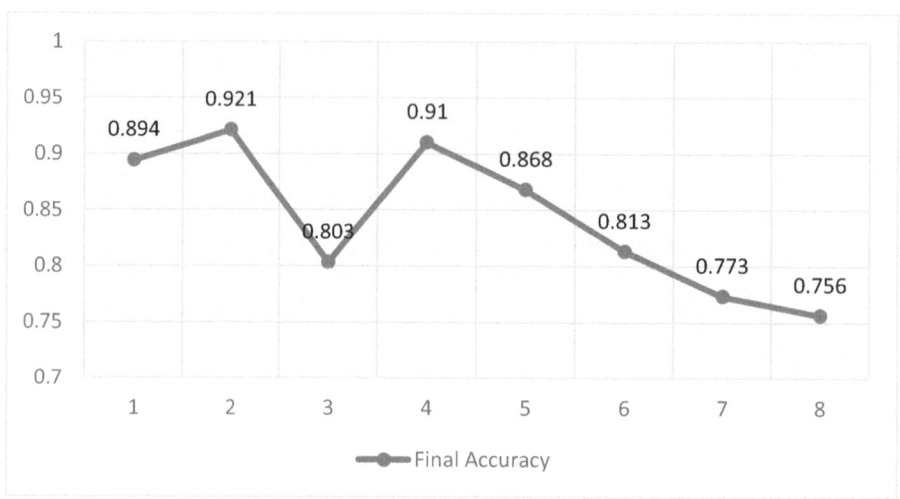

Fig. 6. Jaccard Accuracy and HitRatio.

only, direct processing of requests with no caching, and our proposed Redis method, which combines Bloom filters with the Jaccard similarity algorithm. This caching strategy. Experimental group I, which employs only the Bloom filter caching strategy, is responsible for the initial filtering of requests. Experimental group II employs the Jaccard similarity caching strategy for the exact matching of requests. Experimental group III employs the combination strategy proposed in this paper, namely, a two-step process comprising fast screening by the Bloom filter and subsequent exact matching via the Jaccard similarity algorithm. Experimental group IV does not utilize any caching mechanism, and each request is processed by directly invoking the Large Language Model API. The experimental steps entail implementing the aforementioned strategy for each of the experimental groups, with the objective of processing the same number and type of requests. It is necessary to record the accuracy and average running time of each group of experiments. The data from each group must be analyzed, and the performance differences under different strategies must be compared.

Table 4. Results of Experiment 4.

No.	Experimental Group	Final Accuracy	Time
1	Bloom-only caching	0.61	0.9 s
2	Jaccard-only caching	0.68	3.1 s
3	Bloom-jaccard caching	0.93	1.3 s
4	No-caching	1.00	14.7 s

The results of Experiment 4, as presented in Table 4, demonstrate a significant disparity in the precision and timeliness of the four processing methods. The combined approach of Bloom and Jaccard, in conjunction with similarity matching, represents the optimal solution in terms of precision and timeliness.

By conducting experiments with a combination of different Bloom filter coverage thresholds and Jaccard similarity thresholds, the highest accuracy rate of 92.1% was achieved when the Bloom filter coverage threshold was 0.4 and the Jaccard similarity threshold was 0.6. This finding demonstrates that by carefully selecting the thresholds, the accuracy of text similarity judgment can be effectively improved. Furthermore, the processing time of this method is approximately one to two seconds. Consequently, in contrast to large language models that require seconds or even tens of seconds to generate text in real- time [31], this method can increase the response speed by several times when the text is cached. Consequently, this method can address the issue of high concurrency response in the application of large language models. In conclusion, the experimental results provide valuable reference data for text similarity analysis and demonstrate that the accuracy rate and processing time can be significantly improved by adjusting the threshold value. These findings are highly important for the design of efficient and accurate LLM application systems, particularly when a large amount of data needs to be processed.

5   Conclusion

This study proposes a Redis cache-based similar text matching method to address the challenges of applying large language models in highly concurrent environments. The method achieves fast response and efficient processing of user requests by maintaining a high-frequency request cache and combining the Bloom filter and the Jaccard similarity algorithm. The core goal of this research is to optimize the power load and response speed of the LLM to address the performance and efficiency pressure from large-scale concurrent requests. The preprocessing of text, the utilization of a Bloom filter for initial screening and the application of a Jaccard similarity algorithm for exact matching have been demonstrated to successfully improve the matching accuracy and efficiency. The experimental results indicate that the method significantly improves the processing power and response speed of the system while ensuring high accuracy. In particular, the system shows excellent performance, with an accuracy of 92.1% within a reasonable interval of Bloom filter coverage and the Jaccard similarity threshold. Furthermore, this study examines the significance of caching strategies, including reasonable expiration times and elimination mechanisms, as well as the persistence features of Redis, which play a pivotal role in the stability and reliability of the system. The scalability and high -availability design of the system also facilitates possible future expansion and upgrades.

In summary, this research not only provides an effective technical solution for high concurrency processing of large language models, but also provides valuable experience and insights for researchers in related fields. The implementation of well-designed caching strategies and similarity matching algorithms can significantly enhance the performance of large-scale language models in practical applications, which is highly important for promoting the development of natural language processing technology.

References

1. Bang, F.: Gptcache: an open-source semantic cache for llm applications enabling faster answers and cost savings. In: Proceedings of the 3rd Workshop for Natural Language Processing Open Source Software (NLP-OSS 2023), pp. 212–218 (2023)
2. Cai, H., Xu, B., Jiang, L., Vasilakos, A.V.: IoT-based big data storage systems in cloud computing: Perspectives and challenges. IEEE Internet Things J. **4**(1), 75–87 (2017). https://doi.org/10.1109/JIOT.2016.2619369
3. Chang, K., Xu, S., Wang, C., Luo, Y., Xiao, T., Zhu, J.: Efficient prompting methods for large language models: a survey. arXiv preprint arXiv:2404.01077 (2024)
4. Darwish, T.S., Bakar, K.A.: Fog based intelligent transportation big data analytics in the internet of vehicles environment: motivations, architecture, challenges, and critical issues. IEEE Access **6**, 15679–15701 (2018)
5. Dettmers, T., Zettlemoyer, L.: The case for 4-bit precision: k-bit inference scaling laws. In: International Conference on Machine Learning, pp. 7750–7774. PMLR (2023)
6. Forecast, G., et al.: Cisco visual networking index: global mobile data traffic forecast update, 2017–2022. Update **2017**, 2022 (2019)
7. Frantar, E., Alistarh, D.: Sparsegpt: massive language models can be accurately pruned in one-shot. In: International Conference on Machine Learning, pp. 10323–10337. PMLR (2023)
8. Golrezaei, N., Molisch, A.F., Dimakis, A.G., Caire, G.: Femtocaching and device-to-device collaboration: a new architecture for wireless video distribution. IEEE Commun. Mag. **51**(4), 142–149 (2013)
9. Golzerai, N., Shanmugam, K., Dimakis, A., Molisch, A., Caire, G.: Femtocaching: wireless video content delivery through distributed caching helpers. In: Proceedings of IEEE INFO COM (2012)
10. Guo, D., Wu, J., Chen, H., Yuan, Y., Luo, X.: The dynamic bloom filters. IEEE Trans. Knowl. Data Eng. **22**(1), 120–133 (2009)
11. Ham, J., Choe, Y.J., Park, K., Choi, I., Soh, H.: Kornli and korsts: new benchmark datasets for Korean natural language understanding. arXiv preprint arXiv:2004.03289 (2020)
12. Lee, D., Chu, W.W.: Semantic caching via query matching for web sources. In: International Conference on Information and Knowledge Management (1999). https://api.semanticscholar.org/CorpusID:7024601
13. Leverich, J., Kozyrakis, C.: Reconciling high server utilization and sub-millisecond quality-of-service. In: Proceedings of the Ninth European Conference on Computer Systems, pp. 1–14 (2014)
14. Li, S., Maddah-Ali, M.A., Yu, Q., Avestimehr, A.S.: A fundamental tradeoff between computation and communication in distributed computing. IEEE Trans. Inf. Theory **64**(1), 109–128 (2017)
15. Llorca, J., et al.: Dynamic in-network caching for energy efficient content delivery. In: 2013 Proceedings IEEE INFOCOM, pp. 245–249. IEEE (2013)

16. Luo, L., Guo, D., Ma, R.T., Rottenstreich, O., Luo, X.: Optimizing bloom filter: challenges, solutions, and comparisons. IEEE Commun. Surv. Tutor. **21**(2), 1912–1949 (2018)
17. Maddah-Ali, M.A., Niesen, U.: Fundamental limits of caching. IEEE Trans. Inf. Theory **60**(5), 2856–2867 (2014)
18. Maher, K., Joshi, M.S.: Effectiveness of different similarity measures for text classification and clustering. Int. J. Comput. Sci. Inf. Technol. **7**(4), 1715–1720 (2016)
19. Mikolov, T., Chen, K., Corrado, G., Dean, J.: Efficient estimation of word representations in vector space. arXiv preprint arXiv:1301.3781 (2013)
20. Paschos, G.S., Iosifidis, G., Tao, M., Towsley, D., Caire, G.: The role of caching in future communication systems and networks. IEEE J. Sel. Areas Commun. **36**(6), 1111–1125 (2018)
21. Pennington, J., Socher, R., Manning, C.: GloVe: global vectors for word representation. In: Moschitti, A., Pang, B., Daelemans, W. (eds.) Proceedings of the 2014 Conference on Empirical Methods in Natural Language Processing (EMNLP), pp. 1532–1543. Association for Computational Linguistics, Doha (2014). https://doi.org/10.3115/v1/D14-1162. https://aclanthology.org/D14-1162
22. Shanmugam, K., Golrezaei, N., Dimakis, A.G., Molisch, A.F., Caire, G.: Femtocaching: wireless content delivery through distributed caching helpers. IEEE Trans. Inf. Theory **59**(12), 8402–8413 (2013)
23. Sitikhu, P., Pahi, K., Thapa, P., Shakya, S.: A comparison of semantic similarity methods for maximum human interpretability. In: 2019 Artificial Intelligence for Transforming Business and Society (AITB), vol. 1, pp. 1–4. IEEE (2019)
24. Smith, S., et al.: Using deepspeed and megatron to train megatron-turing nlg 530b, a large-scale generative language model. arXiv preprint arXiv:2201.11990 (2022)
25. Sreekanti, V., et al.: Cloudburst: stateful functions-as-a-service. arXiv preprint arXiv:2001.04592 (2020)
26. Tandon, R., Simeone, O.: Harnessing cloud and edge synergies: toward an information theory of fog radio access networks. IEEE Commun. Mag. **54**(8), 44–50 (2016)
27. Woo, S., Jeong, E., Park, S., Lee, J., Ihm, S., Park, K.: Comparison of caching strategies in modern cellular backhaul networks. In: Proceeding of the 11th Annual International Conference on Mobile Systems, Applications, and Services, pp. 319–332 (2013)
28. Xu, C., Zhou, W., Ge, T., Wei, F., Zhou, M.: Bert-of-theseus: compressing bert by progressive module replacing. arXiv preprint arXiv:2002.02925 (2020)
29. Yang, C., et al.: Large language models as optimizers. arXiv preprint arXiv:2309.03409 (2023)
30. Yang, J.A., Huang, J., Park, J., Tang, P.T.P., Tulloch, A.: Mixed-precision embedding using a cache. arXiv preprint arXiv:2010.11305 (2020)
31. Yang, J., et al.: Harnessing the power of llms in practice: a survey on chatgpt and beyond. ACM Trans. Knowl. Disc. Data (2023)

Research and Application Status of a Tendency-Based Gas Source Localization Strategy via the Active Olfaction Method: A Review

Li Wang[1], Ziyu Ren[1], Shurui Fan[2(✉)], and Lili Xu[1]

[1] School of Electronic Information Engineering, Hebei University of Technology,
Tianjin 300401, China
[2] Innovation and Research Institute, Hebei University of Technology,
Shijiazhuang 050299, China
fansr@hebut.edu.cn

Abstract. The leakage and diffusion of chemical gases pose a serious threat to humans. Rapid and accurate localization of gas sources is currently a research hotspot. On the basis of different tendencies, robot active olfactory localization strategies can be divided into three types: chemical, wind, and information tendencies. The research background and current status of gas source localization are first introduced. Second, the main improvement directions of the three strategies, along with their representative contents, are summarized and compared. The application and existing problems of the gas source localization method, which is based on a tendency strategy, are analyzed in various scenarios with different gas environments, obstacles, and time-varying sources. Finally, future development trends of gas source localization strategies are discussed and compared, providing a reference for further research on gas source localization issues. In this review, the technical bottlenecks and challenges in the gas source localization field are analyzed in depth, which provides a reference for future research and innovation.

Keywords: Gas source localization · Robot active olfaction · Chemotaxis strategy · Anemotaxis strategy · Infotaxis strategy

1 Introduction

The leakage and dispersion of flammable, explosive, toxic gases pose significant safety risks to industrial operations as well as everyday life. Rapidly identifying and locating gas leak sources [1], followed by prompt intervention, are crucial in mitigating these risks. Gas source localization generally refers to the process of "actively" discovering and tracking gas diffusion plumes through sensor networks or mobile robots [2]. This method involves continuously searching to ultimately determine the location and relevant parameters of the gas leakage source [3]. Gas leak source localization based on sensor networks typically requires sensors that have been settled in the search area in advance.

These sensors collect information about gas leaks in the environment [4], which are then used as initial conditions for models to perform source localization [5]. This method has a relatively low cost, but its accuracy is limited because it heavily relies on a predeployed sensor network [6]. Additionally, it fails to integrate well with surrounding environmental information and may even waste some environmental data. However, these issues can be effectively addressed by the robot active olfactory method, which boasts excellent maneuverability.

The robot active olfaction method involves single or multiple robots actively searching for gas leak sources by combining environmental information on plume diffusion. Compared with sensor networks, robot active olfaction methods have advantages such as flexibility, real-time responsiveness, and autonomy [7]. It has become a research hotspot in the field of gas source localization. Common active olfaction localization strategies for robots include the following three types: the chemotaxis strategy [8, 9], the anemotaxis strategy [10], and the infotaxis strategy [11]. The chemotaxis strategy involves localizing gas sources by detecting the chemical composition of diffusing substances [12]. It often relies on determining the gas concentration to determine if a threshold is reached [13], thus confirming whether the robot has arrived at the gas source location. Classic chemotaxis algorithms include the Moth algorithm [14], the Coli algorithm [15, 16] and the Lobster algorithm [17]. The chemotaxis strategy, along with many early olfaction localization algorithms inspired by biological behaviors, evolved from the predatory and mate-seeking behaviors of organisms [18]. This strategy can rapidly and accurately localize gas sources when gas concentration information is abundant. However, in sparse gas environments, where gas concentration gradients are not continuous [19], strategy performance decreases. Building upon the chemotaxis strategy, researchers have proposed the wind-taxis strategy to address this issue [20]. The anemaxis strategy involves robots that combine the gas concentration with the wind direction to locate gas sources by searching against the wind. Common Anemotaxis algorithms include the Silkworm Moth Algorithm [21], Spiral-Surge Algorithm [22], and Plume Center Algorithm. The fundamental idea of the Anemotaxis strategy is to infer paths and speeds of gas diffusion on the basis of gas concentration measurements and wind field information, thereby deducing possible gas source locations [23, 24]. Although the anemotaxis algorithm overcomes the influence of sparse gas environments, turbulent gas characteristics affect the algorithm. In turbulent gases, the shape of gas plumes changes, rendering the combination of upwind searching and concentration ineffective [25]. At this time, the infotaxis strategy is developed. The infotaxis algorithm was first proposed by Massimo et al. [26] in 2007. This method, which is based on Bayesian methods [27], generates probability maps by calculating posterior probabilities to guide robots toward the most likely locations of gas sources. Compared with other types of autonomous search strategies, the infotaxis strategy performs well in autonomously searching tasks, even in sparse gas concentration environments. Although the sensors carried by a robot do not reach their threshold, the robot can still effectively accomplish its autonomous search task. The infotaxis method itself essentially follows the framework of classic infotaxis methods, which include Bayesian estimation and information entropy-based decision-making [28]. However, the infotaxis algorithm faces two issues: computational redundancy and path repetition [29]. Figure 1 illustrates the number of publications on three trend-following strategies in recent years. The graph clearly shows that in recent years, there has been a general upward trend in the use of these three strategies to address gas source localization issues, with the infotaxis strategy being the most widely applied.

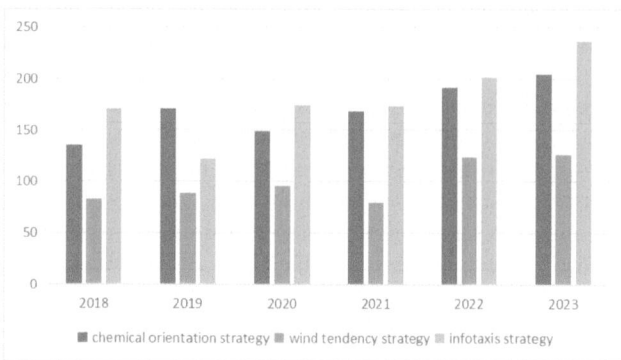

250

200

150

100

50

0

2018   2019   2020   2021   2022   2023

■ chemical orientation strategy ■ wind tendency strategy ■ infotaxis strategy

Fig. 1. Evolutionary trajectories of three directional strategies

At present, there are many gas source location technologies in the robot active olfaction field, which can adapt to different application scenarios with excellent gas source location effects. The latest research achievements, technological advancements, and method applications of typical convergent gas source location strategies are summarized, and technical bottlenecks and gas source location challenges are deeply analyzed to provide directions for further research and innovation. The rest of this paper is organized as follows. Sections 2 to 4 are devoted to comprehensive summaries of these three trend-following strategies and their respective avenues for improvement. In Sect. 5, the advantages and disadvantages of these strategies are discussed. In Sect. 6, conclusions are drawn, and future directions for improvement are outlined. The existing research findings are acknowledged initially, followed by an analysis and synthesis of prior studies to guide future developments.

2   Chemotaxis Strategy

The chemotaxis strategy that originated in early gas source localization stages was inspired by certain biological behaviors. Most bioinspired search algorithms can be taken as series of predefined motion sequences triggered by observations to locate the position of the release source. This method was first proposed by Rozas et al. [30] to capture and measure odor through a hardware system equipped with six different sensors. Although gas sources are located simply and roughly, the positioning accuracy is not ideal in turbulent air, so improvements have been made on the basis of this method, which can be divided into two main directions: chemotaxis and biased random walk.

2.1   Chemotaxis Improvement

The simplest Chemotaxis strategy execution steps are as follows: 1. The measured values between two sensors in different positions or directions are compared. 2. At a certain angle, the next position has a relatively high chemical concentration. 3. Move forward a predetermined distance according to a set step size. The above steps are repeated until the distance to the gas source is less than a certain threshold or the robot-measured gas concentration is greater than a preset threshold. Here, chemotaxis does not refer solely to the use of chemical gradients but can involve algorithms that use certain information measured in a particular way.

Cheng et al. [31] combined the zigzag algorithm with the chemotaxis algorithm and additionally integrated a wireless sensor positioning module into the sensor module, thereby increasing the localization accuracy. Ma et al. [8] set up an artificial olfactory system based on gas sensor arrays that support vector machines (SVMs) on mobile robots. To locate gas sources faster, the zigzag–silkworm algorithm, integrated zigzag algorithm and silkworm algorithm are used to conduct outdoor experiments on locating gas release sources. The experimental results are compared with those obtained via the zigzag algorithm and the silkworm algorithm. The comparison indicates that the combined algorithm demonstrates excellent performance in both efficiency and accuracy. Macedo et al. [32] proposed an improved ANN method based on bacterial inspiration, which uses a neural network to take a set of perceptions as inputs and then outputs the resulting spatial coordinates of attraction. The robot motion is guided by the course of this force and a displacement equal to its instantaneous speed, reducing the reliance on manually defined parameters. From the improvements above, it can be seen that chemotaxis improvements focus mainly on improving mobile robot or sensor performance. Algorithmically, they often involve simple combinations with geometric search methods. However, the specific step size and turning angle significantly affect the accuracy of gas source localization. To address this issue, researchers have proposed the biased random walk method.

2.2 Biased Random Walk Improvement

In each step of the odor source localization process, evaluations are made on step length and turning angle distributions, which are incorporated into some constitutive equations. If both distributions are locally isotropic, the result is a pure random walk model. If one or both probability distributions are biased in the direction of the chemical gradient, then a biased random walk model describing the effects of chemotaxis [33] is identified.

A representative algorithm of biased random walk is the Escherichia coli algorithm, where bacteria can analyze the concentration gradient around them through chemotaxis, aiming to approach nutrients or move away from toxic substances in the shortest time possible [34, 35]. The main steps are as follows: first, the robot measures the concentration information and makes random turns according to the Poisson distribution. Second, the robot advances a certain step, measures the concentration information again, and calculates the concentration difference between the initial position and the current position. Third, the next moving step length is randomly selected from another Poisson distribution according to the concentration difference. The above steps are repeated until the gas source is located.

Although the traditional Escherichia coli algorithm can accurately locate gas sources, the search route is often lengthy. To shorten the search route, Wang et al. [36] designed a fuzzy controller to perceive the environment and adjust trajectory parameters according to the current search situation. A robot can automatically adjust the scale of its search trajectory, allowing adjustments to environmental changes to balance environmental exploration with local search. Optimization of the search path is also reflected in a series of adaptive step size algorithms [37, 38] and path planning methods [39]. Improvements in random biased walking methods mainly aim to balance the global exploration path length with the necessary path length to locate the gas plume and achieve rapid localization with minimal consumption.

3   Anemotaxis Strategy

In nature, some organisms, such as beetles and moths, exhibit a behavior known as upwind or upcurrent movement when foraging or searching for food. Specifically, when they sense odor, they move toward the odor source by using upwind or upcurrent methods. Early researchers were inspired by this and proposed using this upwind behavior of organisms for gas source localization research. This approach includes algorithms such as the zigzag algorithm [40] and the Moth algorithm [41]. Anemotaxis searching is a type of global search, which requires a long search time when the search space is large and often lacks precision in localization. Therefore, it is often used in conjunction with other search strategies.

3.1   Hybrid Strategy Combining Anemotaxis and Chemotaxis Strategies

The wind-chemotaxis strategy, which is simple in structure but tends to be time-consuming, is quite common. Turbulent gas environments pose another challenge for this method. Indoor gas environments are relatively stable, whereas outdoor environments exhibit significant turbulence. In response to different airflow environments indoors and outdoors, the following improvements have been made. Table 1 shows that, in both indoor and outdoor environments, the anemotaxis strategy is preferable for integration with hardware systems because of its simple algorithmic structure. In indoor environments, where the gas environment is stable, environmental parameters are easier to measure, and anemaxis strategies tend to use smaller robots for more precise searches. In outdoor environments, accurately determining the robot position and navigation methods are core issues. Compared with the algorithm itself, there is a greater tendency to improve a robot's sensor modules.

3.2   Hybrid Strategy Combining an Anemotaxis Strategy with Biomimetic Algorithms

Hayes et al. [22] divided the gas source localization problem into three stages: plume discovery, plume tracking, and source confirmation. To ensure efficient exploration of the environment's plume distribution while minimizing the time cost, the anemaxis strategy is often employed at the plume discovery stage to quickly find the gas diffusion plume. At the plume tracking stage, a hybrid strategy combining anemotaxis with bioinspired algorithms is considered to locate the gas source.

Table 1. Improvement methods for the wind-chemotaxis strategy in different airflow environments.

	Improvement methods	Advantages	Disadvantages
Indoor	Improve robot structure by reducing robot size and adding wind wings [42]	Faster sensor response time, higher sensitivity, better obstacle avoidance	Difficult in scaling to multirobot mode
	Add fuzzy system [43, 44]	Adjust the scale of their searching trajectory automatically to adjust environmental changes, resulting in a higher success rate in searching	The algorithm structure is complex and not easy to integrate into systems
	Incorporating physical gas diffusion models [45, 46]	Allowing usage in scenarios where significant airflow disturbances are present with robot	Depends on accurate physical models and prior knowledge of searching environment
Outdoor	Improve sensor structure by adding localization and other sensor modules [48–50]	Simple structure, easily integrated into a system, resulted in shorter search times	High cost
	Combine machine learning and deep learning methods, utilizing DNNs [51], SUV [52]	Source localization accuracy is increased, robots can search sources autonomously and flexibly	Long computation time, preliminary setup work are required
	Incorporate different navigation modes [53]	Beneficial for the development of autonomous systems in complex gas environments	The selection of navigation model to be used is determined based on prior knowledge

Owing to turbulence in the search area, the gas plume distribution is sparse, and the concentration gradients change discontinuously [54]. At the plume tracking stage, the aforementioned issues can be overcome through a hybrid algorithm that combines an anemotoxicity strategy with bioinspired algorithms, which exhibit superior performance.

Common examples include particle swarm optimization (PSO) [3, 55], ant colony optimization (ACO) [56, 57], gray wolf optimization (GWO) [58, 59], and the firefly algorithm [60]. Anemotaxis strategies combined with different bioinspired methods have the following advantages and disadvantages, as shown in Table 2.

The hybrid strategies generally exhibit advantages of fast and accurate localization. However, the local optima issue needs further resolution.

Table 2. Advantages and disadvantages of hybrid strategies.

Names of algorithms	Advantages	Disadvantages
PSO	Fewer parameters are needed, influence of obstacles in the environment is minimal	Prone to falling into local optima, long computation time
ACO	Simple structure, less computing cost, population diversity is enhanced, conduct multirobot source search easier	Improving single-robot source search capability is not considered, may not effectively respond to time-varying gas source information
GWO	Combining plume dispersion models to improve source localization accuracy, facilitating multirobot collaboration	The impact of turbulence on gas dispersion is overlooked and only a constant wind field environment has been considered
Fireworks Algorithm	Adaptive search step size, resulting in fast localization speed	Prior knowledge is deeply relied on

4 Infotaxis Strategy

The infotaxis strategy, which is based on information theory principles and mostly uses Bayesian frameworks [61] for information computation, can overcome the adverse effects of discontinuous, multimodal, and time-varying characteristics [62] of plumes in turbulent gas, making it currently the primary method for gas source localization. It is widely applied because of its broad applicability. The search process involves partitioning grids, and probability maps represented by these grids lead to a significant amount of computational redundancy [63], reducing search efficiency. Moreover, it may lead to local optima and repetitive search paths. The essence of using the infotaxis strategy for gas source localization is guiding the robot to move in the exact direction in which the information uncertainty decreases fastest [64], thereby reducing uncertainty and locating the gas source quickly. Owing to the computational redundancy and path repetition issues inherent in traditional infotaxis algorithms, improvements have focused on two main directions: enhancements related to reward functions and motion modes and improvements based on particle filtering methods.

4.1 Improvements in Reward Functions and Motion Strategies

In the infotaxis algorithm, after the odor particle distribution is calculated via the gas diffusion model, the current information state of the robot position is computed via posterior probability [64]. This information state represents the certainty level of the gas source location, with uncertainty typically quantified via information entropy in standard algorithms [27]. Owing to the robot's tendency to move in the direction where information uncertainty decreases fastest, the current focus of improvement lies in the use of information entropy and a reward function to calculate information uncertainty

[65]. Various traversal strategies are subsequently employed to reach the gas source location.

The preceding horizon method [66] has been applied to information-driven search, aiming to avoid becoming stuck in local optimal states while finding more successful paths for information collection in obstacle-rich urban environments. Song et al. [67] proposed an improved infotaxis algorithm based on a local probability reliability function. This function is related to whether odor particles are captured or not, addressing the issue of suboptimal searching performance in regions with dense local clues or near source areas. However, the above methods have a high time cost. To solve this problem, Song et al. [68] proposed a new searching scheme based on minimum free energy that combines entropy and potential energy to further collect more effective information and improve information utilization efficiency. Park et al. [69] combined the infotaxis method with a Gaussian mixture model to determine candidate actions for sampling the next best information in continuous space by appropriately clustering possible source locations obtained from a particle filter. Additionally, they extended the information-driven strategy from two dimensions to three dimensions by refining the reward function through increasing the types of movement directions. Fan et al. [65] compared 4-direction and 8-direction movement strategies on the basis of the reward function SInfotaxis in 2D scenarios and then applied the reward function SInfotaxis to 3D search. They proposed 6, 14, and 26 directional movement strategies and compared their source-seeking performance. This improvement extends from two dimensions to three dimensions without changing the algorithm structure. Owing to its high universality and simplicity in computation, it has been widely applied. However, the localization accuracy depends on the partition accuracy of the search space grid. This may lead to localization failure when the grid accuracy is too low but may cause computational redundancy when it is too high.

4.2   Improvements Based on Particle Filtering Methods

Infotaxis methods address the challenge of locating gas sources in environments where gas plumes exhibit discontinuous concentration distributions at a macroscopic scale. These methods utilize Bayesian theory to update posterior probability distributions on the basis of likelihood functions and prior information about odor diffusion parameters. By achieving consistency between measurement results and posterior probability density functions, the algorithm estimates parameters of interest through sampling methods [70]. However, the infotaxis algorithm may encounter issues such as becoming trapped in local optima, resulting in repetitive searching paths. To mitigate this, researchers have integrated particle filters (PFs) [59, 62, 71, 72] with information entropy, effectively avoiding the complexity associated with approximating probability maps. The advantages and disadvantages of improvement directions based on PF are summarized in Table 3.

PF algorithms have fewer adjustable parameters, making them easy to implement. However, they often suffer from particle degeneracy issues in the later stages of iteration. Many improvement methods aim to increase particle diversity. Currently, relatively few particle filtering algorithms exist in three-dimensional space, which is a primary direction for future development.

Table 3. Advantages and disadvantages of improvement directions based on PF.

Improvements	Advantages	Disadvantages
Incorporating traversal methods such as octrees [73] and random trees [74]	Further weighting the historical candidate waypoints to avoid repeating searching paths	High computational load, complex algorithm structure, high storage performance of the devices is needed
Combine bioinspired methods [75, 76]	Particle convergence form is optimized, convergence speed is accelerated, computational time is reduced	Prone to local optima issues, hyperparameters have great influence on performance
Improve gas diffusion model [77, 78]	Searching accuracy is improved, so as probability map accuracy	Dependent on prior environment knowledge
Improve resampling methods [79, 80]	Resampled particle number is adjusted intelligently, avoid particle degeneracy issues, computational complexity is reduced during resampling	Particle depletion issue

5   Application Scenarios and Existing Issues

The application scenarios of directional strategies can be categorized on the basis of the gas environment into three forms: those based primarily on gas diffusion, those based primarily on turbulence, and those based on a hybrid of the two [81]. Additionally, they can be classified on the basis of whether obstacles are present in the gas diffusion environment and whether the gas release source is variable. Different directional strategies have their own advantages and disadvantages in various application scenarios.

5.1   Applications in Different Gas Environments

When the three strategies address issues related primarily to the gas environment, gas diffusion predominates in indoor scenarios without airflow. In outdoor environments, turbulence is the primary factor, whereas in indoor environments with weak airflow, a combination of the two methods is observed. In indoor environments without wind, gas diffusion exhibits clear and continuous concentration gradients. Owing to its simple algorithmic structure, the chemotaxis strategy is highly praised for its ease of integration into hardware systems. However, if it is to be extended to multirobot cooperative localization, further improvements are needed. Anemotaxis algorithms are limited in windless environments because of their reliance on wind direction information. The infotaxis strategy, on the other hand, relies on accurate diffusion equations. Although the strategy is more complex, it can achieve precise localization. In outdoor environments, gas plumes are notably sparse, and concentration gradients are discontinuous [19]. Under such conditions, chemotaxis strategies struggle to accomplish gas source localization tasks effectively. Robots find it challenging to accurately determine the correct direction of gradient descent, ultimately leading to significantly increased searching time or failure to localize. Although real-time and dynamic changes in turbulent wind conditions can indeed affect the effectiveness of wind tendency strategies [79], methods that combine a wind tendency strategy with a chemotaxis strategy can still accurately identify the direction of gas plume diffusion. However, for more complex 3D environments with intricate wind fields, further research is necessary. On the other hand, infotaxis strategies can overcome the drawbacks of turbulent environments and are more universally applicable. However, the selection of initial values for the gas diffusion model can significantly impact the final results. In indoor environments with weak airflow, all three strategies are applicable. Zhang et al. [20] compared these three strategies in such an environment, and the experimental results revealed that the infotaxis strategy achieved a balance between exploration and utilization, overcoming the local optima issue, and demonstrated better search performance in this application scenario. However, in the aforementioned gas environments, there has been limited consideration of disturbances caused by robot movement through the search area. Future research could explore this aspect by integrating the disturbance characteristics of different robots.

5.2 Application in Obstacle Environments

In obstacle environments, odor molecules tend to accumulate on the windward side of obstacles [65], forming pseudosources, whereas gas voids may form on the left side, causing interference with gas source localization and restricting the movement of mobile robots. In such environments, single chemotaxis or anemotaxis strategies are no longer applicable, and hybrid strategies, which combine both chemical and wind cues, have become prominent. These strategies leverage terrain information for comprehensive exploration and localization, facilitating collaborative search and localization among multiple robots. However, they often require longer localization times. Information-based strategies can optimize probability maps by combining multiple positive and negative sources [77]. However, in scenarios with irregularly shaped obstacles, this approach may lead to significant computational overhead, necessitating further research.

5.3 Application in a Time-Varying Gas Source Environment

In environments where the gas release rate is not constant during the search process, known as time-varying gas sources, the search efficiency can be increased by combining collaborative wind tendency strategies with bioinspired algorithms. This approach performs well in locating time-varying sources, aiding robots with less time cost on the basis of concentration gradients and the wind direction. However, in such scenarios, each robot's speed is constrained by its airflow, limiting its searching capabilities and increasing its likelihood of becoming stuck in local optima. Infotaxis strategies, which rely on probability maps, require clear information about time-varying source dynamics. To improve localization efficiency, these strategies could be combined with predeployed sensor networks.

6 Discussion

On the basis of existing representative research results, the positioning methods of the chemotaxis strategies, anemaxis strategies, and infotaxis strategies are classified and summarized. The advantages and disadvantages of different methods and research ideas are summarized in Table 4:

Table 4. Advantages and disadvantages of different methods and research ideas.

Gas source localization strategies	Advantages	Disadvantages
Standard Chemotaxis	Easy to understand and implement, low computational load	Localization accuracy is related to the step size
Chemotaxis improvements	Computational time is greatly reduced compared to standard Chemotaxis strategies. Localization accuracy is higher in weak turbulent environment	Depends on the gas environment with continuous significant concentration gradient changes
Biased random walk Improvement	Balance path length of global exploration and path length necessary for positioning after finding the plume. Don't depend on preset parameters	The time cost and accuracy of locating are random
Standard Anemotaxis	The model is simple and has better global searching ability	Application scenarios are limited, computation takes a long time
Wind-Chemotaxis	Easy to integrate with hardware system and has certain adaptability to turbulent environment	Poor adaptability to the search environment with obstacles or time-varying sources
Hybrid strategy combining anemotaxis strategy with biomimetic algorithms	Suitable for all gas environments, high localization accuracy and less time consuming	Easy to fall into local optimal problem, premature convergence
Standard Infotaxis	Wide application scenarios, easy to integrate with hardware systems	Highly relies on prior knowledge of searching environment and accurate gas diffusion models
Improvements to reward functions and motion	Simple calculation, high universality and high information utilization efficiency	Localization accuracy depends on grid division

(*continued*)

Table 4. (*continued*)

Gas source localization strategies	Advantages	Disadvantages
PF	High positioning accuracy, less adjustable parameters, easy to implement	Has particle phenomena problems

As seen from the above table, the chemotaxis strategy can achieve effective gas source location in weakly turbulent environments in general, but the environmental concentration gradient needs to change significantly, and it is not suitable for sparse gas environments. The combination of the anemotaxis strategy with different methods can better complete the gas source location task in a multigas environment, but the location effect is significantly decreased in the search scenario with obstacles. The infotaxis strategy has the most extensive application scenarios. It has the highest utilization rate for various types of information in the search environment. However, it also relies on prior knowledge as well as high requirements for information collection.

7 Conclusion and Outlook

The gas source localization problem is not only important for solving hazardous gas leakage issues but also has potential applications in areas such as earthquake disaster search and rescue, forest fire detection, counterterrorism, and bomb disposal. As research progresses, the advantages of chemotaxis strategies, anemaxis strategies, and infotaxis strategies are increasingly being demonstrated in different application environments. A review of representative research results and improvement directions for these three strategies is provided, starting from their early proposals. Alternative approaches can be used to enhance different strategies when facing various problems, and their advantages and disadvantages can be compared. Finally, the application of directional strategies in different scenarios is discussed, the advantages and disadvantages of different improvement methods are compared, and existing shortcomings and limitations of current technologies, along with prospects, are addressed for future development. Possible future directions include the following:

1. Improving real-time sensor processing capabilities: With the development of sensor technology, the real-time processing capabilities of sensors carried by robots will be further improved. The use of high-performance sensors and advanced data processing algorithms enables robots to quickly acquire gas concentration data, accelerate data transmission, and perform real-time processing and analysis. This will help improve the real-time response capability of gas source localization systems and enable faster and more accurate localization of gas leak sources.
2. Integration of gas source localization systems: By optimizing the structure of sensors and data transmission mechanisms, the rapid acquisition and real-time transmission of sensor data can be achieved, combined with efficient data processing algorithms to integrate real-time responsive gas source localization systems.

3. Application of new materials and technologies: The application of new materials and technologies is also an important development direction for active odor gas source localization strategies for robots. For example, the development of new gas sensors with high sensitivity and selectivity can make robots more sensitive to low concentrations of gas and distinguish between different types of gases. In addition, using machine learning and artificial intelligence technologies to perform deep learning and pattern recognition on sensor data can further improve the performance and intelligence level of gas source localization systems.

References

1. Juffry, Z.H.M., et al.: Application of deep neural network for gas source localization in an indoor environment. Int. J. Comput. Commun. Control **18**(3), 335084 (2023)
2. Tao, J., Zizhen, Y., Qinghao, M.: RAOS: a three-dimensional robot active olfaction simulator. Robot **43**(3), 308–320 (2021)
3. Feng, Q., Cai, H., Li, F., Yang, Y., Chen, Z.: Locating time-varying contaminant sources in 3D indoor environments with three typical ventilation systems using a multirobot active olfaction method. Build. Simul. **11**(3), 597–611 (2018)
4. Yong, Z., Zhang, L.Y., Han, J.F., Zhe, B., Yi, Y.: An indoor gas leakage source localization algorithm using distributed maximum likelihood estimation in sensor networks. J. Ambient. Intell. Humaniz. Comput. **10**(5), 1703–1712 (2019)
5. Lin, S., Zhou, Y.C., Hu, J.H., Sun, Z.J., Zhang, T.Y., Wang, M.: Exploration for a BP-ANN model for gas identification and concentration measurement with an ultrasonically radiated catalytic combustion gas sensor. Sens. Actuators B Chem. **362**, 131733 (2022)
6. Jia, H.Y., Kikumoto, H.: Line source estimation of environmental pollutants using super-Gaussian geometry model and Bayesian inference. Environ. Res. **194**, 110706 (2021)
7. Fu, J., Shen, L., Liu, R.: An indoor odor source locating method for multirobot active olfaction based on improved AEO. Chin. J. Sens. Actuators **34**(10), 1406–1411 (2021)
8. Ma, D., Mao, W., Tan, W., et al.: Emission source tracing based on bionic algorithm mobile sensors with artificial olfactory system. Robotica **40**(4), 976–996 (2022)
9. Terutsuki, D., Uchida, T., Fukui, C., et al.: Electroantennography-based biohybrid odor-detecting drone using silkmoth antennae for odor source localization. J. Vis. Exp (174), e62895 (2021)
10. Jiang, M., Liao, Y., Guo, X., et al.: A comparative experimental study of two multirobot olfaction methods: toward locating time-varying indoor pollutant sources. Build. Environ. **207**, 108560 (2022)
11. Ji, Y., Chen, F., Chen, B., et al.: Multi-robot collaborative source searching strategy in large-scale chemical clusters. IEEE Sens. J. **22**(18), 17655–17665 (2022)
12. Fan, S., Hao, D., Sun, X., Sultan, Y.M., Li, Z., Xia, K.: A study of modified algorithms in 2D and 3D turbulent environments. Comput. Intell. Neurosci. **2020**, 4159241 (2020)
13. Russell, R.A.: Comparing search algorithms for robotic underground chemical source location. Auton. Robots **38**(1), 49–63 (2015)
14. Gao, B., Li, H., Li, W., et al.: 3D moth-inspired chemical plume tracking and adaptive step control strategy. Adapt. Behav. **24**(1), 52–65 (2016)
15. Cremer, J., Honda, T., Tang, Y., et al.: Chemotaxis as a navigation strategy to boost range expansion. Nature **575**(7784), 658–663 (2019)
16. Purnamadjaja, A.H., Russell, R.A.: Pheromone communication: implementation of necrophoric bee behaviour in a robot swarm. In: IEEE Conference on Robotics, Automation and Mechatronics, Piscataway, USA, pp. 638–643. IEEE (2004)

17. Grasso, F.W., Basil, J.A.: How lobsters, crayfishes, and crabs locate sources of odor: current perspectives and future directions. Curr. Opin. Neurobiol. 12(6), 721–727 (2002)
18. Bartumeus, F., Campos, D., Ryu, W.S., et al.: Foraging success under uncertainty: search tradeoffs and optimal space use. Ecol. Lett. 19(11), 1299–1313 (2016)
19. Chen, X.X., Huang, J.: Odor source localization algorithms on mobile robots: a review and future outlook. Robot. Auton. Syst. 112, 123–136 (2019)
20. Zhang, S., Cui, R., Xu, D.: Performance analysis on the algorithm for searching in dilute environments. Robot 35(4), 432–438 (2013)
21. Russell, R.A., Bab-Hadiashar, A., Shepherd, R.L., et al.: A Comparison of reactive robot chemotaxis algorithms. Robot. Auton. Syst. 45(2), 83–97 (2003)
22. Hayes, A.T., Martinoli, A., Goodman, R.M.: Distributed odor source localization. IEEE Sens. J. 2(3), 260–271 (2002)
23. Huang, X.: Improved algorithm-based cooperative multi-USV pollution source search approach in lake water environment. Symmetry 12(4), 549 (2020)
24. Zhang, J., et al.: PSO-based sparse source location in large-scale environments with a UAV swarm. IEEE Trans. Intell. Transp. Syst. 24(5), 5249–5258 (2022)
25. Murlis, J., Elkinton, J.S., Carde, R.T.: Odor plumes and how insects use them. Annu. Rev. Entomol. 37(1), 505–532 (1992)
26. Vergassola, M., Villermaux, E., Shraiman, B.I.: 'Lévy flights' as a strategy for searching without gradients. Nature 445(25), 406–409 (2007)
27. Wang, R., Chen, B., Qiu, S., et al.: Hazardous source estimation using an artificial neural network, particle swarm optimization and a simulated annealing algorithm. Atmosphere 9(4), 119 (2018)
28. Cheng, S.: Research on odor source search of robots based on tracing. Ph.D. thesis, Northwestern Polytechnical University, Xi'an (2020)
29. Deng, S., Fan, S., Zhang, Y.: Research on location method of research algorithm on gas diffusion model. Electron. Meas. Technol. 45(12), 58–65 (2022)
30. Rozas, R., Morales, J., Vega, D.: Artificial smell detection for robotic navigation. In: Fifth International Conference on Advanced Robotics, 'Robots in Unstructured Environments', pp. 1730–1733. IEEE (1991)
31. Cheng, L., Zhang, D., Liu, B., Wu, H.Y., Wang, Y.J.: Design of gas leak source localization robot based on wireless sensor networks. Transducer Microsyst. Technol. 34(2), 85–91 (2015)
32. Macedo, J., Marques, L., Costa, E.: Evolving neural networks for multi robot odor search. In: IEEE International Conference on Autonomous Robot Systems and Competitions, Braganca, Portugal, pp. 288–293. IEEE (2016)
33. Anderson, M.J., Sullivan, J.G., Horiuchi, T.K., et al.: A bio-hybrid odor-guided autonomous palm-sized air vehicle. Bioinspir. Biomim. 16(2), 026002 (2020)
34. Dhariwal, A., Sukhatme, G.S., Requicha, A.A.: Bacterium-inspired robots for environmental monitoring. In: IEEE International Conference on Robotics and Automation, vol. 2, pp. 1436–1443. IEEE (2004)
35. Naeem, W., Sutton, R., Chudley, J.: Chemical plume tracing and odour source localisation by autonomous vehicles. J. Navig. 60(2), 173–190 (2007)
36. Wang, L., Pang, S.: Robotic odor source localization via adaptive bio-inspired navigation using fuzzy inference methods. Robot. Auton. Syst. 147, 103914 (2021)
37. Scase, M., Hewitt, R.: Unsteady turbulent plume models. J. Fluid Mech. 697, 455–480 (2012)
38. Celani, A., Villermaux, E., Vergassola, M.: Odor landscapes in turbulent environments. Phys. Rev. X 4(4), 041015 (2014)
39. Wang, J., Zhao, H.M.: Variable step-size odor tracing and source localization algorithm for mobile robot. Comput. Eng. Appl. 45(2), 243–245 (2009)

40. Ishida, H., Suetsugu, K., Nakamoto, T., et al.: Study of autonomous mobile sensing system for localization of odor source using gas sensors and anemometric sensors. Sens. Actuators A Phys. **45**(2), 153–157 (1994)
41. Kuwana, Y., Shimoyama, I., Sayama, Y., et al.: Synthesis of pheromone-oriented emergent behavior of a silkworm moth. In: IEEE/RSJ International Conference on Intelligent Robots and Systems (IROS'96), Osaka, Japan, pp. 1722–1729. IEEE (1996)
42. Anderson, M.J., Sullivan, J.G., Horiuchi, T.K., Fuller, S.B., Daniel, T.L.: A bio-hybrid odor-guided autonomous palm-sized air vehicle. Bioinspir. Biomim. **16**(2), 026002 (2022)
43. Wang, L., Pang, S.: Robotic odor source localization via adaptive bio-inspired navigation using fuzzy inference methods. Robot. Auton. Syst. **147**, 103914 (2022)
44. Wang, L., Pang, S., Li, J.: Olfactory-based navigation via model-based reinforcement learning and fuzzy inference methods. IEEE Trans. Fuzzy Syst. **29**(10), 3014–3027 (2021)
45. Wiedemann, T., Schaab, M., Gomez, J.M., Shutin, D., Scheibe, M., Lilienthal, A.J.: Gas source localization based on binary sensing with a UAV. In: IEEE International Symposium on Olfaction and Electronic Nose (ISOEN 2022), Aveiro, Portugal, 29 May–01 June 2022. IEEE (2022)
46. Gan, L.X., Lu, T.F., Shu, Y.Q.: Diffusion and superposition of ship exhaust gas in port area based on Gaussian puff model: a case study on Shenzhen port. J. Mar. Sci. Eng. **11**(2), 330 (2023)
47. Terutsuki, D., Uchida, T., Fukui, C., Sukekawa, Y., Okamoto, Y., Kanzaki, R.: Electroantennography-based bio-hybrid odor-detecting drone using silkmoth antennae for odor source localization. J. Vis. Exp. **174**, e62895 (2022)
48. Anderson, M.J., Sullivan, J.G., Talley, J.L., Brink, K.M., Fuller, S.B., Daniel, T.L.: The "Smellicopter," a bio-hybrid odor localizing nano air vehicle. In: IEEE/RSJ International Conference on Intelligent Robots and Systems (IROS 2019), Macau, China, 04–08 November 2019, pp. 6077–6082. IEEE (2019)
49. Terutsuki, D., Uchida, T., Fukui, C., Sukekawa, Y., Okamoto, Y., Kanzaki, R.: Real-time odor concentration and direction recognition for efficient odor source localization using a small bio-hybrid drone. Sens. Actuators B Chem. **339**, 129770 (2021)
50. Wang, L., Yin, Z., Pang, S.: Learn to trace odors: robotic odor source localization via deep learning methods with real-world experiments. In: IEEE SoutheastCon Conference, Orlando, FL, 01–16 April 2023, pp. 524–531. SoutheastCon (2023)
51. Horibe, J., Ando, N., Kanzaki, R.: Odor-searching robot with Insect-behavior-based olfactory sensor. Sens. Mater. **33**(12), 4185–4202 (2021)
52. Golov, Y., Benelli, N., Gurka, R., Harari, A., Zilman, G., Liberzon, A.: Open-source computational simulation of moth-inspired navigation algorithm: a benchmark framework. MethodsX **8**, 101529 (2021)
53. Zhou, X., Wang, F., Yang, L., Gu, M.: Locating indoor time-variant contaminant sources based on nelder-mead algorithm using robot active olfaction method. J. Tongji Univ. Nat. Sci. **50**(6), 812–820 (2022)
54. Feng, Q., et al.: Experimental study on a comprehensive particle swarm optimization method for locating contaminant sources in dynamic indoor environments with mechanical ventilation. Energy Build. **196**, 145–156 (2019)
55. Meng, Q., Yang, W., Wang, Y., et al.: Multirobot odor-plume tracing in indoor natural airflow environments using an improved ACO algorithm. In: IEEE International Conference on Robotics and Biomimetics, Tianjin, China, 14–18 December 2010, pp. 110–115. IEEE (2010)
56. Meng, Q., Yang, W., Wang, Y., et al.: Adapting an ant colony metaphor for multi-robot chemical plume tracing. Sensors **12**(4), 4737–4763 (2012)
57. Shen, X.Y., Yuan, J.: Robot plume tracking method based on grey wolf optimization algorithm. Sci. Technol. Eng. **21**(11), 4498–4505 (2021)

58. Jin, X.Q., Zhang, X., Jiang, H., Tian, J.: Hybrid strategy improved grey wolf optimization algorithm for plume tracking and localization method in indoor weak wind environment. IEEE Access **10**, 100976–100986 (2022)
59. Miao, Y., Ma, X., Jin, X., et al.: Mobile robot odor source localization based on modified FWA. In: IEEE 8th Annual International Conference on CYBER Technology in Automation, Control, and Intelligent Systems (CYBER 2018), Tianjin, China, 11 April 2019
60. Hutchinson, M., Liu, C., Chen, W.H.: Information-based search for an atmospheric release using a mobile robot: algorithm and experiments. IEEE Trans. Control Syst. Technol. **27**(6), 2388–2402 (2019)
61. Jones, C.D.: On the structure of instantaneous plumes in the atmosphere. J. Hazard. Mater. **7**(2), 87–112 (1983)
62. Park, M., Oh, H.: Cooperative information-driven source search and estimation for multiple agents. Inf. Fus. **54**, 72–84 (2020)
63. Loisy, A., Eloy, C.: Searching for a source without gradients: how good is infotaxis and how to beat it. Proc. R. Soc. A Math. Phys. Eng. Sci. **478**(2262), 20220118 (2022)
64. Karpas, E.D., Shklarsh, A., Schneidman, E.: Information socialtaxis and efficient collective behavior emerging in groups of information-seeking agents. Proc. Natl. Acad. Sci. U.S.A. **114**(22), 5589–5594 (2017)
65. Fan, S., Hao, D., Sun, X., et al.: A study of modified infotaxis algorithms in 2D and 3D turbulent environments. Comput. Intell. Neurosci. **2020**, 4159241 (2020)
66. Park, M., Ladosz, P., Kim, J., Oh, H.: Receding horizon-based infotaxis with random sampling for source search and estimation in complex environments. IEEE Trans. Aerosp. Electron. Syst. **59**(1), 591–609 (2023)
67. Song, C., He, Y., Yang, P., Lei, X.: An infotaxis strategy for seeking a dispersion source using local probabilistic reliability. J. Northwest. Polytech. Univ. **34**(5), 843–850 (2016)
68. Song, C., He, Y., Lei, X.: Autonomous searching for a diffusive source based on minimizing the combination of entropy and potential energy. Sensors **19**(11), 2465 (2019)
69. Park, M., An, S., Seo, J., Oh, H.: Autonomous source search for UAVs using Gaussian mixture model-based infotaxis: algorithm and flight experiments. IEEE Trans. Aerosp. Electron. Syst. **57**(6), 4238–4254 (2021)
70. Stone, L.D., Streit, R.L., Corwin, T.L., et al.: Bayesian Multiple Target Tracking. Artech House, Fitchburg (2013)
71. Zhu, H., Wang, Y., Du, C., Zhang, Q., Wang, W.: A novel odor source localization system based on particle filtering and information entropy. Robot. Auton. Syst. **132**, 103619 (2020)
72. Hutchinson, M., Oh, H., Chen, W.H.: Entrotaxis as a strategy for autonomous search and source reconstruction in turbulent conditions. Inf. Fus. **42**, 179–189 (2018)
73. Wang, Y., Zhu, H., Wang, W.: Octree-based repetitive pose detection of large-scale cyclic environments. In: 2018 3rd International Conference on Robotics and Automation Engineering (ICRAE), pp. 60–64. IEEE (2018)
74. An, S., Park, M., Oh, H.: Receding-horizon RRT-infotaxis for autonomous source search in urban environments. Aerosp. Sci. Technol. **120**, 107276 (2022)
75. Tian, M., Liu, L., Chen, Z., Fang, Y.: Firefly algorithm optimized particle filter based on spring mechanism. Control Decis. **39**(2), 420–428 (2024)
76. Chen, Z., Bo, Y., Wu, P., Zhu, K., Yin, M.: Novel landscape adaptive particle filter algorithm based on convergent particle swarm and its application. J. Nanjing Univ. Sci. Technol. **36**(5), 861–868 (2012)
77. Wang, Y.: Using Information Entropy to Search and Localize a Gas Source in a Cluttered Scene. Harbin Inst. Technol., Harbin (2019)
78. Farrell, J., Pang, S., Li, W., et al.: Plume mapping via hidden Markov methods. IEEE Trans. Syst. Man Cybern. B Cybern. **33**(6), 850–863 (2003)

79. Wang, Z., Lu, W., Chang, Z.: Joint inverse estimation of groundwater pollution source characteristics and model parameters based on an intelligent particle filter. J. Hydrol. **625**, 129965 (2023)
80. Zafar, T., Tariq, M., Alam, A., Rasheed, H.: Hybrid resampling scheme for particle filter-based inversion. IET Sci. Meas. Technol. **14**(4), 396–406 (2020)
81. Li, F.: Multi-Robot Odor-Source Localization in Turbulence Dominated Airflow Environments. Tianjin Univ., Tianjin (2010)

Exploring Named Entity Recognition in Medical Knowledge Graphs with Pre-trained Language Models and Attention Mechanism

Junsong Zhang[1,2], Askar Hamdulla[1,2], and Turdi Tohti[1,2(✉)]

[1] School of Computer Science and Technology, Xinjiang University,
Urumqi 830017, China
`zhangjunsong@stu.xju.edu.cn`, {`askar,turdy`}`@xju.edu.cn`
[2] Xinjiang Key Laboratory of Signal Detection and Processing,
Urumqi 830017, China

Abstract. Named entity recognition plays a crucial role in natural language processing and directly impacts the performance of downstream tasks. Pre-trained models have become a groundbreaking advancement in artificial intelligence and are commonly employed for NER tasks. However, when dealing with medical-related tasks, utilizing a pre-trained model trained on general corpora may lead to a significant decline in performance. To address this issue, we propose a neural network called the "Named Entity Recognition Model Integrated with Medical Knowledge Graph." This model builds upon previous research in medical named entity recognition tasks. In the knowledge representation phase, our model incorporates a knowledge graph by employing relative position encoding. This inclusion enhances the pre-trained model's ability to capture domain-specific knowledge and mitigates the loss of semantic and structural information that typically occurs when incorporating triplets. Additionally, a multi-head attention layer aligns the knowledge graph with semantic features, enabling the model to learn the semantic associations between the two and aiding the model in making accurate annotation decisions. Experimental results demonstrate that our model outperforms other mainstream named entity recognition models. On the diabetes and CCKS2017 datasets, our model achieves an absolute increase of 4.8% and 2.13% in the F1 value compared to the baseline model, confirming the superior performance of our proposed model. Furthermore, we conduct corresponding ablation experiments to showcase the effectiveness of our innovation.

Keywords: Named entity recognition · Pre-trained models · Medical knowledge graph

H. Yu et al. (Eds.): CCF NCCA 2024, CCIS 2274, pp. 167–182, 2024.
https://doi.org/10.1007/978-981-97-9671-7_11

1   Introduction

Named Entity Recognition (NER) plays a crucial role in natural language processing (NLP) and is an essential component of information extraction. It involves identifying entities within text and classifying them into predefined categories. NER can be broadly categorized into general NER and domain-specific NER, as discussed in reference [1]. General NER identifies universal entities like personal names, organizations, and geographical locations. In contrast, domain-specific NER focuses on recognizing entity types that are specific to a particular field. For example, in the context of diabetes research, entities such as specific diseases, causes, clinical symptoms, and pharmaceutical names are often identified. These specialized entity types differentiate domain-specific NER from its general counterpart and are particularly important in the medical field.Pre-trained models have revolutionized NLP, demonstrating superior performance in various tasks such as named entity recognition, sentiment analysis, and reading comprehension. Fine-tuning these pre-trained models on NER datasets has become the prevailing approach. However, there are significant differences in language usage and text characteristics between general corpora and domain-specific texts. Research [2] has shown that the coverage rate of domain-specific vocabularies in different domains can be as low as 12.7%. When dealing with domain-specific texts, laypeople often rely on context alone to understand sentence meanings, while domain experts draw on their specialized knowledge for deeper comprehension. Although pre-trained models improve their understanding through continuous learning from open-domain corpora, the overlap in knowledge between these open domains and specific fields is minimal. As a result, pre-trained models often underperform in entity recognition tasks within certain domains. Building domain-specific pre-trained models is a direct solution to this problem. However, this approach requires significant time and expensive computational resources, making it unfeasible for most researchers. Furthermore, in more specialized domains, the scarcity of data further hampers the feasibility of this approach.

Moreover, although pre-trained models can learn domain knowledge during pretraining, this learning process is more costly and inefficient [3]. For instance, to enable a model to acquire knowledge about how chronic obstruction can cause renal hydronephrosis, a significant occurrence of both "chronic obstruction" and "renal hydronephrosis" in the pretraining corpus is required, which places high demands on the training data and makes it difficult to achieve [4]. Leveraging knowledge graphs to supplement domain knowledge offers a promising solution to address this issue. Several domain-specific knowledge graphs have been constructed, such as the Traditional Chinese Medicine knowledge graph [5] and the China tourist attraction knowledge graph [6]. Integrating knowledge graphs with pre-trained models can provide the necessary domain knowledge, thereby enhancing their performance on specific domain tasks.

However, there are two challenges in integrating knowledge graphs into pre-trained models. Firstly, the entities in medical domain texts and knowledge graphs are obtained differently, resulting in inconsistent vector spaces. Secondly, incorporating knowledge graphs may lead to a deviation from the original sentence's intended meaning. Figure ?? illustrates the entity recognition model proposed in this paper and the main contributions of this paper are as follows:

(1) Introducing medical knowledge graphs into existing general pre-trained models: By encoding the text and entities using the pre-trained model, the inconsistency in vector spaces between them is avoided, effectively enhancing the model's ability to encode medical knowledge.

(2) Designing a relative position encoding to alleviate the loss of semantic structural information caused by the introduction of knowledge graphs. This encoding annotates the positions of text and entities, preventing semantic ambiguity resulting from knowledge insertion.

(3) Utilizing an attention layer to combine the heterogeneous features formed by the original text and the fused text. This assists the model in learning the semantic correlation between the two and guides the model in effectively aligning and merging the features.

2   Related Work

Early Chinese named entity recognition methods were initially inspired by English tasks, which are easier to handle due to the absence of word segmentation challenges. Lample et al. developed an end-to-end model for word embedding and entity classification using a bidirectional LSTM (Long Short-Term Memory) + CRF (Conditional Random Fields) structure, specifically designed for modeling English words [7]. To improve entity recognition in Chinese, Peng et al. combined Chinese word segmentation with named entity recognition, leveraging shared information and joint learning [8]. Liu et al. proposed the LM-LSTM-CRF model architecture, which utilizes bidirectional LSTM to model character-level representations of words. This information is then passed to the input interfaces of the language model (LM) and named entity recognition task, jointly training the language model to enhance overall model performance [9].

Compared to traditional methods, deep neural network-based named entity recognition methods offer significant advantages, including efficient expressive power and automatic capture of latent features. Representation learning is crucial for effectively utilizing deep neural networks in named entity recognition [10]. Initially, researchers widely used static word vector representations, such as Word2vec [11]. However, since the introduction of BERT (Bidirectional Encoder Representations from Transformers) by Google in 2018, named entity recognition performance has significantly improved. BERT leverages its powerful and efficient text embedding technique to provide dynamic word vectors, which can be fine-tuned through supervised learning [12]. In optimizing the training of pre-trained models to enhance entity recognition performance, Baidu-ERNIE and BERT-wwm adopted full-word masking instead of single-word masking during BERT pre-training on Chinese corpora. This modification has proven effective in improving the training process and enhancing the model's representation of text [13,14]. Additionally, references [15] proposed an extension to BERT by masking consecutive random spans and introducing a span boundary objective. RoBERTa optimized BERT's pre-training in three ways: removing the next sentence prediction task, dynamically changing the masking strategy, and using more and longer sentences for training [16]. These methods have enhanced the training process of pre-trained models and improved the representation of text. However, they still do not fully account for the differences between general text and medical field text.

In addition to optimizing the training process of pre-trained models, incorporating additional knowledge into the models is another approach to improve medical entity recognition performance [17–19]. In Chinese named entity recognition, the Lattice-LSTM model is a typical approach where the input text is represented as a directed acyclic graph, with each node representing an input word and edges representing dependencies between words [20]. The Lattice-LSTM model utilizes forward and backward propagation to update hidden states, taking into account the dependencies between nodes and edges. This bidirectional propagation mechanism helps the model better understand contextual information in the text, thus improving its performance.

The MECT architecture incorporates convolutional neural networks to encode Chinese character structural information while integrating lexical information [21]. The Flat model proposes a dual-stream Transformer architecture to establish connections between external lexical knowledge and external character structural understanding. However, this structure lacks deep integration of domain knowledge and has many parameters with low fusion efficiency [22]. Sun et al. consider the polyphonic nature of Chinese characters and use convolutional neural networks to encode the composition of Chinese characters' pinyin. They fuse encoded data from different font styles of character images and employ a fusion layer to combine three types of information, forming rich semantics for model training [23]. These approaches enhance entity recognition performance by incorporating lexical information as external knowledge, aligning with the characteristics of the Chinese language. However, when applied to medical named entity recognition tasks, they still face certain limitations due to the complexity of medical vocabulary, which includes a plethora of abbreviations and intricate term structures.

In terms of integrating pre-trained models with knowledge graphs, THU-ERNIE is a pioneer in this direction. It enhances language representation by automatically detecting entities and incorporating them as additional context. However, it does not consider the relationships between entities and does not fully exploit the learning potential of the knowledge graph, leading to suboptimal performance [24]. COMET uses triples from a knowledge graph as a corpus to train GPT for common-sense learning. However, the effectiveness is limited due to the relatively weak association between the triples [25]. Prior to the rise of pre-trained models, a model introduced in the literature effectively captured the compositional structure of textual relationships, optimizing jointly for entities, knowledge base integration, and textual relationship representations [27].

Another study combines cross-lingual word alignment techniques and distant supervision mechanisms to automatically capture the correspondence between languages, jointly learning representations of entities and vocabulary from large-scale text resources [28]. However, these methods did not integrate with pre-trained models. While joint representation techniques bring entities and words closer together in vector space, there may still be a bias or imbalance in their respective representations. Additionally, when applied to large-scale knowledge graphs containing millions of entities, these methods require substantial computational power, resulting in high training expenses [29,30].

3   Method

Our model consists of four main components: the knowledge fusion layer, the BiLSTM-IDCNN layer, the feature fusion layer, and the decoding layer. The model first incorporates the triples from the knowledge graph into the original text through the knowledge fusion layer. It then utilizes a pre-trained model to obtain vector representations for original and fused texts. These representations are fed into the BiLSTM-IDCNN (Incremental Dilated Convolution Neural Networks) layer for semantic encoding. The feature vectors are further weighted using an attention mechanism and passed to the CRF decoding layer. Finally, the model outputs the predicted labels for each token. As shown in the following Fig. 1.

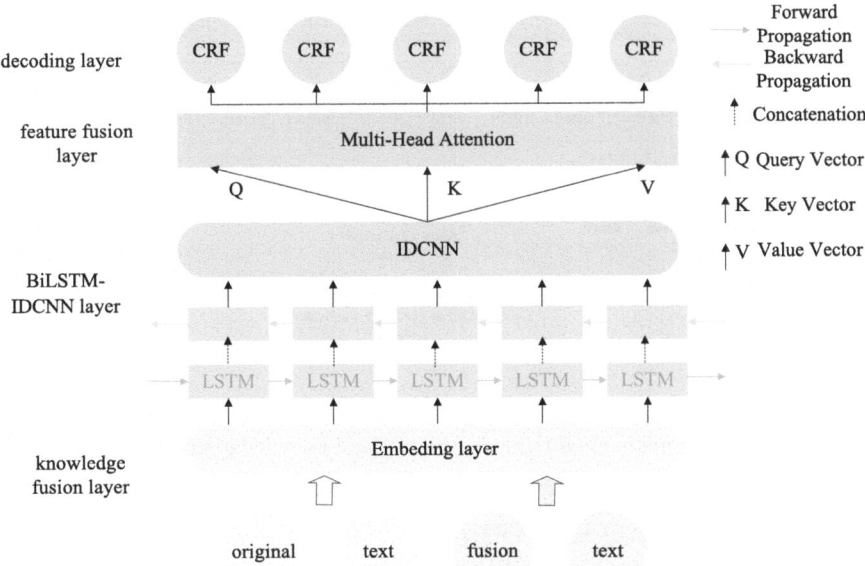

Fig. 1. Model in this article.

3.1   Knowledge Fusion Layer

In the knowledge fusion layer, medical domain knowledge is injected into the text by inserting corresponding triples from the medical knowledge graph into the input text. This enriches the domain knowledge within the text. A relative positional encoding is proposed to mitigate the potential disruption of the text's semantic structure caused by injecting domain knowledge. Here, h_i and t_i represent the names of the head and tail entities, respectively, and r_i represents the relationship between the two entities. The process of obtaining the fused text can be divided into two steps: domain knowledge query and domain knowledge injection. The original text $s = \{W0, W1, W2, ..., Wi, ..., Wn\}$ is inputted in the domain knowledge query step. The entity name h_i is matched with the original text, and if there exists $c = (h_i, r_i, t_i)$ in the knowledge graph, where h_i is the same as Wi, the relationship r_i and the tail entity t_i from the triple are inserted into the corresponding position after Wi in the original text, achieving the injection of domain knowledge. This results in the fused text $s = \{W0, W1, W2, ..., Wi, \{r_i, t_i\}, ..., Wn\}$, which contains the fused medical domain knowledge. To address the inconsistency between the vector spaces of text and entity embeddings in triples, our approach leverages the pre-trained BERT model for semantic encoding of both the original and fusion text information. This strategy effectively fills the gap in integrating domain knowledge, ensuring a more coherent representation across textual and entity data.

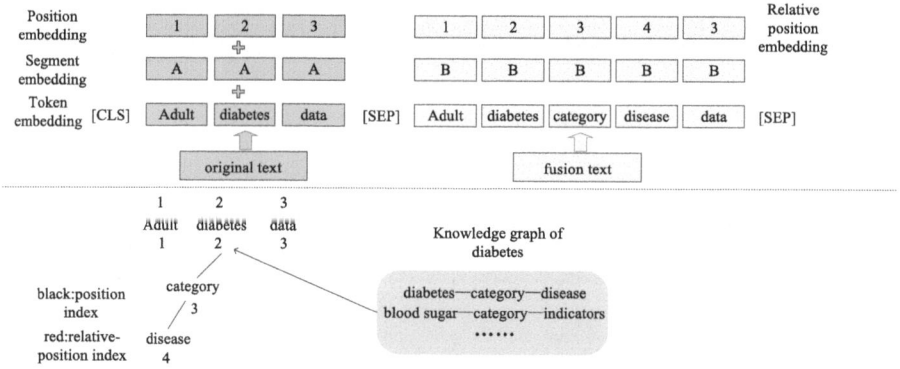

Fig. 2. Knowledge fusion layer. Knowledge is stored in triples in the knowledge graph, where $\epsilon = (h_i, r_i, t_i)$ represents a triple in the medical knowledge graph K, with $\epsilon \in K$.

While obtaining fused text, our method directly concatenates the triples after the entities in the text. This direct insertion of triples disrupts the syntactic structure of the original text and alters the positional encoding of the text. When BERT represents the fused text $s = \{W0, W1, W2, ..., Wi, \{r_i, t_i\}, ..., Wn\}$, we aim to minimize the semantic loss caused by inserting the triple $\{r_i, t_i\}$ and the resulting issue of unclear word order. To achieve this, relative positional encoding is introduced to complement the structural information of the fused text. The original text and the fused text constitute a set of heterogeneous features. The original text is retained to enrich the model's input information. By concatenating the original and fused texts as the input, the model can leverage both characteristics to provide more comprehensive and rich data. Compared to the relative positional encoding in the fused text, the absolute positional encoding in the original text can provide auxiliary information for the fused text. Utilizing these heterogeneous features can enhance the model's understanding and prediction capability of medical entities, thereby improving the performance and accuracy of the model.

In the knowledge fusion layer, we annotate both absolute and relative positions within the integrated text. Absolute positioning sequences the characters' locations within the sentence, while relative positioning maintains the original order of characters before the knowledge graph is incorporated. When integrating triples from the knowledge graph, their relative positions are established according to the segments of the corresponding entities. The rest of the text is kept intact, which minimizes disruption to the textual sequence and preserves the structural coherence of the fused text, as depicted in Fig. 2. This approach enables BERT to effectively capture the interrelations between texts by computing attention mechanisms in the correct sentence order.

3.2 BiLSTM-IDCNN Layer

To further capture contextual information and spatial dependencies in the sequence, we introduce the BiLSTM-IDCNN structure. The BiLSTM model can model the contextual dependencies in the text sequence, while the IDCNN model captures the spatial dependencies in the text sequence. By increasing the receptive field of the convolutional kernel, the IDCNN model reduces information loss caused by pooling operations, thereby improving the accuracy of sequence modeling. Therefore, combining BiLSTM and IDCNN can better extract semantic features from the text sequence and enhance the model's performance. The specific definitions are as follows:

$$i_t = \delta(W_{ix}x_t + W_{ih}h_{t-1} + W_{ic}c_{t-1} + b_i) \tag{1}$$

$$f_t = \delta(W_{fx}x_t + W_{fh}h_{t-1} + W_{fc}c_{t-1} + b_f) \tag{2}$$

$$c_t = f_{t-1}c_{t-1} + i_t tanh(W_{cx}x_t + W_{ch}h_{t-1} + b_c) \tag{3}$$

$$o_t = \delta(W_{ox}x_t + W_{oh}h_{t-1} + W_{oc}c_t + b_0) \tag{4}$$

$$h = o_t tanh(c_t) \tag{5}$$

$$h_t = h^l \oplus h^r \tag{6}$$

where i_t is the output of the input gate of the model, representing important feature information input to the model, ϵ is an activation function. f_t is the output of the forget gate, representing the critical feature information that the model retains for the current sequence. o_t is the output of the output gate, meaning the selected output features based on the relevance dependencies in the modeling. h_{t-1} represents the hidden layer features at time step $t-1$, and c_{t-1} represents the cell state at time step $t-1$. w_* denote the corresponding trainable parameters, randomly initialized at the beginning of model training and continuously optimized during training. b_* represents the related bias parameters, which are not trainable and randomly initialized. h represents the output of the LSTM. Since BiLSTM is a bidirectional model, its feature output is bidirectional, denoted as h^l and h^r, representing the output feature vectors in the forward and backward directions of the BiLSTM model, respectively. $\oplus$ represents the vector concatenation operation, which concatenates the two feature vectors along a particular feature dimension. After obtaining the feature vector $H = [h_1, h_2, \cdots, h_t, \cdots, h_n]$ containing dependencies in the given text sequence through the BiLSTM model, it is fed into the IDCNN model to capture the spatial dependency relationships in the given text sequence.

IDCNN comprises several dilated convolution blocks of the same size, allowing the model to capture a broader range of contexts and exhibit good generalization capabilities. As shown in Fig. 3, the input feature vector is h_t, D_ϵ^j represents the expansion convolutional layer with expansion width ϵ in layer j, $C_t^{(j)}$ represents the features obtained through the addition convolutional network in layer j, and r represents the $ReLU$ activation function. Multiple expansion convolutional layers are cascaded to create an expansion convolutional block, which is denoted by $B()$, and the input k is iterated times to obtain $b_t^{(k)}$. Finally, use the parameter W_o to perform a linear fitting of formula (10) on the output $b_t^{(k)}$ of the last convolutional block, and obtain the output $o_t^{(k)}$ of the module. The IDCNN process is as follows.

$$C_t^{(0)} = D_1^{(0)} h_t \tag{7}$$

$$C_t^{(j)} = r(D_\epsilon^{(j-1)} C_t^{(j-1)}) \tag{8}$$

$$b_t^{(k)} = B(b_t^{(k-1)}) \tag{9}$$

$$O_t^{(k)} = W_o b_t^{(k)} \tag{10}$$

Both employing a kernel size of 3, are compared. Notably, both configurations consist of two convolutional layers. However, the dilated convolution boasts a larger context size of 7, in contrast to the traditional convolution's context size of 5. The illustration depicts dilated convolutions with dilation rates of 1 and 2, demonstrating how dilated convolutions augment the model's receptive field by incorporating gaps within the convolutional kernel.

3.3 Feature Fusion Layer and Decoding Layer

The feature fusion layer receives its input from the IDCNN layer, which encapsulates semantic aspects of the text, captures long-range dependencies within the text sequence, and retains local segments of the sequence. In this layer, a

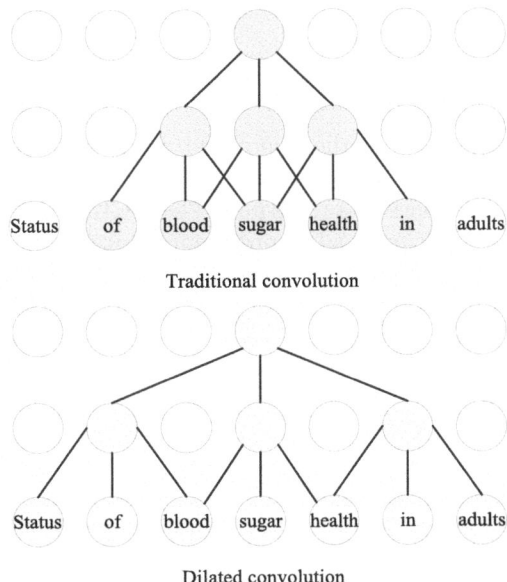

Fig. 3. Traditional and dilated convolutions.

multi-head attention layer is employed to integrate these elements, enabling the model to concentrate on various sections of the input independently through distinct query, key, and value weight matrices. This allows the model to extract additional features and relationships, highlighting significant features while diminishing the focus on less relevant parts. As a result, a unified feature vector sequence is produced, blending the original text with the fused text. Following the multi-head attention output, this sequence is then fed into the CRF model for the decoding process.

$$Loss = -\frac{1}{T}\sum_{t=1}^{T}logP(y^{(t)}|x^{(t)}) \tag{11}$$

T represents the number of training samples, y^t and x^t represent the output and input of the training data, respectively, and $P(y^{(t)}|x^{(t)})$ denotes the probability of output x^t given input y^t. The logarithmic likelihood loss function (11) is used to compare the predicted values of the CRF with the ground truth values, defining the loss as the negative maximization of the likelihood function.

Random variables $x = \{x_1, x_2, \cdots, x_t\}$ and $y = \{y_1, y_2, \cdots, y_t\}$ represent the observation and label sequences, respectively, and $P(y|x)$ denotes the conditional probability, $f(y_{t-1}, y_t, x)$ is used to calculate the transition scores from y_{t-1} to y_t and the score of y_t. The specific calculation is as follow.

$$P(y|x) = \frac{\sum_{t=1}^{T} e^{f(y_{t-1}, y_t, x)}}{\sum_{y}^{Y(x)}\sum_{t=1}^{T} e^{f(y_{t-1}, y_t, x)}} \tag{12}$$

4   Experiments

4.1   Datasets

For this research, we employed the annotated diabetes data from the Tianchi competition, along with the publicly accessible CCKS2017 electronic medical record dataset as our experimental corpus. The CCKS2017 dataset comprises four categories of electronic medical records: general items, medical history characteristics, diagnostic and therapeutic processes, and discharge summaries. The training set consists of 960 clinical records, while the testing set includes 120 clinical records, covering five distinct types of entities. The CCKS2017 dataset corresponds to the Medical Concept Knowledge Graph [32], which includes five categories of triples: symptoms, body parts, diseases, examinations, and treatments. The Medical Concept Knowledge Graph comprises 13,864 triples (Table 1).

Table 1. Entity classification of CCKS2017.

number	Entity classification labels
1	*Body*
2	*Check*
3	*Disease*
4	*Signs*
5	*Treatment*

The diabetes dataset was characterized by titles, abundant superfluous line breaks, punctuation, random characters, and excessive whitespace, which led to sentence fragmentation and posed challenges for sentence segmentation. Consequently, a thorough cleaning process was essential to eliminate these extraneous elements, followed by a reannotation and recounting of the dataset. Subsequently, the dataset was randomly shuffled and divided into training and test subsets in a ratio of 8:2, respectively, based on the length of the text. The training subset comprises 70,085 entities and the test subset consists of 23,366 entities. The diabetes-specific knowledge graph, DiaKG [31], corresponding to the diabetes dataset, encompasses 22,050 entities and 6,890 relationships. It stands out as the pioneering diabetes-related knowledge graph in the Chinese language domain. The dataset encompasses 15 distinct entity classification labels (Table 2).

4.2   Experimental Settings

Experimental environment configuration: GPU-NVIDIA GeForce RTX 3090Ti, Ubuntu (18.04.1), CUDA (11.3), Torch (1.12.1+cu113), Python (3.8). The learning rate is set to 1e-5, batch size is 32, maximum sentence length is 128, dropout

Table 2. Entity classification of diabetes.

number	Entity classification labels
1	*Disease*
2	*Reason*
3	*Symptom*
4	*Test*
5	*Test_Value*
6	*Lowest Level Heading*
7	*Drug*
8	*Amount*
9	*Method*
10	*Treatment*
11	*Operation*
12	*ADE*
13	*Anatomy*
14	*Level*
15	*Duration*

is 0.3, and the AdamW optimizer is used for optimization. Additionally, an early stopping mechanism is implemented during the training process, with an early stopping patience of 5 to prevent overfitting of the model.

4.3 Comparative Experiment

(1) BiLSTM-CRF: A traditional model for named entity recognition utilizes word2vec word embeddings as its input representation, while employing a BiLSTM-CRF architecture for efficient encoding and accurate prediction.

(2) IDCNN-CRF: This model also utilizes pre-trained word2vec word embeddings as input representation, with the semantic encoding part using an IDCNN model.

(3) BERT-CRF: Only employs the pre-trained BERT model for text embedding.

(4) BERT-BiLSTM-CRF: By leveraging a pre-trained BERT model to extract rich semantic features, the resulting representations are subsequently fed into a BiLSTM network. The final step involves applying a CRF layer to refine the labeling process, culminating in the production of the final set of labels.

(5) CRFCW-BiLSTM-CRF [33]: The model ingests input features including word2vec word embeddings, word boundary information, and part-of-speech tags, and it captures sentence dependency relationships using self-attention mechanisms.

(6) MC-BERT [34]: Leveraging a pre-trained medical model that shares the same architecture as the reference baseline model.

(7) Our (TransE): We train the TransE model on the triplets, aligning the triplet features with the text features within the same vector space, while preserving the integrity of other structural elements.

(8) Our (f): The knowledge fusion layer retains the fusion text while maintaining the integrity of the remaining structural components.

(9) Our (nkg): We have omitted the knowledge graph component from our model's architectural design.

(10) Our: The complete model proposed in this paper. In this study, model (4) was selected as the baseline model, and the comparative experimental results are shown in Table 5.

Table 3. Comparative experimental results in CCKS2017.

model	P	R	F1
(1) BiLSTM-CRF	80.83	81.82	81.32
(2) IDCNN-CRF	80.46	82.44	81.44
(3) BERT-CRF	91.50	89.94	90.71
(4) BERT-BiLSTM-CRF	91.64	90.98	91.31
(5) CW-BiLSTM-CRF	/	/	/
(6) MC-BERT	92.03	93.39	93.00
(7) Our(TransE)	93.14	92.07	92.60
(8) Our(f)	93.51	92.86	93.18
(9) Our(nkg)	91.92	90.86	91.38
(10) Our	93.76	93.14	93.44

The table shows that the proposed model in this paper outperforms the baseline model BERT-BiLSTM-CRF with performance improvements of 5.05% and 2.13% on the two datasets, respectively. On the diabetes dataset, CW-BiLSTM-CRF[33] utilizes lexical information but does not use pre-trained models, resulting in limited text representation capability and a decrease of 1.03% in F1 score compared to the baseline model. On the CCKS2017 dataset, the proposed model shows a 0.44% improvement over the medical pre-trained model MC-BERT[34], confirming the effectiveness of introducing a medical knowledge graph. Furthermore, we compared the traditional knowledge representation model TransE, and the proposed model achieves performance improvements of 3.74% and 0.84% on the two datasets. Additionally, compared to the model Our(f), which only uses fused text, the proposed model shows performance improvements of 0.34% and 0.26% on the two datasets, demonstrating the effectiveness of retaining the original text in the knowledge fusion layer. Our(nkg) model does not incorporate knowledge graph information, yet it achieves F1 score improvements of

2.28% and 0.07% compared to the baseline model on the two datasets, proving our work's effectiveness even without including medical domain knowledge (Tables 3 and 4).

Table 4. Comparative experimental results in Diabetes dataset.

model	P	R	F1
(1) BiLSTM-CRF	75.60	69.48	72.41
(2) IDCNN-CRF	75.81	70.32	72.96
(3) BERT-CRF	79.49	75.74	77.57
(4) BERT-BiLSTM-CRF	82.62	80.34	81.46
(5) CW-BiLSTM-CRF	86.84	74.91	80.43
(6) MC-BERT	/	/	/
(7) Our(TransE)	83.24	81.83	82.52
(8) Our(f)	86.78	85.09	85.92
(9) Our(nkg)	85.04	83.46	84.24
(10) Our	87.13	85.39	86.26

4.4 Ablation Experiment

(1) Our(kg): The complete model.
(2) BERT(kg)-BiLSTM-IDCNN-CRF: Removing the attention layer from the our model.
(3) BERT(kg)-BiLSTM-ATT-CRF: Removing the IDCNN module from the our model.
(4) BERT(kg)-BiLSTM-CRF: Removing the attention layer and IDCNN module in our model.
(5) Our(n): Removing the relative positional encoding from the our model.

Table 5. Comparative experimental results for two datasets.

model	Diabetes dataset			CCKS2017		
	P	R	F1	P	R	F1
(1) Our(kg)	87.13	85.39	86.26	93.76	93.14	93.44
(2) BERT(kg)-BiLSTM-IDCNN-CRF	86.54	84.11	85.31	92.04	91.36	91.69
(3) BERT(kg)-BiLSTM-ATT-CRF	86.71	84.44	85.56	92.87	92.10	92.48
(4) BERT(kg)-BiLSTM-CRF	83.57	82.80	83.18	91.83	91.42	91.62
(5) Our(n)	84.12	82.76	83.43	92.17	90.12	92.14

The table shows that the proposed model in this paper outperforms the baseline model BERT-BiLSTM-CRF with performance improvements of 5.05% and 2.13% on the two datasets, respectively. On the diabetes dataset, CW-BiLSTM-CRF[33] utilizes lexical information but does not use pre-trained models, resulting in limited text representation capability and a decrease of 1.03% in F1 score compared to the baseline model. On the CCKS2017 dataset, the proposed model shows a 0.44% improvement over the medical pre-trained model MC-BERT[34], confirming the effectiveness of introducing a medical knowledge graph. Furthermore, we compared the traditional knowledge representation model TransE, and the proposed model achieves performance improvements of 3.74% and 0.84% on the two datasets. Additionally, compared to the model Our(f), which only uses fused text, the proposed model shows performance improvements of 0.34% and 0.26% on the two datasets, demonstrating the effectiveness of retaining the original text in the knowledge fusion layer. Our(nkg) model does not incorporate knowledge graph information, yet it achieves F1 score improvements of 2.28% and 0.07% compared to the baseline model on the two datasets, proving our work's effectiveness even without including medical domain knowledge.

5   Conclusion

This paper introduces a named entity recognition approach that integrates a medical knowledge graph. We enhance the pre-trained model's capability by incorporating domain-specific medical knowledge through the insertion of knowledge graph triplets into the text. To preserve semantic structural information and facilitate the model's understanding, we employ relative positional encoding and a BiLSTM-IDCNN layer, which jointly capture contextual and spatial dependencies. The attention mechanism fuses the original and enriched text, enhancing vector representation. Our method achieves effective entity recognition without altering the pre-trained model's architecture or requiring retraining. However, the efficacy of this method hinges upon the robustness and accuracy of the knowledge graph. Additionally, addressing noise within the integrated triplets represents a significant challenge that warrants further investigation.

Acknowledgments. This work has been supported by the National Natural Science Foundation of China (62166042), and Natural Science Foundation of Xinjiang, China (2021D01C076).

References

1. Ahmed, R., Berntsson, P., Skafte, A., et al.: EasyNER: a customizable easy-to-use pipeline for deep learning-and dictionary-based named entity recognition from medical text (2023)
2. Gururangan, S., Marasović, A., Swayamdipta, S., et al.: Don't stop pretraining: adapt language models to domains and tasks. In: Process of the 58th Annual Meeting of the Association for Computational Linguistics, pp. 8342–8360 (2020)

3. Liu, Z., Huang, D., Huang, K., et al.: Finbert: a pre-trained financial language representation model for financial text mining. In: Process of the Twenty-Ninth International Conference on International Joint Conferences on Artificial Intelligence, pp. 4513–4519 (2021)
4. Liu, W., Zhou, P., Zhao, Z., et al.: K-bert: enabling language representation with knowledge graph. In: Process of the AAAI Conference on Artificial Intelligence, vol. 34, no. 03, pp. 2901–2908 (2020)
5. Fu, L., Cao, Y., Bai, Y., et al.: Development status and prospects of knowledge graphs in domestic vertical fields. Appl. Res. Comput./Jisuanji Yingyong Yanjiu **38**(11) (2021)
6. Yochum, P., Chang, L., Gu, T., et al.: Linked open data in location-based recommendation system on tourism domain: a survey. IEEE Access **8**, 16409–16439 (2020)
7. Lample, G., Ballesteros, M., Subramanian, S., et al.: Neural architectures for named entity recognition. In: Process of the Conference of the North American Chapter of the Association for Computational Linguistics: Human Language Technologies 2016, pp. 260–270 (2016)
8. Peng, N., Dredze, M.: Improving named entity recognition for Chinese social media with word segmentation representation learning. In: Process of the 54th Annual Meeting of the Association for Computational Linguistics, vol. 2: Short Papers, pp. 149–155 (2016)
9. Liu, L., Shang, J., Ren, X., et al.: Empower sequence labeling with task-aware neural language model. In: Process of the AAAI Conference on Artificial Intelligence, vol. 32, no. 1 (2018)
10. Li, Z., Liu, X., Wang, X., et al.: Transo: a knowledge-driven representation learning method with ontology information constraints. World Wide Web **26**(1), 297–319 (2023)
11. Allen, C., Hospedales, T.: Analogies explained: towards understanding word embeddings. In: International Conference on Machine Learning, pp. 223–231. PMLR (2019)
12. Kenton, J.D.M.W.C., Toutanova, L.K.: Bert: pre-training of deep bidirectional transformers for language understanding. In: Process of naacL-HLT, pp. 1–2 (2019)
13. Yu, F., Tang, J., Yin, W., et al.: Ernie-vil: knowledge enhanced vision-language representations through scene graphs. In: Process of the AAAI Conference on Artificial Intelligence, vol. 35, no. 4, pp. 3208–3216 (2021)
14. Cui, Y., Che, W., Liu, T., et al.: Pre-training with whole word masking for Chinese bert. IEEE/ACM Trans. Audio Speech Lang. Process. **29**, 3504–3514 (2021)
15. Joshi, M., Chen, D., Liu, Y., et al.: Spanbert: improving pre-training by representing and predicting spans. Trans. Assoc. Comput. Linguist. **8**, 64–77 (2020)
16. Liu, Z., Lin, W., Shi, Y., et al.: A robustly optimized BERT pre-training approach with post-training. In: China National Conference on Chinese Computational Linguistics, pp. 471–484. Springer, Cham (2021)
17. Budi, I., Suryono, R.R.: Application of named entity recognition method for Indonesian datasets: a review. Bull. Electr. Eng. Inf. **12**(2), 969–978 (2023)
18. Chen, B., Xu, G., Wang, X., et al.: Aishell-ner: named entity recognition from Chinese speech. In: ICASSP 2022-2022 IEEE International Conference on Acoustics, Speech and Signal Processing (ICASSP), pp. 8352–8356. IEEE (2022)
19. Zhu, Z., Zhang, D., Li, L., et al.: Knowledge-guided multi-granularity GCN for ABSA. Inf. Process. Manag. **60**(2), 103223 (2023)
20. Zhang, Y., Wang, Y., Yang, J.: Lattice LSTM for Chinese sentence representation. IEEE/ACM Trans. Audio Speech Lang. Process. **28**, 1506–1519 (2020)

21. Wu, S., Song, X., Feng, Z.: MECT: multi-metadata embedding based cross-transformer for Chinese named entity recognition. In: The Joint Conference of the 59th Annual Meeting of the Association for Computational Linguistics and the 11th International Joint Conference on Natural Language Processing (ACL-IJCNLP 2021) (2021)

22. Li, X., Yan, H., Qiu, X., et al.: FLAT: Chinese NER using flat-lattice transformer. In: Process of the 58th Annual Meeting of the Association for Computational Linguistics, pp. 6836–6842 (2020)

23. Sun, Z., Li, X., Sun, X., et al.: Chinesebert: Chinese pretraining enhanced by glyph and pinyin information. arXiv preprint arXiv:2106.16038 (2021)

24. Wang, X., Gao, T., Zhu, Z., et al.: KEPLER: a unified model for knowledge embedding and pre-trained language representation. Trans. Assoc. Comput. Linguist. **9**, 176–194 (2021)

25. Malaviya, C., Bhagavatula, C., Bosselut, A., et al.: Commonsense knowledge base completion with structural and semantic context. In: Process of the AAAI Conference on Artificial Intelligence, vol. 34, no. 03, pp. 2925–2933 (2020)

26. Radford, A., Narasimhan, K., Salimans, T., et al.: Improving language understanding by generative pre-training (2018)

27. Toutanova, K., Chen, D., Pantel, P., et al.: Representing text for joint embedding of text and knowledge bases. In: Process of the 2015 Conference on Empirical Methods in Natural Language Processing, pp. 1499–1509 (2015)

28. Cao, Y., Hou, L., Li, J., et al.: Joint representation learning of cross-lingual words and entities via attentive distant supervision. Association for Computational Linguistics (2018)

29. Sung, M., Jeong, M., Choi, Y., et al.: BERN2: an advanced neural biomedical named entity recognition and normalization tool. Bioinformatics **38**(20), 4837–4839 (2022)

30. Ma, Y., Zhang, Y., Sangaiah, A.K., et al.: Active learning for name entity recognition with external knowledge. ACM Trans. Asian Low-Res. Lang. Inf. Process. (2023)

31. Chang, D., Chen, M., Liu, C., Liu, L., Li, D., Li, W., Kong, F., Liu, B., Luo, X., Qi, J., Jin, Q., Xu, B.: DiaKG: an annotated diabetes dataset for medical knowledge graph construction. In: Qin, B., Jin, Z., Wang, H., Pan, J., Liu, Y., An, B. (eds.) CCKS 2021. CCIS, vol. 1466, pp. 308–314. Springer, Singapore (2021). https://doi.org/10.1007/978-981-16-6471-7_26

32. Ye, M., Cui, S., Wang, Y., et al.: Medpath: augmenting health risk prediction via medical knowledge paths. In: Proceedings of the Web Conference 2021, pp. 1397–1409 (2021)

33. Houchang, Z., Chengliang, L.: Chinese medical named entity recognition using embedded word features. Chin. J. Med. Libr. Inf. Technol. **30**(9), 42–49 (2022)

34. Chen, P., Zhang, M., Yu, X., et al.: Named entity recognition of Chinese electronic medical records based on a hybrid neural network and medical MC-BERT. BMC Med. Inf. Decis. Mak. **22**(1), 1–13 (2022)

A Text-Oriented Transformer with an Image Aesthetics Assessment Fusion Network for Visual-Textual Sentiment Analysis

Ziyu Liu and Zhonglin Zhang(⊠)

School of Electronic and Information Engineering, Lanzhou Jiaotong University,
Lanzhou 730070, China
zhangzl@mail.lzjtu.cn

Abstract. The rapid advancement of social networks has significantly altered how people convey their emotions, increasingly through a mix of images and text on social media platforms. Visual-textual sentiment analysis has garnered considerable attention because it incorporates visual data into textual sentiment analysis. Moreover, most current visual-textual sentiment analysis approaches underperform because of their limited exploitation of the correlations between these two modalities. Furthermore, current methods for visual analysis tend to focus excessively on extracting image features while neglecting the aesthetic aspects of images. To address these issues, this study introduces a text-oriented transformer with an image aesthetics assessment fusion network mechanism, ter633133_1_En_12_Chaptermed ToTIAN. This approach comprises two main components: aesthetics-oriented visual feature extraction and a text-oriented transformer. It integrates textual information with image aesthetics—which include emotional cues—via a multiattention mechanism, resulting in a comprehensive representation enriched with emotional cues. Extensive experiments on two publicly available datasets confirmed the superior efficacy of the proposed ToTIAN approach compared with the prevalent unimodal and multimodal methods.

Keywords: Text orientation · Image aesthetics assessment · Visual-textual sentiment analysis

1 Introduction

With the rapid development of digital technologies and social networking platforms, individuals increasingly opt to express their emotions on social media through the use of both images and texts, a phenomenon that significantly enhances the diversity and complexity of emotional expressions. Therefore, the ability to understand and analyze affective states in multimodal data is not only a prominent topic in the field of affective computing but also a key technological challenge across various fields, including social media analysis [1], human–computer interaction [2], and mental health monitoring [3]. Visual-textual sentiment analysis aims to examine the integration of sentiment information in images and texts and, on this basis on this, abstracts and infers the emotional intent expressed by humans.

© The Author(s), under exclusive license to Springer Nature Singapore Pte Ltd. 2024
H. Yu et al. (Eds.): CCF NCCA 2024, CCIS 2274, pp. 183–200, 2024.
https://doi.org/10.1007/978-981-97-9671-7_12

As is widely acknowledged, images and texts serve as two distinct carriers of information, each harboring rich and unique semantic and emotional content [4]. Images convey visual emotions through color, composition, aesthetics, and other content features, whereas texts directly present the semantic and emotional tones of language. In addition to accurately extracting sentiment information from each modality, the efficient realization of fusion between them is critical to visual-textual sentiment analysis.

Aesthetic features of images play an important role in contextual understanding [5]. Most traditional visual-textual sentiment analysis methods overlook the importance of assessing aesthetics, thereby failing to identify the emotional cues conveyed by aesthetic features. This omission hampers the accurate capture of complex affective responses triggered by visual imagery, consequently undermining the accuracy of sentiment analysis.

Extracting, understanding, and fusing semantic information in both visual and textual media is crucial for visual-textual sentiment analysis. Some prior studies have focused separately on extracting features from visual and textual content via techniques such as low-rank multimodal fusion (LMF) [6], interactive typical correlation analysis (ICCN) [7], and modal invariant and specific representation (MISA) [8]. These approaches often undervalue the performance impact of each modality individually, instead assessing through gradient updates and each modality's contribution level to the outcome. This approach results in text, which holds the densest semantic content, being equally weighted in the construction of multimodal features, potentially deteriorating the effectiveness of multimodal sentiment analysis in capturing emotional content.

To address these challenges, this paper presents a text-oriented transformer with an image aesthetic assessment fusion network (ToTIAN). ToTIAN comprises two primary components: aesthetics-oriented visual feature extraction (AoV) and the text-oriented transformer fusion network (ToT). The AoV adopts image aesthetics evaluation in its visual feature extraction to pinpoint and elaborate on the emotional cues embodied within those aesthetic features. ToT emphasizes incorporating semantic details into textual representation to bolster the fused representation. Through rigorous validation on two extensively utilized datasets, ToTIAN shows superior performance in visual-textual sentiment analysis, outperforming existing avant-garde methods. This research not only enhances image content feature extraction and image representation accessibility but also underscores the importance of integrating image aesthetic attributes with textual content to increase model efficacy.

Our contributions can be summarized as follows:

1) We propose ToTIAN, a model that adeptly bridges the semantic gap between visual and textual data, utilizing a sophisticated visual-textual fusion mechanism for enhanced multimodal information integration.

2) To reinforce the integration of textual semantics into visual features, we craft a text-oriented transformer fusion network that emphasizes textual information. Concurrently, we introduced an aesthetics-oriented transformer equipped with an aesthetics-oriented visual feature extraction module to grasp the compositional and color aesthetics of images, thereby offering more nuanced emotional cues.

3) After conducting rigorous experiments and thorough analyses on two public datasets, our findings reveal that the ToTIAN approach outperforms existing unimodal

and multimodal models in terms of performance. This underscores its potent and efficient capabilities in tackling visual-textual sentiment analysis tasks.

The remainder of this paper is organized as follows: Sect. 2 reviews prior work in visual-textual sentiment analysis and image aesthetic evaluation. In Sect. 3, we introduce the proposed ToTIAN model. Sect. 4 elaborates on the experimental investigations and analysis of the ToTIAN model across two datasets. Sect. 5 discusses the model's limitations, and Sect. 6 concludes with the study's key findings.

2 Related Work

2.1 Visual-Textual Sentiment Analysis

Numerous multimodal emotion classification approaches have been developed to integrate various modalities. These approaches are classified into three distinct categories: early fusion, intermediate fusion, and late fusion.

Early fusion transforms all features from each modality into the same format and combines them into a single feature for input into the classification algorithm. Xu et al. [9] examined visual-text interactions and proposed a common memory network to iteratively model the interactions between visual and textual content. Peng et al. [10] designed a cross-modal complementary network for multimodal sentiment classification to integrate features from different modalities, thereby reducing the performance degradation that could be caused by integrating irrelevant modalities. Zhang et al. [11] extracted image features after eliminating noise in the text and employed symmetry to discern the internal features of text and images through the attention mechanism.

Intermediate fusion transforms different modal data into high-dimensional feature representations before fusing them in the intermediate layer of the model. Zhou et al. [12] emphasized cross-modal consistency and correlation, extracting semantic and affective interactions between images and text hierarchically to address the noise issue in visual-textual sentiment analysis. Xu et al. [13] designed a progressive dual attention module to capture the correlation between modalities, combining it with rich social data to enhance multimodal sentiment analysis performance. Zhu et al.[14] introduced a cross-modal alignment module to capture region-word correspondences, fusing the multimodal features through an adaptive gating module and integrating individual modal features for enhanced sentiment prediction accuracy. Tashu et al. [15] employed sequential joint attention and weighted modal fusion for sentiment classification in art. Ortis et al. [16] extracted objective textual descriptions of images, combined them with visual features, and inferred sentiment polarity via supervised support vector machines. Yang et al. [17] utilized an updating memory network for deep semantic image-text features and designed a multimodal sentiment analysis model with a multiview attention network, comprising feature mapping, interactive learning, and fusion components. Yadav et al. [18] designed a deep multilevel attention network to generate dual-attention visual maps, enhancing spatial and channel-dimensional representations within CNNs. Yang et al. [19] developed an emotion-aware multichannel graph neural network for image-text emotion detection, encoding modalities and learning multimodal representations, culminating in deep fusion achieved with a multihead attention mechanism.

Late fusion processes classify the features of each modality independently and then fuse the classification results to form a final decision vector, which subsequently generates sentiment predictions. Zhang et al. [20] designed a hybrid fusion network to extract intra- and intermodal features, used multihead visual attention guided with visual features to obtain explicit semantic sentiment features from the text, and finally trained a baseline classifier to learn discriminative knowledge from different modalities to complete decision fusion. Kumar et al. [21] performed image sentiment scoring of regions via SentiBank and SentiStrength scores with R-CNN, obtained text sentiment scores via a novel context-aware blending technique, and finally aggregated independently processed sentiment scores via an optical character recognizer.

2.2 Assessment of Image Aesthetics

The assessment of image aesthetics seeks to evaluate the aesthetic quality of an image by appraising its visual appeal. Owing to its significant potential application in pattern recognition and computer vision, this field is emerging as a prominent research area, garnering increasing attention. A principal task of aesthetic evaluation involves extracting the pertinent aesthetic features. Efforts in feature extraction, ranging from initial hand-designed characteristics to contemporary deep neural networks for automated feature extraction, have culminated in considerable enhancements in aesthetic evaluation performance. Tong et al. [22] identified global low-level features of an image to construct a model of the image's aesthetics, thereby categorizing photographs for both photographers and general users. Datta et al. [23] utilized support vector machines and classification trees to develop automatic classifiers and employed linear regression on feature polynomials to deduce numerical aesthetic ratings, thereby investigating the relationship between emotion and image content. Ke et al. [24] employed color distribution, contrast, and brightness as image features to distinguish between the perceptual factors of professional photographs and snapshots and formulated high-level semantic features to quantify perceptual differences. Liu et al. [25] modeled human visual perception and proposed a semisupervised algorithm designed to optimally eliminate noise and redundant low-level image features.

A myriad of deep learning methods for image representation have emerged in recent years. Krizhevsky and colleagues[26] made significant strides with a CNN using ReLU activation functions and efficient GPU implementation for image classification. Wang and team[27] created a brain-inspired deep network for aesthetic ratings on the basis of the associations among various features. The A-LAMP architecture, proposed by Ma et al. [28], focuses on learning from both fine-grained and global image aspects. Tailoring CNNs, work by Talebi et al. [29] introduced a fully connected layer to better predict the range of human aesthetic preferences. Liu et al. [30] explored image partitioning into local regions to compute aesthetic features, revealing their interconnectivity through graph convolutional networks. In terms of optimizing aesthetics, Zhao et al. [31] deployed gated units to marry compositional and aesthetic image traits. Finally, Li et al. [32] designed a dual-path network that combines manually extracted and convolutionally learned features to assess an image's aesthetic value.

The previously described methodologies do not account for color composition or spatial configuration, both of which are critical for image aesthetics. Addressing this

gap, Lyu et al. [33] leverage user interactions for image ranking and utilize deep reinforcement learning for aesthetic evaluation, thereby creating personalized assessments
aligned with diverse tastes. Chambe et al. [34] explored the generalization capabilities of
a common aesthetic computational model and refined it with professional photographs
to improve accuracy. Sheng et al. [35] introduce a novel multiblock aggregation approach for aesthetic image evaluation that efficiently trains the model in an end-to-end
manner using solely aesthetic labels. Furthermore, they present targets that incorporate
three common attention mechanisms (average, minimum, and adaptive) and assess their
influence through the AVA (Aesthetic Visual Analysis) benchmark. Yi et al. [36] unveil
a new style-specific art evaluation network that adeptly harnesses both style-driven and
generalized aesthetic data to critique images. Finally, Chen et al. [37] present an emotion-
aware multibranch network that enriches aesthetic traits with emotional insights to render
more nuanced aesthetic verdicts.

3 Method

In this section, we detail the ToTIAN methodology, which is composed of two principal components: the Aesthetics-oriented Visual feature extraction (AoV) and the Text-
oriented Transformer Fusion Network (ToT). Figure 1 illustrates the comprehensive
architecture underpinning the ToTIAN framework.

3.1 Model Overview

Our ToTIAN approach encompasses two integral modules: aesthetics-oriented visual
feature extraction (AoV) and the text-oriented transformer fusion network (ToT). The
AoV module is designed to distinguish aesthetic attributes from visual inputs, thus facilitating sentiment analysis through the evaluation of compositional factors, the extraction
of color features via color moments, and the synthesis of aesthetic captions drawn on
pretrained models.

Conversely, the ToT module accentuates textual elements by leveraging a text-
oriented multicentered attention mechanism. This innovation ensures heightened sensitivity to contextually pertinent information and adeptly captures the nuanced emotional
undertones and perspectives conveyed within accompanying text, acknowledging both
the sentiment and the 'emotional color' indicated by the language used.

3.2 Aesthetics-Oriented Visual Feature Extraction

Content Understanding Network
Composition is a fundamental element in photography, painting, design, and various
other visual arts and plays a pivotal role in image aesthetics. A good composition not
only directs the viewer's gaze but also effectively transmits emotions and amplifies the
overall aesthetic appeal of the work. In this work, we input an image into the ResNet
model, which was pretrained on ImageNet, after resizing it to 224 × 224 pixels. Considering that content and color impart a more pronounced effect on an image's emotional

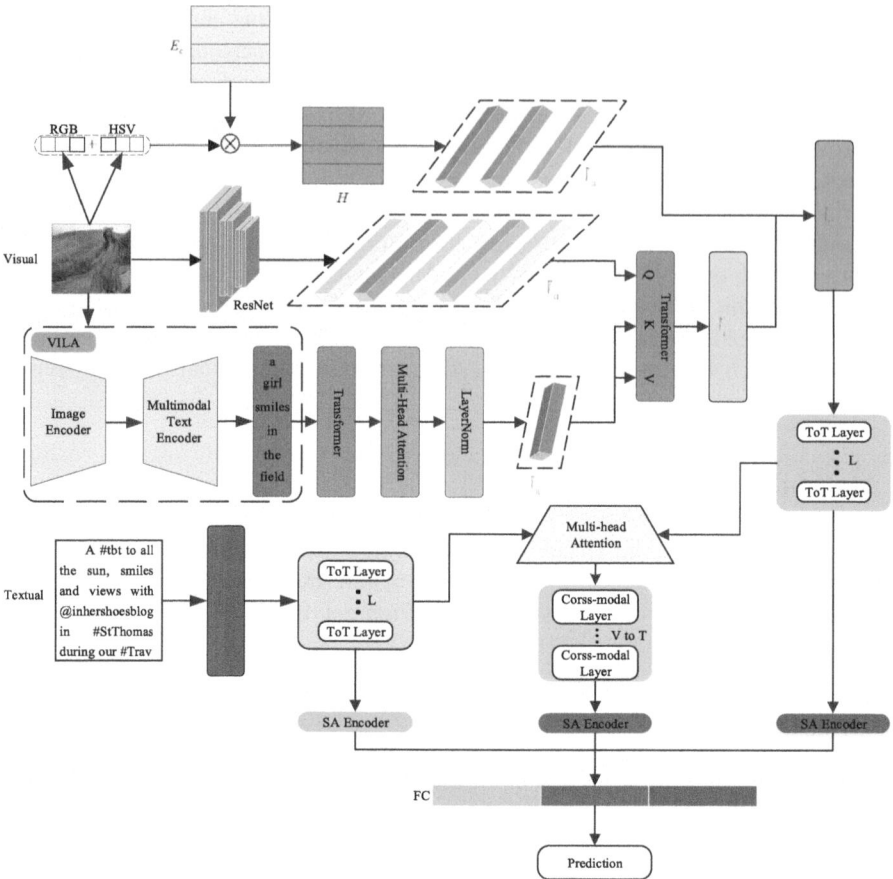

Fig. 1. Overview of the Text-oriented Transformer with the Image Aesthetic assessment fusion Network (ToTIAN)

resonance than spatial factors do, we employed the feature vectors from the pooling layer to represent the content of the image I:

$$V'_{cu} = \text{Resnet}(I) \tag{1}$$

where $V'_{cu} \in \mathbb{R}^{2048}$, 2048 indicates the dimensionality of the feature vector extracted from the image content.

After a linear transformation function is used on the content representation V'_{cu}, the final content features are obtained $V_{cu} = W^T V_{cu'}$, where $W \in \mathbb{R}^{2048 \times d}$ is the trainable parameter matrix.

Color-aware Network (CAN)

Color can not only enhance the aesthetic appeal of images but also communicate deeper meanings and emotions, thereby amplifying visual effects and expressiveness. To augment the semantic representation of colors in images, which have a significant relationship with emotions, inspired by 38 and others, this study initially characterizes color information via low-dimensional vectors and subsequently employs embedding methods to further enhance the representation of emotional information conveyed by colors.

Color moments constitute a technique in digital image processing and analysis for describing and characterizing the color distribution within an image. On the basis of statistical principles, color moments capture the essential properties of color through the calculation of the first-order (mean), second-order (variance), and third-order moments (skewness) pertaining to an image's color distribution. The first-order moments (means) depict the average brightness or intensity for each color channel (e.g., red, green, blue) within an image. Second-order moments (variance) quantify the dispersion within the color distribution of an image, namely, the deviation of each color channel's values from its mean value. Third-order moments (skewness) reveal the asymmetry within an image's color distribution, indicating the extent to which the distribution leans toward the brighter or darker side. Extracting and comparing an image's color moments enables the effective identification and differentiation of image content, as distinct images frequently exhibit substantial variations in the statistical properties related to their color distribution. Furthermore, color moments serve as an invaluable tool in practical applications owing to their simplicity in computation, efficiency, and independence from image size and rotation.

Among the color spaces, the red, green, blue (RGB) color space aligns with intuitive human perception and has a broad range of applications. Furthermore, the hue, saturation, value (HSV) space more closely mirrors people's perception of color and provides an intuitive means to convey the hue, saturation, and brightness of colors. Consequently, we employ the color moment method for collecting image color features from both the RGB and HSV color spaces:

$$c = \text{ColorMoment}(I) \tag{2}$$

To augment the emotional semantic content of the color features identified, we employ the embedding technique to convert the color feature vector c into the color feature matrix H, thus circumventing the issue of semantic scarcity inherent to low-dimensional color feature vectors:

$$H = c \otimes E_c \tag{3}$$

where $E_c \in \mathbb{R}^{18 \times d}$ is the color embedding matrix and where d is the dimension of each color factor. $\otimes$ denotes elemental multiplication. These features are then fed into the Softmax layer to obtain the final high-level color feature V_{ca}:

$$V_{ca} = \text{softmax}(H) \tag{4}$$

Aesthetic Fusion Network

We employ a pretrained image aesthetic learning framework, VILA[39], for generating aesthetic captions pertinent to each image. Subsequently, RoBERTa[40] is utilized to obtain the contextual feature representation $V_t = \left(V_{t_1}, V_{t_2}, ..., V_{t_n}\right)$. Which is further fed into the transformer encoder to obtain the textual aesthetic feature $V_{ta} = \left(V_{ta_1}, V_{ta_2}, ..., V_{ta_n}\right)$

as follows:

$$\mathrm{ATT}^i(V_t, V_t, V_t) = \mathrm{softmax}\left(\frac{\left[W_q^i V_t\right]^\top \left[W_k^i V_t\right]}{\sqrt{d/h}}\right)\left[W_v^i V_t\right]^\top \tag{5}$$

$$\widetilde{V_t} = W_h\left[\mathrm{ATT}^1(V_t, V_t, V_t); ...; \mathrm{ATT}^h(V_t, V_t, V_t)\right]^\top \tag{6}$$

$$\overline{V_t} = LayerNorm(\widetilde{V_t} + V_t) \tag{7}$$

$$V_{ta} = LayerNorm(FFN(\overline{V_t}) + \overline{V_t}) \tag{8}$$

where h denotes the number of heads in multihead attention (MHA). $\left\{W_q^i, W_k^i, W_v^i\right\} \in \mathbb{R}^{d/h \times d}$ denotes the i-th learnable weight matrix of Query, Key, and Value, respectively. The parameter matrix of MHA is denoted as $W_h \in \mathbb{R}^{d \times d}$. *Layernorm* and *FFN* refer to the layer normalization and feedforward networks, respectively.

We consider the content feature V_{cu} as Q and the textual aesthetic feature V_{ta} as K and V. Then, these feature representations are processed through the transformer encoder to obtain the aesthetic feature $V_a \in \mathbb{R}^{d \times 49}$.

The high-level color feature V_{ta} and the aesthetic feature V_a are connected to obtain the final feature I_v of the image as follows:

$$I_v = V_{ca} \oplus V_a \tag{9}$$

3.3 Text-Oriented Transformer Module

Word Embedding
BERT (bidirectional encoder representations from transformers) is a cutting-edge pre-trained deep learning model developed in 2018 by researchers at Google. The defining feature of BERT is its bidirectional context encoding capability, offering rich semantic information for textual models. We leverage BERT as a text encoder; for a given original sentence $S = \{w_1, w_2, \ldots, w_n\}$ consisting of n words, the sequence is formed by concatenating S with special tokens [CLS] and [SEP], which is then input into the encoder. This process yields a sequence representation $I_t \in \mathbb{R}^{T_t \times d_t}$ enriched with contextual information, which serves as the input to the textual model:

$$I_t = \mathrm{BERT}(S, \theta_{BERT}) \tag{10}$$

Text-oriented Transformer
Motivated by the Mult framework, we design a Text-oriented Transformer (ToT) module to encode both visual and textual data. ToT computes attention weights between visual and textual elements, facilitating an exchange of information and emotional content between these modalities. The crucial role of text in multimodal applications is well acknowledged. Text, by expressing complex concepts, emotions, and information in a

highly dense form, serves not only as a direct source of information but also as a crucial bridge connecting other modal data, thus enhancing the system's comprehension and processing abilities. Consequently, image aesthetic features often lack emotion-related information. To mitigate this issue, we enhance the interaction between image aesthetic features and text through the multihead attention mechanism utilized by ToT.

The result of the multiattention mechanism, denoted as $output = Multihead(Q, K, V)$, is formulated as follows:

$$output = concat(head_1, head_2, \ldots, head_{h_n})W^O \tag{11}$$

$$head_i = Attention(Q, K, V) \tag{12}$$

$$Attention(Q, K, V) = \text{softmax}(QK^T \lambda)V \tag{13}$$

where $Q \in \mathbb{R}^{T_m \times d_f}$, $K \in \mathbb{R}^{T_m \times d_f}$, $V \in \mathbb{R}^{T_m \times d_g}$ is the number of 'heads' in the multiple attention mechanism.

To obtain the mapping between visual-textual modalities, we stack the L layers of TOT for the $i = 1, 2, ..., L$ layers:

$$I_{v \to t} = I_v \tag{14}$$

$$I_{v \to t}^i = \text{MHA}_{v \to t}^i(\text{LN}(I_{v \to t}^{i-1}), c_a) + \text{LN}(I_{v \to t}^{i-1}) \tag{15}$$

$$I_{v \to t}^i = \text{FFN}(\text{LN}(I_{v \to t}^i)) + \text{LN}(I_{v \to t}^i) \tag{16}$$

where $I_{v \to t} \in \mathbb{R}^{T_m \times d}$, $I_{v \to t}^i \in \mathbb{R}^{T_m \times d_g}$. $\text{MHA}_{v \to t}^i$ is the multihead attention from vision to text in the i-th layer. LN() denotes layer normalization, and FFN() denotes the feedforward network.

By concatenating the outputs from the final layer of the transformer for each modality, we obtain a modality-specific representation. This representation is then input into the transformer encoder to achieve self-attention. The representations of all the modalities are ultimately concatenated and processed through a fully connected layer for sentiment analysis.

$$y = FC(SA(I_v); SA(I_t); SA(I_{v \to t})) \tag{17}$$

ToT leverages the fact that the textual modality contributes most significantly to sentiment analysis among all modalities. It enhances the role of text and integrates textual information with visual features to achieve a multimodal representation enriched with sentiment-related information.

3.4 Classifier

The Softmax activation function necessitates the use of cross-entropy as the loss function to maintain a consistent probability distribution of the results:

$$L = -\frac{1}{M} \sum_{M}^{i=0} y_i \ln \hat{y}_i \tag{18}$$

where M denotes the number of sentiments and where y_i and $\hat{y}_i$ represent the true and predicted (TP) values of i-th sentiments, respectively.

4 Experiments

To validate the performance of our proposed method, we conducted experiments and benchmarked it against 11 other methods across 2 datasets. Our experimental setup is detailed first, followed by a presentation of the comparative results. The significance of two key modules, the AoV and ToT, is subsequently elucidated through an ablation study.

4.1 Datasets

We conduct experiments on two datasets. Table 1 gives detailed statistics for these datasets.

Twitter: The Twitter 100k dataset, proposed by Hu et al. [41], was selected for its coverage of approximately one quarter of image-text pair data, totaling 19,878 samples after the removal of garbled and erroneous entries.

Flickr: Samples with text lengths greater than 100 or fewer than 5 were selected from the Flickr CC dataset, as presented by Borth et al. [42], which includes 106,897 samples.

Table 1. Statistics of the datasets

Dataset	Positive	Negative	Total
Twitter	11835	8043	19878
Flickr	75087	31810	106897

4.2 Compared with Baselines

SPAN [43]: A model for sequence annotation tasks in the field of natural language processing. In sequence annotation tasks, the model needs to annotate each token in the input sequence with predefined labels, such as named entity recognition and lexical annotation. The core idea of the Span model is to annotate by recognizing consecutive segments of tokens in the sequence rather than independent tokens. This approach captures the contextual information between tokens and helps improve the performance of the model on the annotation task.

RoBERTa [40]: a pretrained language representation model proposed by Facebook AI, which has been refined and enhanced on the basis of the BERT model. The text representation is then input directly into the transformer encoder.

BART [44]: A pretrained sequence-to-sequence model devised by Facebook AI that leverages the transformer architecture and integrates the concepts of both the autoencoder and decoder.

DMAF [45]: Two distinct unimodal attention models have been developed to create proficient sentiment classifiers for visual and textual modalities, and subsequently, a fusion-centric multimodal attention model has been introduced to harness the inherent interconnections among various modalities.

AMGN [46]: A visual semantic attention model has been designed to learn attentional visual features for each word. To efficiently synthesize emotional data from both image and text modalities, a modality-gated LSTM has been developed to acquire multimodal insights by selectively prioritizing the modality with more pronounced emotional content. Finally, a semantic self-attention model was introduced to focus autonomously on discriminative attributes for emotion classification.

TEMMA [47]: Investigation into long short-term memory (LSTM) networks and transformer encoders, leveraging multimodal multihead attention, aims at depicting complex temporal sequences. The use of early, late, and model fusion techniques enhances performance by harnessing complementary modal data.

DTRN [48]: Broadens the scope of affective dynamics into multimodal contexts and introduces a dialog transformer tasked with mapping both intramodal and intermodal emotional dynamics. It is designed so that intramodal dynamics capture temporal sequences and meet context-specific needs within each modality. Moreover, cross-modal dynamics strive to address spatial dependencies of varied granularity across modalities.

VLP2MSA [49]: This method produces distinctive textual cues derived from video content to bolster textual representations and synchronizes unimodal video and text features via video-textual contrastive loss.

ICU [40]: Decomposes the visual-linguistic challenge into two distinct stages: the visual-linguistic component generates image captions in English, and the multilingual language model (MLM) employs these captions for alternative text, facilitating cross-language understanding. This approach relocates the multilingual processing load from the visual-linguistic component to the MLM.

CMMT [51]: This method integrates a module with dual auxiliary tasks to elucidate an intramodal representation of emotional aspects and launches a text-directed cross-modal mechanism that dynamically modulates the visual data's impact on word representations during intermodal exchanges.

4.3 Implementation Details

The experiments utilized an NVIDIA RTX 4070Ti. Image preprocessing was set to 224×224 dimensions as input for ResNet50, which underwent fine-tuning. Consequently, the text underwent transformation into word vectors, attaining lengths of up to 36 and a vector size of 200 each. The learning rate across datasets was set at 4e-5. The hyperparameters α and β maintained fixed values of 0.1 and 0.7 across the dataset. The hidden dimension d, warming ratio, and quantity of attention heads h were 768, 0.1, and 12, respectively. Training spanned 40 epochs, with the model architecture built on the PyTorch framework.

4.4 Evaluation Metrics

We utilize accuracy and F1 scores as principal metrics for evaluating model performance, identified as Acc and F1.

$$F1 = \frac{2TP}{2TP + FN + FP} \tag{19}$$

$$Acc = \frac{TP + TN}{TP + TN + FP + FN} \tag{20}$$

TP (true positive): refers to the number of positive samples that the classifier correctly determines as positive. TP indicates that the model correctly identifies true positive examples.

TN (true negative): refers to the number of negative samples that the classifier correctly determines to be negative. TN indicates that the model correctly excludes true negative samples.

FP (false positive): refers to the number of negative samples that the classifier incorrectly determines to be positive cases. FP indicates that the model incorrectly misclassified true negative cases as positive cases.

FN (false negative): refers to the number of positive case samples that the classifier incorrectly determines as negative cases. FN indicates that the model incorrectly omits real positive cases as negative cases.

4.5 Results and Analysis

Table 2 presents the experimental outcomes of the proposed model. As depicted in Table 2, the ToTIAN model outperforms the baseline in terms of accuracy and F1 scores across both datasets.

Several conclusions can be drawn from the analysis of the results presented in Table 2. Initially, the RoBERTa and BART models demonstrate superior performance within the single-text model domain, surpassing even various multimodal methodologies. This observation provides substantial evidence supporting the benefits conferred by large-scale pretrained language models.

Second, the annotation of data samples in the Twitter and Flickr datasets is predominantly text dependent, demonstrating the most robust relational correlation with the corresponding sentiment labels. Owing to the ToT module, which leverages the predominant contribution of the textual modality in sentiment analysis, the ToTIAN model significantly outperforms conventional text-based approaches by reinforcing textual dominance and integrating textual evidence within visual features. This outcome implies that transmuting visual content into a textual domain as ancillary textual evidence is advantageous for our model. Furthermore, a comparative examination of model performance illuminates an intriguing phenomenon: while theoretically, multimodal frameworks should augment text comprehension via visual information integration, this theoretical advantage is not consistently evident in empirical implementations. This discrepancy may stem from interference of the internal attention mechanisms within multimodal models by visual data when processing purely textual information, thereby impinging upon the models' textual processing capabilities.

Table 2. Performance on Twitter and Flickr

Model	Twitter		Flickr	
	Acc	F1	Acc	F1
SPAN	58.54	72.27	66.11	77.55
RoBERTa	75.24	80.07	80.27	85.82
BART	90.29	92.61	85.33	90.42
DMAF	79.64	81.79	82.06	88.52
AMGN	86.06	88.42	83.88	88.96
TEMMA	90.95	92.99	85.77	90.80
DTRN	91.01	92.60	86.39	90.50
VLP2MSA	92.58	93.65	86.54	90.99
ICU	94.71	94.88	87.05	91.54
CMMT	96.57	95.69	88.04	92.04
ToTIAN (Ours)	**97.80**	**97.68**	**88.91**	**92.57**

Third, the ToTIAN model manifests marked improvements in comparison with a majority of extant multimodal feature fusion methodologies. With respect to the quantifiable scores, ToTIAN registers incremental gains of 0.32 and 0.08 on the Twitter dataset, as well as 0.33 and 0.28 on the Flickr dataset. The ascendancy of ToTIAN over foundational multimodal methodologies is ascribable to several contributory factors: (1) In contrast to techniques that naively amalgamate visual and textual features, the ToTIAN framework proficiently mediates visual-textual schism by diminishing inter-modal disparities via a more granulated mechanism. The gating mechanism deployed for multimodal granular integration within ToTIAN is posited to be more efficient and cogent than the baseline model's approach to intermodal affective dynamics. (2) Most baseline multimodal avenues fail to capture the unspoken emotions encapsulated by the aesthetic attributes of imagery, whereas ToTIAN inaugurated a novel means of encasing aesthetic indicators to enhance visual content representation. Employing the AoV module to steer the framework in assimilating and efficaciously merging aesthetic data from images—a process exceptionally valuable for proffering visual context, aiding the clarification of emotions, and seizing delicate emotional subtleties—is intrinsically beneficial. (3) Models such as the ICU, which harness pretrained algorithms to fabricate image captions subsequently utilized as surrogate text for transcultural semantic comprehension by a multilingual language processor, confront an issue wherein the generated captions are bereft of latent emotional elements. Unlike their counterparts, the aesthetically infused captions synthesized by ToTIAN are imbued with copious sentiment cues, considerably fortifying its ability to perform sentiment analysis.

4.6 Ablation Studies

In this section, a series of ablation studies are conducted to assess the effectiveness of individual components within the ToTIAN model. More specifically, the impact of the following subcomponents is investigated: (1) "w/o AoV" refers to the removal of the aesthetics-guided visual feature extraction module, whereby only visual features are extracted from the image via ResNet50, leading to the extracted visual features lacking aesthetic information. (2) "w/o ToT" indicates the exclusion of the transformer module, which omits textual guidance and classifies visual and textual features by simple concatenation through a fully connected layer. In this scenario, visual and textual modalities are treated equivalently, without the assurance of adequate incorporation of emotional information. (3) "w/o CAN" signifies the removal of the color-aware network, which fails to recognize the RGB and HSV features of the image, focusing solely on its compositional features. The experimental findings are presented in Table 3.

Table 3. Results of Ablation Studies on Twitter and Flickr

Model	Twitter		Flickr	
	Acc	F1	Acc	F1
ToTIAN (Ours)	**97.80**	**97.68**	**88.91**	**92.57**
w/o AoV	92.36	89.81	83.45	84.21
w/o ToT	89.01	82.11	74.57	80.13
w/o CAN	94.85	94.27	84.64	81.66

Effects of AoV
Upon exclusion of the AoV module, the performance of the ToTIAN model decreases on both datasets. Notably, the Flickr dataset demonstrates a more pronounced decline in performance than does the Twitter dataset. This observation is attributed to the AoV branch being specifically designed for detecting implicit emotional cues in images. Furthermore, the prevalence of samples with high similarity in the Flickr dataset exceeds that in the Twitter dataset; therefore, its absence results in a more substantial performance deterioration in datasets characterized by a higher concentration of emotionally charged samples.

Effects of ToT
Table 3 shows that the exclusion of the ToT module leads to the worst performance on the two datasets. The reason is that text can explicitly express specific meanings, concepts, and affective attitudes. This explicit expression of text illuminates the semantic level of language and furnishes detailed background information, enriching the understanding of multimodal interactions. Additionally, the textual modality functions not only as a carrier of information but also as a bridge that connects the other modalities. Via association learning, the model can simultaneously process and comprehend information from various modalities, explore the intrinsic connections between distinct modalities, and foster an in-depth semantic understanding and sentiment analysis. Such capabilities

are crucial for the development of comprehension and inference models, particularly in applications demanding profound parsing of content meanings and sentiment analyses.

Effects of CAN

The performance deteriorated across both datasets upon the removal of the color-aware network. Color is universally recognized as a potent medium for conveying emotions, playing an indispensable role in the complexity of emotional expression through the revelation of nuances within emotional dimensions. As a primary visual attribute, color embodies features at the perceptual level. It possesses a unique capacity to deliver additional cues and subtle distinctions, which are crucial to revealing and comprehending emotional information in images. Overlooking the utility of color in sentiment analysis not only precludes the identification of cues pivotal to the emotional accessibility of images but also impedes the attainment of a holistic understanding of emotional expression in social media.

5   Limitations

As observed from the experiments we conducted, the ToTIAN model demonstrates considerable adaptability to visual-textual sentiment analysis tasks. However, no method can achieve 100% accuracy, and our method is not exempt from limitations. For example, individuals with "sunny depression" may conceal their genuine emotions, presenting a cheerful disposition, while the content posted on social platforms might display positive textual cues and negative visual cues. Furthermore, the use of irony and sarcasm can involve combining textual content of a neutral or seemingly positive nature with visual representations that contrast with the text, limiting the model's analytical effectiveness in these cases. This insight underscores the need to integrate more granular semantic knowledge into the model in subsequent research.

6   Conclusion

Our study introduces ToTIAN, a new visual-textual sentiment analysis model that ingeniously integrates aesthetics-oriented visual feature extraction with a text-oriented transformer, revealing the emotional nuances within multimodal content. With a focus on the aesthetic aspects of imagery, ToTIAN seamlessly weaves these elements with text semantics, exhibiting exceptional capabilities for the visual-textual sentiment analysis task. Experimentally validated on two public datasets, it outperforms numerous baseline models, notably in discerning subtle emotional cues in visual data. Further analysis reveals the critical contributions of each ToTIAN component, with the amalgamation of visual aesthetics and text significantly amplifying the model's efficacy. However, despite its robustness in multimodal sentiment analysis, ToTIAN faces challenges in processing scenes that conceal true emotions or content laced with irony and sarcasm. Such insights and hurdles pave the way for further research, including the assimilation of more nuanced semantic knowledge to refine ToTIAN's interpretation of complex semantics. In summary, ToTIAN heralds a potent method for decoding and inspecting multimodal content, particularly in the social media sphere, with its pioneering fusion of visual and textual elements sparking fresh avenues for ensuing research and practical implementations.

Acknowledgments. This project was supported by the National Natural Science Foundation of China (Grant No. 61662043) and the Phased Research Results of Gansu Philosophy and Social Sciences Planning Project (20YB056).

Conflict of Interest. The authors declare that they have no conflicts of interest.

References

1. Shah, R.R.: Multimodal analysis of user-generated content in support of social media applications. In: Proceedings of the 2016 ACM on International Conference on Multimedia Retrieval, 423–426 (2016)
2. Azofeifa, J.D., Noguez, J., Ruiz, S., et al.: Systematic review of multimodal human–computer interaction[C]//Informatics. MDPI 9(1), 13 (2022)
3. Garcia-Ceja, E., Riegler, M., Nordgreen, T., et al.: Mental health monitoring with multimodal sensing and machine learning: A survey[J]. Pervasive Mob. Comput. **51**, 1–26 (2018)
4. Jindal, K., Aron, R.: A novel visual-textual sentiment analysis framework for social media data[J]. Cogn. Comput. **13**, 1433–1450 (2021)
5. Miao, H., Zhang, Y., Wang, D., et al.: Multioutput learning based on multimodal GCN and coattention for image aesthetics and emotion analysis[J]. Mathematics **9**(12), 1437 (2021)
6. Liu, Z., et al.: Efficient low-rank multimodal fusion with modality-specific factors[J]. arXiv preprint arXiv:1806.00064 (2018)
7. Sun, Z., et al.: Learning relationships between text, audio, and video via deep canonical correlation for multimodal language analysis. In: Proceedings of the AAAI Conference on Artificial Intelligence. **34**(05), 8992–8999 2020
8. Hazarika, D., Zimmermann, R., Poria, S.: Misa: modality-invariant and-specific representations for multimodal sentiment analysis.In: Proceedings of the 28th ACM International Conference on Multimedia, 1122–1131 (2020)
9. Xu, N., Mao, W., Chen, G.: A comemory network for multimodal sentiment analysis. In: The 41st International ACM SIGIR Conference on Research & Development in Information Retrieval, 929–932 2018
10. Peng, C., Zhang, C., Xue, X., et al.: Cross-modal complementary network with hierarchical fusion for multimodal sentiment classification[J]. Tsinghua Sci. Technol. **27**(4), 664–679 (2021)
11. Zhang, K., Geng, Y., Zhao, J., et al.: Sentiment analysis of social media via multimodal feature fusion[J]. Symmetry **12**(12), 2010 (2020)
12. Zhou, T., Cao, J., Zhu, X., et al.: Visual-textual sentiment analysis enhanced by hierarchical cross-modality interaction[J]. IEEE Syst. J. **15**(3), 4303–4314 (2020)
13. Xu, J., Li, Z., Huang, F., et al.: Social image sentiment analysis by exploiting multimodal content and heterogeneous relations[J]. IEEE Trans. Industr. Inf. **17**(4), 2974–2982 (2020)
14. Zhu, T., et al.: Multimodal sentiment analysis with image-text interaction network[J]. IEEE Trans. Multimedia (2022)
15. Tashu, T.M., Hajiyeva, S., Horvath, T.: Multimodal emotion recognition from art using sequential coattention[J]. J. Imaging **7**(8), 157 (2021)
16. Ortis, A., Farinella, G.M., Torrisi, G., et al.: Exploiting objective text description of images for visual sentiment analysis[J]. Multimedia Tools Appl. **80**(15), 22323–22346 (2021)
17. Yang, X., Feng, S., Wang, D., et al.: Image-text multimodal emotion classification via multiview attentional network[J]. IEEE Trans. Multimedia **23**, 4014–4026 (2020)

18. Yadav, A., Vishwakarma, D.K.: A deep multilevel attentive network for multimodal sentiment analysis. ACM Trans. Multimed. Comput. Commun. Appl. **19**(1), 1–19 (2023)
19. Yang, X., et al.: Multimodal sentiment detection based on multichannel graph neural networks.In: Proceedings of the 59th Annual Meeting of the Association for Computational Linguistics and the 11th International Joint Conference on Natural Language Processing vol. 1: Long Papers, 328–339 (2021)
20. Zhang, S., Li, B., Yin, C.: Cross-modal sentiment sensing with visual-augmented representation and diverse decision fusion. Sensors **22**(1), 74 (2021)
21. Kumar, A., Garg, G.: Sentiment analysis of multimodal twitter data[J]. Multimedia Tools Appl. **78**, 24103–24119 (2019)
22. Tong, H.H., Li, M.J., Zhang, H.J., He, J.R., Zhang, C.S.: Classification of digital photos taken by photographers or home users. In: Advances in Multimedia Information Processing-PCM 2004. Berlin, Heidelberg: Springer Berlin Heidelberg, 198–205 (2004)
23. Datta, R., Joshi, D., Li, J., Wang, J.Z.: Studying aesthetics in photographic images using a computational approach. In: Computer Vision-ECCV 2006. Berlin, Heidelberg: Springer Berlin Heidelberg, 288–301 2006
24. Ke, Y., Tang, X.O., Jing, F.: The design of high-level features for photo quality assessment. In: 2006 IEEE Computer Society Conference on Computer Vision and Pattern Recognition. New York, NY, USA, IEEE, 419–426 (2006)
25. Liu, Z., Wang, Z., Yao, Y., Zhang, L., Shao, L.: Deep active learning with contaminated tags for image aesthetics assessment. IEEE Trans. Image Proce. **1** (2018)
26. Krizhevsky, A., Sutskever, I., Hinton, G.E.: ImageNet classification with deep convolutional neural networks. Commun. ACM **60**(6), 84–90 (2017)
27. Wang, Z.Y., Chang. S.Y., Dolcos, F., Beck, D., Liu, D., Huang. T.S.: Brain-inspired deep networks for image aesthetics assessment (2016). arXiv:1601.0415. https://arxiv.org/abs/1601. 04155
28. Ma, S., Liu, J., Chen, C.W.: A-lamp: adaptive layout-aware multipatch deep convolutional neural network for photo aesthetic assessment. In: 2017 IEEE Conference on Computer Vision and Pattern Recognition. Honolulu, HI, USA, IEEE, 722–731 (2017)
29. Talebi, H., Milanfar, P.: NIMA: neural image assessment. IEEE Trans. Image Process. **27**(8), 3998–4011 (2018)
30. Liu, D, Puri, R, Kamath, N, Bhattacharya, S.: Composition-aware image aesthetics assessment. In: 2020 IEEE Winter Conference on Applications of Computer Vision. Snowmass, CO, USA, IEEE, 3558–3567 (2020)
31. Zhao, L., Shang, M., Gao, F., Li, R., Huang, F., Yu, J.: Representation learning of image composition for aesthetic prediction. Comput. Vis. Image Underst. **199**, 103024 (2020)
32. Li, X., Li, X., Zhang, G., Zhang, X.: A novel feature fusion method for computing image aesthetic quality. IEEE Access, 863043–63054 (2020)
33. Lyu, P., et al.: User-guided personalized image aesthetic assessment based on deep reinforcement learning (2021)
34. Chambe, M., Cozot, R., Le Meur, O.: Behavior of recent aesthetics assessment models with professional photography (2019)
35. Sheng, K.K., Dong, W.M., Ma, C.Y., Mei, X., Huang, F.Y., Hu, B.G.: Attention-based multipatch aggregation for image aesthetic assessment. In: Proceedings of the 26th ACM International Conference on Multimedia. Seoul Republic of Korea, New York, NY, USA (2018)
36. Yi, R., et al.: Toward artistic image aesthetics assessment: a large-scale dataset and a new method. In: Proceedings of the IEEE/CVF Conference on Computer Vision and Pattern Recognition, 22388–22397 (2023)
37. Chen, H., et al.: Image Aesthetics Assessment with Emotion-Aware Multi-Branch Network. IEEE Trans. Instrum. Meas. (2024)

38. Stricker, M.A., Orengo, M.: Similarity of color images[C]//Storage and retrieval for image and video databases III. SPiE **2420**, 381–392 (1995)
39. Ke, J., et al.: Vila: learning image aesthetics from user comments with vision-language pre-training. In: Proceedings of the IEEE/CVF Conference on Computer Vision and Pattern Recognition, 10041–10051 (2023)
40. Liu, Y., et al.: Roberta: A robustly optimized bert pretraining approach. arXiv preprint arXiv: 1907.11692 (2019)
41. Hu, Y., Zheng, L., Yang, Y., et al.: Twitter100k: a real-world dataset for weakly supervised cross-media retrieval. IEEE Trans. Multimedia **20**(4), 927–938 (2017)
42. Borth, D., et al.: Large-scale visual sentiment ontology and detectors using adjective noun pairs. In: Proceedings of the 21st ACM International Conference on Multimedia, 223–232 (2013)
43. Hu, M., et al.: Open-domain targeted sentiment analysis via span-based extraction and classification. arXiv preprint arXiv:1906.03820 (2019)
44. Yan, H., et al.: A unified generative framework for aspect-based sentiment analysis. arXiv preprint arXiv:2106.04300 (2021)
45. Huang, F., Zhang, X., Zhao, Z., et al.: Image–text sentiment analysis via deep multimodal attentive fusion. Knowl.-Based Syst. **167**, 26–37 (2019)
46. Huang, F., et al.: Attention-based modality-gated networks for image-text sentiment analysis. ACM Trans. Multimedia Comput. Commun. Appl. (TOMM), **16**(3), 1–19 2020
47. Cai, C., et al.: Multimodal sentiment analysis based on recurrent neural network and multimodal attention. In: Proceedings of the 2nd on Multimodal Sentiment Analysis Challenge, 61–67 (2021)
48. Mao, Y., et al.: Dialoguetrm: exploring the intra-and intermodal emotional behaviors in the conversation. arXiv preprint arXiv:2010.07637 (2020)
49. Yi, G., Fan, C., Zhu, K., et al.: Vlp2msa: expanding vision-language pretraining to multimodal sentiment analysis. Knowl.-Based Syst. **283**, 111136 (2024)
50. Wu, G.: ICU: Conquering Language Barriers in Vision-and-Language Modeling by Dividing the Tasks into Image Captioning and Language Understanding. arXiv preprint arXiv:2310. 12531 (2023)
51. Yang, L., Na, J.C., Yu, J.: Cross-modal multitask transformer for end-to-end multimodal aspect-based sentiment analysis. Inf. Process. Manage. **59**(5), 103038 (2022)

Coverage Path Planning for Aircraft Skin Inspection UGV under Curvature Constraints

Minnan Piao[1], Mingze Sun[1], Haifeng Li[1(✉)], and Yonghui Xie[2]

[1] College of Computer Science and Technology, Civil Aviation University of China, Tianjin 300300, China
hfli@cauc.edu.cn

[2] Intelligent Manufacturing Department, CISDI Information Technology Co., Ltd., Chongqing 401147, China

Abstract. This paper presents a coverage path planning method tailored for the automated inspection of aircraft skin using an Unmanned Ground Vehicle (UGV). Initially, by employing a dual sampling approach, the viewpoint space satisfying various constraints such as photogrammetry, safety distance, and equipment status is computed for each target detection cell across the aircraft's surface. Subsequently, adhering to the principle of minimizing the UGV waypoints, suitable viewpoints are selected from the viewpoint space based on metrics of image quality. In order to extend the coverage range of the onboard detection equipment, the UGV is usually equipped with an automatic lifting device, which leads to a higher center of gravity of the UGV, increasing the risk of tilting during motion. Therefore, it is necessary to strictly limit the curvature of the UGV's path. Considering this, based on the waypoints of the UGV, a coverage path satisfying constraints including curvature and obstacle avoidance is further planned using the Bezier curve method. Finally, the effectiveness of the proposed method is verified through simulations.

Keywords: Aircraft Skin Inspection · Coverage Path Planning · Photogrammetry Constraint · Curvature Constraint

1 Introduction

Aircraft skin damage, such as dents, cracks, and lightning strikes, poses significant safety hazards to flight. To ensure the continuous airworthiness of aircraft, it is essential to conduct Aircraft Skin Inspection (ASI). Currently, approximately 90% of aircraft skin damage detection tasks rely predominantly on manual visual inspection, which presents challenges including high labor intensity, low operational efficiency, and inability to guarantee full coverage. The rapid development of robotics technology offers a promising solution to address these challenges [1–3]. Unmanned Ground Vehicles (UGVs) have strong payload capacity and endurance, enabling them to simultaneously carry various detection equipment such as high-definition array cameras, active infrared thermal imagers, and laser scanners. As a result, the UGV has emerged as an important platform for ASI (as can be seen in Fig. 1).

© The Author(s), under exclusive license to Springer Nature Singapore Pte Ltd. 2024
H. Yu et al. (Eds.): CCF NCCA 2024, CCIS 2274, pp. 201–213, 2024.
https://doi.org/10.1007/978-981-97-9671-7_13

Coverage Path Planning (CPP) is the core technology for ASI robots, enabling automatic coverage of specified areas by planning the set of camera poses (also called "viewpoints") and their traversal sequence. The CPP algorithm for ASI needs to satisfy various constraints. Specifically, to ensure accurate identification and measurement of small damages, strict limitations on the poses of the camera during data collection are necessary, such as the camera's shooting distance and shooting angle. Due to the restricted height and angles of the lifting device and the camera, as well as the constraint of safe operation distance, the reachable viewpoint space is confined, further increasing the difficulty of viewpoint generation. Due to the augmentation of an automatic lifting device, the center of gravity of the UGV becomes higher, increasing the risk of tilting during motion. Therefore, it is also necessary to strictly limit the curvature of the UGV's path, and also restrict the height of the lifting device during the UGV's motion.

Existing research on aircraft skin CPP primarily focuses on Unmanned Aerial Vehicle (UAV) platforms, and the goal is to achieve high coverage ratios or algorithmic efficiency, whether in model-based or model-free scenarios. In [4] and [5], viewpoints are generated using an adaptive primal sampling technique, directing the search towards regions with low accuracy and coverage. A graph-based search algorithm is then employed to determine the path. In [6], due to potential unavailability of the aircraft model, a guided next best view approach coupled with waypoint prediction is proposed to expedite CPP exploration while maintaining high coverage. In contrast, [7] initially constructs a coarse model of the aircraft by guiding a UAV-RGB-D camera system along a predefined path. Subsequently, viewpoints are generated using the primal sampling method, and the CPP problem is tackled utilizing the Monte Carlo tree search algorithm. To enhance the CPP efficiency, the inspection environment is subdivided into multiple subspaces at various resolutions. High-level and low-level CPP algorithms are then developed to plan paths among and within these subspaces, respectively [8].

Compared to the UAV, the CPP based on UGVs faces more constraints, such as the constraints on the curvature of the UGV's path and the height of the lifting device during the UGV's motion, yet there is currently limited research in this area. Therefore, we have developed a dedicated CPP method for the UGV, which can effectively address the multiple practical constraints above mentioned. To be specific, by using the dual sampling approach, the viewpoint space satisfying various constraints such as photogrammetry, safety distance, and equipment status is computed for each target detection cell across the aircraft's surface. Subsequently, adhering to the principle of minimizing the UGV waypoints, suitable viewpoints are selected from the viewpoint space based on metrics of image quality. Then, a coverage path satisfying constraints such as curvature and obstacle avoidance is planned using the Bezier curve method [9].

The remainder of this paper is organized as follows. In Sect. 2, the problem is formulated. In Sect. 3, the CPP method is presented. Simulations are performed in Sect. 4. In Sect. 5, we give the concluding remarks.

2 Problem Description

A UGV is used to achieve automatic inspection of a specified Region of Interest (ROI) on aircraft fuselage surface. The UGV carries a lifting device and a PTZ camera to expand the visible space. The pose of the UGV, the height of the lifting device, and the PTZ

Fig. 1. A UGV for ASI developed by us at Civil Aviation University of China.

of the camera can be controlled to reach a desirable viewpoint ("viewpoint" refers to "camera pose" in this paper), which is denoted as

$$T_i = \left(x_i, y_i, \varphi_i, h_i, \mathcal{P}_i, \mathcal{T}_i, \mathcal{Z}_i\right) \tag{1}$$

where (x_i, y_i) and φ_i are the position and the yaw angle of the UGV, h_i is the height of the camera, $(\mathcal{P}_i, \mathcal{T}_i)$ are the pan and tilt rotational angles of the Pan Tilt Unit (PTU), and $\mathcal{Z}_i$ is the zoom magnification. $P_{i,i+1}$ is used to denote the path of the UGV between (x_i, y_i, φ_i) and $(x_{i+1}, y_{i+1}, \varphi_{i+1})$. Subsequently, the effects of various constraints on CPP are described in detail.

Constraints on Photography: To capture high-quality images and ensure accurate identification and measurement of minor surface damages, it is necessary to restrict the camera pose when taking photos, such as the shooting distance and shooting angle. Therefore, the shooting angle θ (the angle between optical axis and surface normal vector) and the shooting distance d should satisfy

$$0 \le |\theta| \le \theta_{\max}, \ d_{\min 1} \le d \le d_{\max 1}, \ d_{\min 2} \le d \le d_{\max 2} \tag{2}$$

where $\theta_{\max}$ is the maximum allowable value of θ. Due to the obstruction of the wings and engines, the UGV needs to move to a farther position for shooting the fuselage above the wings. Therefore, zooming is used to ensure resolution accuracy in long-distance shooting. In this paper, $\mathcal{Z}_n$ is used for shooting close-range fuselage, while $\mathcal{Z}_d$ is used for shooting distant fuselage. For $\mathcal{Z}_n$, the minimum and maximum shooting distances are $d_{\min 1}$ and $d_{\max 1}$, and for $\mathcal{Z}_d$, the minimum and maximum shooting distances are $d_{\min 2}$ and $d_{\max 2}$.

Constraints on Reachable Viewpoint Space: Due to the physical limitations of the lifting device, the PTU, the camera, as well as the UGV, the reachable viewpoint space is very confined, which can be represented by

$$
\begin{aligned}
&h_{\min} \le h_i \le h_{\max}, \ \mathcal{P}_{\min} \le \mathcal{P}_i \le \mathcal{P}_{\max}, \\
&\mathcal{T}_{\min} \le \mathcal{T}_i \le \mathcal{T}_{\max}, \ D\left(x_i, y_i\right) \ge d_s
\end{aligned}
\tag{3}
$$

where the subscripts "min" and "max" denote the minimum and the maximum values, respectively, $D(x_i, y_i)$ is the distance between the UGV and the nearest obstacle, and d_s is the required safety distance.

Constraints on UGV' Motion: To ensure the stability and safety of the entire aircraft skin image collection system during operation, the lifting device is restricted to maintain its minimum height, and that is $h = h_{min}$ while the UGV is in motion. In addition, define the curvature of a point on the path as κ:

$$\kappa = 1/r \tag{4}$$

where r is the turning radius, and we have the following restriction:

$$K_{min} \leq \kappa \leq K_{max} \tag{5}$$

where K_{min} and K_{max} are the minimum and maximum values of κ.

The objective of the CPP algorithm is to achieve full coverage of specified areas of the aircraft fuselage while satisfying all the aforementioned constraints. The outputs of the algorithm are the sequence of viewpoints $\{T_1, T_2, \cdots, T_N\}$ and the path of the UGV $\{P_{1,2}, P_{2,3}, \cdots, P_{N-1,N}\}$.

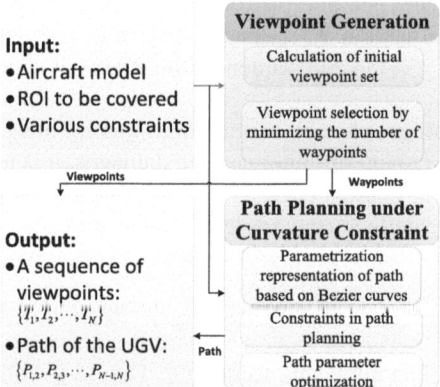

Fig. 2. Overview of the proposed algorithm.

3 Proposed Method

To ensure the lifting device maintains its minimum height while the UGV is in motion, a motion mode of column-wise scanning is adopted. First, the UGV moves to the planned waypoint for a specific column of cells, adjusting the lifting height and sequentially reaching multiple viewpoints with the PTZ to achieve image coverage of that column of cells. Then, the lifting height is lowered to the minimum, and the UGV proceeds to the next waypoint corresponding to another column of cells. This process is repeated until the coverage of the target detection area is complete. The waypoint set of the UGV is obtained by two steps. First, initial viewpoint set for each cell is generated to satisfy various constraints such as photogrammetry, safety distance, and equipment status, and then the viewpoint is selected by minimizing the number of waypoints to improve the inspection efficiency. Based on the waypoints, a coverage path satisfying constraints such as curvature and obstacle avoidance is planned using the Bezier curve method. Overview of the proposed algorithm is presented in Fig. 2.

3.1 Viewpoint Generation

(1) Calculation of initial viewpoint set

The ROI to be covered is first decomposed into unit cells by using the method in [3]. To meet the constraints of the shooting angle and shooting distance, the four side planes passing through the four edges of the unit cell and forming the maximum shooting angle with the unit cell are obtained, and the four hemispheres are established with the unit cell center as the sphere center and the minimum/maximum shooting distance as the radius. The viewpoint space satisfying the shooting constraints is enclosed by the four planes and four hemispheres, and it is represented as follows:

$$(s - v_j)^T n_j \geq 0, j = \{1, 2, 3, 4\}$$
$$d_{\min 1} \leq |s - O| \leq d_{\max 1} \tag{6}$$
$$d_{\min 2} \leq |s - O| \leq d_{\max 2}$$

where s denotes the position coordinates of the viewpoint, O represents the center point of the unit cell, v_j ($j = 1, 2, 3, 4$) are the vertices of the unit cell, and n_j is the normal vector of the jth side plane.

Further, selection of viewpoints satisfying the safety distance within the viewpoint space is needed. Since the algorithm complexity is primarily determined by the size of the viewpoint set, only the viewpoints whose horizontal projections lie on the exterior outline S are selected to reduce computational burden. S is obtained by inflating the outline of the projection of the aircraft on the horizontal plane with the safety distance d_s. Let N_S represent the number of points on S, and S_k ($k = 1, 2, \cdots, N_S$) denote a point on S. For each S_k, calculate the height of the viewpoint that can satisfy the front-view constraint as much as possible:

$$h_k = O(3) + \frac{n_0(3)\sqrt{((O(1) - S_k(1))^2 + (O(2) - S_k(2))^2)}}{\sqrt{(1 - n_0^2(3))}} \tag{7}$$

where n_0 represents the normal vector of the unit cell, which can be obtained using principal component analysis. h_k is constrained by $h_{\min}$ and $h_{\max}$ as

$$h_k = \begin{cases} h_{\min}, & h_k \leq h_{\min} \\ h_{\max}, & h_k \geq h_{\max} \\ h_k, & \text{else} \end{cases} \tag{8}$$

If $V_k = \left[S_k(1), S_k(2), h_k \right]$ lies within the viewpoint space defined by (6), it is added to the initial viewpoint set; otherwise, it is discarded.

Next, the local potential field method is employed to calculate the orientation $\mathcal{P}_k$ and $\mathcal{T}_k$ at V_k. First, the normalized vectors between all points on the unit cell U and V_k are obtained, and the mean of all vectors is calculated as

$$w_k = \frac{1}{N_U} \sum_{m=1}^{N_U} (P_m - V_k) / |P_m - V_k| \tag{9}$$

where P_m denotes a point on U, N_U represents the number of points on U, and w_k denotes the normalized directional vector. Once w_k is obtained, the angles of the gimbal can be calculated as

$$\begin{cases} \mathcal{P}_k = \operatorname{atan2}\big(w_k(1),\ w_k(2)\big) - 3\pi/2 - \varphi_i \\ \mathcal{T}_k = -\arctan\left(w_k(3) \Big/ \sqrt{w_k(1)^2 + w_k(2)^2} \right) \end{cases} \tag{10}$$

where φ_i is subsequently calculated based on adjacent UGV waypoints. The actual commands for $\mathcal{P}_k$ and $\mathcal{T}_k$ are given by:

$$\mathcal{P}_k = \begin{cases} \mathcal{P}_{\min}, & \mathcal{P}_k \le \mathcal{P}_{\min} \\ \mathcal{P}_{\max}, & \mathcal{P}_k \ge \mathcal{P}_{\max}, \\ \mathcal{P}_k, & \text{else} \end{cases} \quad \mathcal{T}_k = \begin{cases} \mathcal{T}_{\min}, & \mathcal{T}_k \le \mathcal{T}_{\min} \\ \mathcal{T}_{\max}, & \mathcal{T}_k \ge \mathcal{T}_{\max} \\ \mathcal{T}_k, & \text{else} \end{cases} \tag{11}$$

$\mathcal{Z}_k$ can be obtained by the following equation:

$$\mathcal{Z}_k = \begin{cases} \mathcal{Z}_n, & d_{\min 1} \le |V_k - O| \le d_{\max 1} \\ \mathcal{Z}_d, & d_{\min 2} \le |V_k - O| \le d_{\max 2} \end{cases} \tag{12}$$

In fact, viewpoints whose horizontal projections lie on or outside the outline S in the viewpoint space can also simultaneously satisfy the safety distance and shooting constraints. However, considering all the possible viewpoints would lead to a significant amount of occlusion culling computations in the subsequent steps, thereby reducing the efficiency of the algorithm. Apart from reducing computational load, selecting viewpoints with horizontal projections lying on S can also minimize the shooting distance and increase the resolution.

Another basic constraint for a viewpoint is to achieve complete coverage of a specified cell, whose validation includes two main calculation steps, frustum culling and occlusion culling, respectively. Frustum culling is to extract all the voxels within the camera's view frustum, while occlusion culling further culls the voxels obstructed by the airplane itself or other obstacles. Based on the method in [3], the coverage ratio of a viewpoint for the corresponding unit cell is derived. If the coverage ratio is 1, the viewpoint is added to the initial viewpoint set; otherwise, it is discarded.

After the above calculations, for the *ith* unit cell, we can derive its feasible viewpoint set $VPSet_i$, whose component is $T_k = \big(x_k, y_k, \varphi_k, h_k, \mathcal{P}_k, \mathcal{T}_k, \mathcal{Z}_k\big)$, and the index

k is added to the set *IndexSet$_i$*. Let *WayPoint$_k$* $= [x_k, y_k]$ represent the UGV position of T_k, and add *WayPoint$_k$* to the set *WPSet$_i$*. The distance and angle between each T_k ($k \in$ *IndexSet$_i$*) and the corresponding *ith* unit cell are calculated, and then the distance and angle are weighted to obtain a comprehensive performance metric:

$$J_k = |V_k - I_k| + c \arccos(w_k n_0) \tag{13}$$

where I_k denotes the intersection point between w_k and the *ith* unit cell, and c is the weighting factor, which can be adjusted according to actual shooting requirements.

(2) Viewpoint selection by minimizing the number of waypoints

Due to the adoption of a column-scanning motion pattern, a common UGV waypoint is computed for each column of unit cells by selecting an appropriate viewpoint for each unit cell. Firstly, compute the intersection of *WPSet$_i$* for all the cells in each column, which is denoted as *Inter_WPSet*. For each waypoint in *Inter_WPSet*, the sum of the comprehensive performance metrics of all the corresponding viewpoints is calculated as *TJ*. Finally, we select the waypoint with the minimum *TJ* as the common waypoint of that column. The corresponding viewpoints are also selected for the respective unit cells.

Some columns may have UGV waypoints with small distances between them, which can be merged to further improve the inspection efficiency. Therefore, we compute the set of waypoints whose distances between each other are less than r_{min}, and the centroid of this set is calculated as S_o, which represents the new horizontal projection of the viewpoints. Based on S_o, the above computations are repeated. If the newly generated viewpoints still satisfy the various constraints, then the viewpoint adjustment is effective, and merging can be performed; otherwise, merging is not conducted.

3.2 Path Planning Under Curvature Constraint

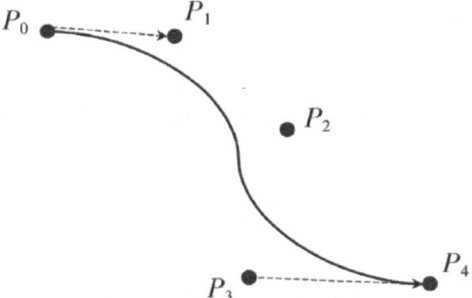

Fig. 3. A quartic Bezier curve.

To further plan a smooth path for the UGV under the curvature constraint, the quartic Bezier curve method is used in this section.

(1) Parametrization representation of path based on Bezier curves

As shown in Fig. 3, five control points $(P_0, P_1, P_2, P_3, P_4)$ can determine a quartic Bezier curve. The curve starts at the first control point P_0 and ends at the last control point P_4. The curve usually does not pass through the control points in the middle (P_1, P_2, P_3), but these points affect how much and in which direction the curve bends. The parameterized expression of the quartic Bezier curve is

$$P(t) = P_0(1-t)^4 + 4P_1(1-t)^3 t + 6P_2(1-t)^2 t^2 + 4P_3(1-t)t^3 + P_4 t^4, \quad t \in [0, 1] \tag{14}$$

where t is a variable that varies within the closed interval $[0,1]$, controlling the curve from the starting point P_0 to the ending point P_4. Let $P(t) = [x(t), y(t)]$ and $P_i = [x_i, y_i]$, where $x(t)$, $y(t)$, x_i, and y_i represent the plane coordinates of $P(t)$ and P_i, respectively, and then we have

$$x(t) = x_0(1-t)^4 + 4x_1(1-t)^3 t + 6x_2(1-t)^2 t^2 + 4x_3(1-t)t^3 + x_4 t^4,$$
$$y(t) = y_0(1-t)^4 + 4y_1(1-t)^3 t + 6y_2(1-t)^2 t^2 + 4y_3(1-t)t^3 + y_4 t^4, \quad t \in [0, 1] \tag{15}$$

The curvature is the rate of rotation of the tangent direction angle with respect to arc length at a certain point on the curve, and its parameterized expression is [9]

$$\kappa(t) = \frac{x'(t)y''(t) - y'(t)x''(t)}{(x'^2(t) + y'^2(t))^{\frac{3}{2}}} \tag{16}$$

(2) Constraints in path planning
(3) Constraints on initial and terminal states

Let the initial state of the UGV be $X_0 = [x_0, y_0, \varphi_0, \kappa_0]$, where (x_0, y_0) is the plane coordinate of P_0, φ_0 is the initial heading angle, and κ_0 is the initial curvature. By setting $d_1 - |P_0 P_1|$, P_1 can be obtained according to the characteristics of the Bezier curve as

$$P_1 = \begin{bmatrix} x_1 \\ y_1 \end{bmatrix} = \begin{bmatrix} x_0 + d_1 \cos(\varphi_0) \\ x_0 + d_1 \sin(\varphi_0) \end{bmatrix} \tag{17}$$

κ_0 can be expressed by

$$\kappa_0 = \frac{3}{4} \times \frac{|(P_1 - P_0)(P_2 - P_1)|}{|P_1 - P_0|^3} \tag{18}$$

Then, P_2 can be obtained under the constraint of κ_0 as

$$P_2 = \begin{bmatrix} x_2 \\ y_2 \end{bmatrix} = \begin{bmatrix} x_2 \\ \dfrac{4\kappa_0(d_1^2 + y_1^2)^{\frac{3}{2}}}{3d_1} + \dfrac{y_1(x_2 - x_1)}{d_1} + 2y_1 \end{bmatrix} \tag{19}$$

Let the terminal state be $X_T = [x_T, y_T, \varphi_T, \kappa_T]$, where (x_T, y_T) is the plane coordinate of P_4, φ_T is the terminal heading angle, and κ_T is the terminal curvature. By setting $d_4 = |P_3P_4|$, P_3 can be obtained as

$$P_3 = \begin{bmatrix} x_3 \\ y_3 \end{bmatrix} = \begin{bmatrix} x_4 - d_4\cos(\varphi_T) \\ x_4 - d_4\sin(\varphi_T) \end{bmatrix} \tag{20}$$

For the local path between two waypoints, P_0 and P_4 are just the waypoints. Besides, φ_0 and φ_T can be calculated according to the adjacent waypoints. To be specific, φ_0 of the current path is the same with φ_T of the last path, and φ_T of the current path is determined according to the end waypoints of the current path and the next path. φ_0 of the first path is calculated by the first two waypoints. Similarly, κ_0 of the current path is the same with κ_T of the last path. κ_0 of the first path can be obtained by

$$\kappa = \frac{1}{r} = \frac{\tan\vartheta}{L} \tag{21}$$

where r is the turning radius, L is the wheelbase, and ϑ is the steering angle. Except for the above variables, there are three remaining variables (d_1, d_4, x_2) to be optimized to determine this quartic Bezier curve.

2) Bounded curvature constraint

Substituting (15) into (16) can obtain

$$\kappa(t) = \frac{At^4 + Bt^3 + Ct^2 + Dt + E}{(Ft^6 + Gt^5 + Ht^4 + It^3 + Jt^2 + Kt + L)^{\frac{3}{2}}} \tag{22}$$

where

$$A = b_xa_y - a_xb_y, \ B = 2(c_xa_y - a_xc_y), \ C = c_xb_y - b_xc_y + 3d_xc_y - 3a_xd_y,$$
$$D = 2(d_xb_y - b_xd_y), \ E = d_xc_y - c_xd_y, F = a_x^2 + a_y^2, \ G = 2(a_xb_x + a_yb_y),$$
$$H = b_x^2 + 2a_xc_x + b_y^2 + 2a_yc_y, \ I = 2(a_xd_x + b_xc_x + a_yd_y + b_yc_y), \tag{23}$$
$$J = c_x^2 + 2b_xd_x + c_y^2 + 2b_yd_y, \ K = 2(c_xd_x + c_yd_y), \ L = d_x^2 + d_y^2,$$

$$a_x = 4x_0 - 16x_1 + 24x_2 - 16x_3 + 4x_4, \ bx = -12x_0 + 36x_1 - 36x_2 + 12x_3,$$
$$c_x = 12x_0 - 24x_1 + 12x_2, \ d_x = -4x_0 + 4x_1,$$
$$a_y = 4y_0 - 16y_1 + 24y_2 - 16y_3 + 4y_4, \ b_y = -12y_0 + 36y_1 - 36y_2 + 12y_3, \tag{24}$$
$$c_y = 12y_0 - 24y_1 + 12y_2, \ d_y = -4y_0 + 4y_1$$

The parameters from A to L can be derived from the above three variables (d_1, d_4, x_2). To ensure motion safety, $\kappa(t)$ $(t \in [0, 1])$ must be constrained by

$$K_{\min} \leq \kappa(t) \leq K_{\max} \tag{25}$$

3) Obstacle avoidance constraint

To achieve obstacle avoidance, each point on the Bezier curve is judged to determine whether it is in the polygon defined by the contour points in S:

$$P(t) \notin \text{Poly}(S), \quad t \in [0, 1] \tag{26}$$

(3) Path parameter optimization

The path planning problem can be characterized as a nonlinear optimization problem as

$$\text{minimize: } J(X_0, X_T, p) = \kappa_{\max}(t_1) - \kappa_{\min}(t_2), \quad t_1, t_2 \in [0, 1]$$
$$\text{s.t. } \kappa_{\max}(t) \leq K_{\max}, \quad \kappa_{\min}(t) \geq K_{\min}, \quad d_1 > 0, d_4 > 0, \tag{27}$$
$$P(t) \notin \text{Poly}(S), \quad t \in [0, 1]$$

where t_1 and t_2 are the values of t that make the path curvature maximum and minimum, respectively. The physical meaning of the objective function is to optimize the parameter $p = [d_1, d_4, x_2]$ to minimize the difference between the maximum and the minimum curvature, so as to obtain a smoother path. In this paper, Sequential Quadratic Programming (SQP) method [10] is used to solve the problem. In each optimization iteration, the maximum and minimum curvature can be obtained by calculating the extreme value of $\kappa(t)$. The core of the SQP algorithm is that at the current iteration point, a quadratic programming is constructed by using the quadratic approximation of the objective function and the first approximation of the constraint function, and the quadratic programming problem is solved to obtain the next better iteration point. If the iteration point satisfies the convergence criterion, it is considered to be the solution of the nonlinear quadratic programming, otherwise continue to iterate.

4 Simulation Results

In this paper, Boeing 737-300 is selected as the object to be inspected, and the inspection area is the side of the fuselage whose height is between 1.5m and 4m. The main parameters of the CPP algorithm are

$$\theta_{\max} = 60°, \ d_{\min 1} = 2\text{m}, d_{\max 1} = 3\text{m}, d_{\min 2} = 9\text{m}, d_{\max 2} = 10\text{m},$$
$$\mathcal{Z}_n = 1, \ \mathcal{Z}_d = 4, \ h_{\min} = 2.35\text{m}, \ h_{\max} = 1.45\text{m}, \ \mathcal{P}_{\min} = 0°, \ \mathcal{P}_{\max} = 360°,$$
$$\mathcal{T}_{\min} = -15°, \ \mathcal{T}_{\max} = 90°, \ d_s = 2\text{m}, \ c = 0.2, \ K_{\min} = -0.187, \ K_{\max} = 0.187 \tag{28}$$

A total of 66 unit cells are obtained by decomposing the ROI for inspection, and the final viewpoint planning results are shown in Fig. 4. The coordinate system composed of three different colored axes represents the camera coordinate system at the planned viewpoint. Each point on the outer contour S represents the UGV waypoint corresponding to each column of cells. Among them, the blue UGV waypoints are obtained by merging waypoints, the circular UGV waypoints indicate that their corresponding viewpoints require a 4x zoom ($\mathcal{Z}_d = 4$), and the remaining star-shaped UGV waypoints indicate that their corresponding viewpoints require a 1x zoom ($\mathcal{Z}_n = 1$). According to the calculation, the planned viewpoints can achieve complete coverage of the specified fuselage area. The computer platform for running the CPP algorithm is ThinkPad X1 (Windows 10 system, Core i7 processor, 32GB memory), and the mathematical software is Matlab 2021b. The algorithm running time is 30.34 s, which can meet the real-time requirements well.

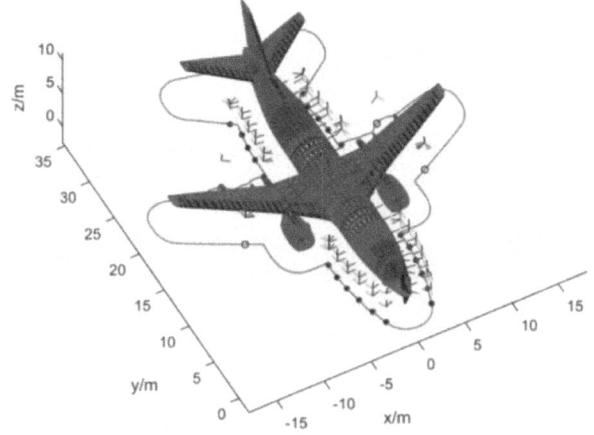

Fig. 4. Viewpoint planning results.

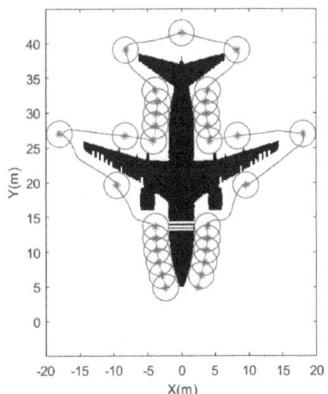

Fig. 5. Planned path based on A* and TEB.

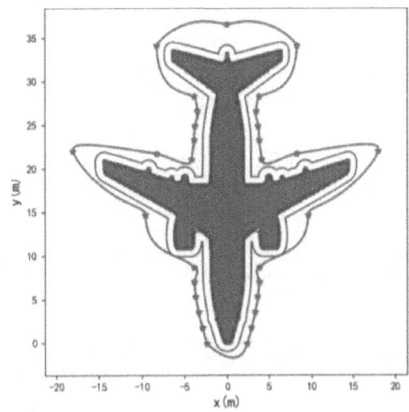

Fig. 6. Planned path with our algorithm.

To demonstrate the advantages of our path planning method, the path generated based on A* and TEB (Timed Elastic Band)algorithms [11] is also presented. Figure 5 shows the path planned by A* and TEB algorithms. Figure 6 shows the path formed by using the Bezier curve algorithm under the curvature constraints ($\kappa_0 = 0.1$). According to Fig. 5, it can be seen that the path based on A* and TEB algorithms cannot make a smooth transition at the waypoint, especially in the path of the wing part, where the turn is too sharp, which may cause the tipping of the UGV. As can be seen from Fig. 6, the path formed by Bezier curve algorithm can make a smooth transition at waypoints, and the curvature constraint is realized in the entire path. In Fig. 5, the maximum curvature at the waypoint reaches 13, which is too large and exceeds the maximum curvature constraint. The paths in Fig. 6 are all generated under curvature constraints, which can meet the requirements of curvature constraints.

5 Conclusion

We have developed a dedicated CPP method tailored for UGVs that effectively addresses multiple practical constraints. Specifically, utilizing the dual sampling approach, the viewpoint space for each target detection cell on the aircraft's surface was computed, considering constraints such as photogrammetry, safety distance, and equipment status. Following this, we employed the principle of minimizing UGV waypoints to select suitable viewpoints based on the image quality metrics. Furthermore, by utilizing the waypoints of the UGV, a coverage path satisfying constraints including curvature and obstacle avoidance using the Bezier curve method was planned. This comprehensive approach ensures efficient and effective aircraft skin inspection while navigating the challenges posed by real-world constraints.

Acknowledgement. The authors acknowledge the financial support from the Research Project of Tianjin Municipal Education Commission under Grant 2021KJ043.

References

1. Jovancevic, S., Larnier, J.J., Orteu, J.B., et al.: Automated exterior inspection of an airplane with a pan-tilt-zoom camera mounted on a mobile robot. J. Electron. Imaging **24**(6), 061110, 1–15 (2015)
2. Leca, D., Cadenat, V., Sentenac, T., et al.: Sensor-based obstacle avoidance using spiral controllers for an aircraft maintenance inspection robot. In: 18th European Control Conference, Naples, Italy, pp. 2083–2089 (2019)
3. Piao, M., Li, H., Zhang, F., et al.: Viewpoint generation for coverage inspection of airplane fuselage surface based on unmanned ground vehicles. In: 42nd Chinese Control Conference, Tianjin, China, pp. 4413–4418 (2023)
4. Almadhoun, R., Taha, T., Gan, D., et al.: Coverage path planning with adaptive viewpoint sampling to construct 3D models of complex structures for the purpose of inspection. In: Proceedings of the 2018 IEEE/RSJ International Conference on Intelligent Robots and Systems (IROS), Madrid, Spain, pp. 7047–7054 (2018)
5. Silberberg, P., Leishman, R. C.: Aircraft inspection by multirotor UAV using coverage path planning. In: Proceedings of the 2021 International Conference on Unmanned Aircraft Systems (ICUAS), Athens, Greece, pp. 575–581 (2021)
6. Almadhoun, R., Taha, T., Seneviratne, L., et al.: Multi-robot hybrid coverage path planning for 3D reconstruction of large structures. IEEE Access **10**, 2037–2050 (2022)
7. Sun, Y., Ma, O.: Automating aircraft scanning for inspection or 3D model creation with a UAV and optimal path planning. Drones **6**(4), 87 (2022)
8. Cao, C., Zhang, J., Travers, M., et al.: Hierarchical coverage path planning in complex 3D environments. In: Proceedings of the 2020 IEEE International Conference on Robotics and Automation (ICRA), Paris, France, pp. 3206–3212 (2020)

9. Chen, C., He, Y., Bu, C., et al.: Quartic Bezier curve based trajectory generation for autonomous vehicles with curvature and velocity constraints. In: Proceedings of the 2014 IEEE International Conference on Robotics and Automation (ICRA), Hong Kong, China, pp. 6108–6113 (2014)
10. Nocedal, J., Wright, S.J.: Numerical Optimization. Springer, New York (2006). https://doi.org/10.1007/978-0-387-73301-2_5
11. Rösmann, C., Feiten, W., Wösch, T., et al.: Efficient trajectory optimization using a sparse model. In: Proceedings of the European Conference on Mobile Robots, Barcelona, Spain, pp. 138–143 (2013)

Less Hallucination and More Factuality: Human Values Alignment in Text Summarization

Zihan Qiu and Yumei Xu[✉]

School of Information Science and Technology, Beijing Foreign Studies University,
Beijing 100089, China
xuyuemei@bfsu.edu.cn

Abstract. Text summarization task aims to condense lengthy documents into shorter texts while preserving the essential information. However, related work mostly focused on preserving semantic information of documents and ignored the human values alignment between documents and summaries. What's more, summaries generated by LLMs are often found to contain hallucinations and lack factuality. In this paper, we introduce a two-stage extract-generate summarization framework to address the above issues. In the first stage, sentences containing both semantic and human values are extracted through a diffusion model. Then in the second stage, overall summaries are generated via LLMs. By extracting important sentences before summarization, we can shift LLMs' attention to preserve more human values information from the documents, hence achieving the goal of human values alignment. Specifically, we follow Schwartz's Theory of Human Values for human values definition due to its wide adoption as a value framework in the realm of NLP(Natural Language Processing). To verify the performance of our framework, we designed several LLM-only methods including in-context methods as baselines to compare with. Experiments conducted on CNN/DM and Reddit summarization datasets demonstrate improvements on both human values alignment and factuality achieved by our framework, indicating the importance of an extractor before leveraging LLM for summarization.

Keywords: Human Values Alignment · Text Summarization · Large Language Model

1 Introduction

Text summarization task aims to generate a short summary from the source document while preserving important information in an easy-to-read and grammatically correct way [1,2]. Existing approaches can be categorized into two categories: extractive approaches [3,4] and abstractive approaches [5,6]. Extractive

This work was supported by the National Social Science Foundation (No.24CYY107).

approaches directly extract sentences from document to form summarization [7], while abstractive approaches paraphrase the main contents of the document [8].

Both extractive and abstractive approaches exhibit limitations. Extractive approaches fail to maintain coherency, particularly in terms of the transition between sentences [9]. Abstractive approaches suffer from factual errors and inaccuracies in grammatical structure since they involve language comprehension and generation [8]. With the advent of large language models (LLMs), abstractive summarization via LLMs has witnessed rapid growth in recent research [10–15]. Research has shown that LLM-generated summaries not only achieved improved coherency and correctness [12,13] but also gained more human preferences compared to traditional automatic metrics like fine-tuning models including BART and T5 [14,15]. For example, it's shown that human evaluators favor summaries generated by GPT-3 over a fine-tuned BART model 70% of the time [15].

However, in LLM-based abstractive summarization approaches, two crucial challenges exist. First, LLMs suffer from hallucination, a phenomenon where the model lies and fabricates non-existent facts, which may lead to generate unfaithful or inaccurate summaries [16,17]. The second challenge arises from the position bias in LLMs, where they tend to focus on the beginning or the end of the source documents [18,19]. This bias steers LLMs to generate summaries unfairly biased towards certain positions within source documents, leading to a lack of comprehensive consideration, particularly noticeable with lengthy documents [20].

Existing research on summarization task mainly focused on improving metrics of coherency, informativeness and factuality [21–24]. Human values alignment, a crucial dimension, is absent from the current text summarization research, mainly due to the traditional extractive or abstractive approaches will not cause human values deviations between documents and summaries. However, the issues of hallucination and position bias that occur in LLMs make it significant to prioritize human values alignment in LLM-based abstractive summarization methods.

Many studies have utilized Schwartz's Theory of Basic Human Values [25] to explore the human values behind LLMs [26,27]. These studies have revealed that LLMs often exhibit human values bias towards specific countries, including USA, as well as some European and South American countries [26]. Such bias may result in deviations in human values between the source documents and summaries. Recent research has made effort to align human values of LLMs with human beings for generating human preferred content [28,29]. However, to the best of our knowledge, we are the first to consider human values alignment of LLMs in text summarization task, regarding it as a part of essential factual consistency.

In this work, we propose a novel two-stage extract-generate summarization framework for human values alignment, while alleviating both the hallucination and position bias issues inherent in LLMs. In the first stage, we extract sentences that capture both semantic information and human values from the documents, with the aim of preserving these values. In the second stage, these selected sentences serve as input of LLMs for generating overall summaries. This framework ensures that LLMs pay more attention to the content of the selected sentences rather than being driven by position bias or inherent human values bias to ensure the human values alignment. The contributions of this paper are as follows:

- This paper presents a first step towards human values alignment in the field of text summarization task. The integrated consideration of human values and summaries enables LLMs to generate summaries that not only accurately capture the essence of the document but also the human values conveyed within it.
- We proposed an extract-generate framework for achieving the human values alignment goal in summarization task while alleviating the hallucination and position bias issues inherent in LLMs. This two-stage approach allows us to extract important sentences from documents for guiding LLMs in generating summaries.
- We conducted experiments on CNN/DM [30] and Reddit [31] summarization datasets, covering both news and social media domains known for conveying human values. The experiment results show that our method can improve the performance both on human values similiarity and FactCC while preserving the performance on ROUGE.

2   Related Work

2.1   Research on Human Values

Human values are beliefs and motivations guiding people's actions and decision-making, including curiosity, hedonism, and etc. [32,33]. In the field of NLP, research concerning human values can be broadly divied into two main categories: human values identification and human values alignment.

Human Values Identification. Research on identifying human values in texts focused on word-level or sentence-level values identification. For example, Chen et al. [34] applied Linguistic Inquiry and Word Count (LIWC) metric to analyze individuals' word use in Reddit posts and examined the correlation between word usage and user's underlying human values. Boyd et al. [35] further explored the relationship between words and human values by employing the Meaning Extraction Method (MEM) for topic modeling on Facebook posts. Their research proved the relevance between word use and human values.

However, these works primarily focused on word-level values identification. To address the issue, Kiesel et al. [36] explored to identify human values behind arguments, thereby facilitating sentence-level value identification. Task 4 (ValueEval) of the SemEval 2023 competition [37] built upon this work, attracting numerous researchers to develop more effective systems for identifying human values behind arguments. Therefore, we leveraged the best-performing system in the competition, as proposed by Schroter et al. [38] for values identification in texts.

To the best of our knowledge, there has been no prior research focusing on identifying summarization-level human values. Given that human values can reflect the author's intent, we consider them as crucial information to preserve during summarization process. By employing the computational methods for values identification, we can compare human values within source documents with those within summaries, thereby assessing the ability of applied summarization models to preserve human values.

Human Values Alignment. With the rapid development of LLMs, human values alignment in the Artificial Intelligence Generated Content (AIGC) by LLMs has attracted much attention [26,27,39,40], for the purpose of achieving responsible AI technology.

Previous research has shown the human values bias towards specific countries exhibited in LLMs [26], indicating the significance of human values alignment for generating equitable and unbiased contexts across different culture. Recent study primarily focused on human values alignment with human intent. For instance, Kang et al. [39] injected specific value distributions into LLMs to predict human opinions. Liu et al. [41] proposed a reinforcement learning based method to align human values of LLMs with humans. In this work, instead of aligning human values with human intent, we focus on human values alignment between summaries and source documents, proposing a newly human values alignment task.

2.2   Research on Text Summarization

There are mainly two approaches of text summarization: extractive and abstractive. Extractive approaches, pioneered by Luhn [42], select sentences directly from documents to form summaries. They often suffer from overall coherency issues due to missing context information [43]. In response, abstractive approaches were introduced to generate summaries by rewriting key ideas from the documents. Abstractive approaches typically leverage transformer-based encoder-decoder models like T5 [44], Bart [45], and Pegasus [46]. Yet, these models struggle with long documents due to the high quadratic self-attention complexity [47] and may produce summaries lacking faithfulness and containing factual errors [46,48]. To address these challenges, a hybrid approach called extract-generate framework has been proposed [49], taking the advantages of both extractive and abstractive approaches for improving summarization performance.

The extract-generate framework enables an extractive model to guide an abstractive model in generating comprehensive summaries, making it particularly effective for handling lengthy documents [49]. This approach enhances overall summaries by achieving higher ROUGE and improving readability [50–52]. Hsu et al. [53] used an extraction model to determine the possibility of each sentence's inclusion, serving as the sentence-level attention, which then modulates word-level attention in the abstractive model to generate summaries with higher ROUGE. Chen et al. [49] bridged these stages using a sentence-level policy gradient method optimized through reinforcement learning. DYLE jointly trained extractive and abstractive models while keeping the extracted text as latent variables [54]. Zhang et al. [55] employed extract-generate framework using ChatGPT for both extractive and generation stage, reducing the hallucination issues in LLM-generated summaries.

Our work follows the extract-generate framework to leverage both the advantages of extractive and abstractive summarization approaches, but different from the previous work, we place particular emphasis on preserving human values during the extractive stage to guide the overall LLM-generated summarization.

3 Preliminaries and Task Formulation

3.1 Preliminaries

Human values are beliefs and motivations guiding people's behaviors in daily lives. In this work, we leverage the refined version of Schwartz's theory of basic human values [25] to give definition of human values. Schwartz's theory contains 19 distinct types of human personality, as listed in Table 1. To detect the underlying human values of given documents, we applied the state-of-the-art human values detection model, revealed in the ValueEval Task of SemEval 2023 competition [37].

Table 1. Human values definition

Id	Value	Definition
1	**Self-direction: thought**	Freedom to cultivate ideas and abilities
2	**Self-direction: action**	Freedom to determine actions
3	**Stimulation**	Excitement, novelty, and change
4	**Hedonism**	Pleasure and sensuous gratification
5	**Achievement**	Success according to social standards
6	**Power: dominance**	Control over people
7	**Power: resources**	Control resources
8	**Face**	Maintain public image
9	**Security: personal**	Safety in immediate environment
10	**Security: societal**	Safety in the wider society
11	**Tradition**	Maintain cultural, family, or religious traditions
12	**Conformity: rules**	Adherence to regulations and obligations
13	**Conformity: interpersonal**	Avoidance of upsetting or harming other people
14	**Humility**	Recognize a modest view of one's importance
15	**Benevolence: caring**	Being a reliable ingroup member
16	**Benevolence: dependability**	Devotion to the welfare of ingroup members
17	**Universalism: concern**	Advocacy for equality and justice
18	**Universalism: nature**	Preservation of the natural environment
19	**Universalism: tolerance**	Inclusivity and empathy

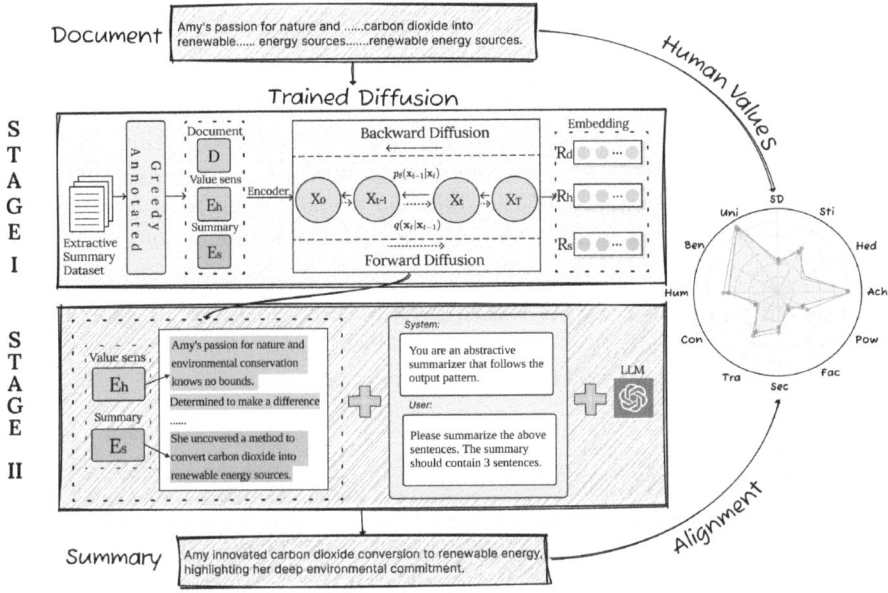

Fig. 1. Two stage summarization framework for human values preservation

3.2 Task Formulation

Our task aims to align human values between the generated summary and source document. Given source document $D = (d_1, d_2, ..., d_m)$ consisting of m sentences, we aim to generate a summary $S = (s_1, s_2, ..., s_n)$ consisting of $n(n < m)$ sentences, where D and S should exhibit consistent human values. Given the predefined l dimension human values, the human values of D and S are denoted as $V_d = (v_1, v_2, ..., v_l)$ and $V_s = (v_1, v_2, ..., v_l)$, respectively. We use the Euclidean distance metric [56] to measure the distance between V_d and V_s, as shown in below:

$$\text{Distance}_{V_d, V_s} = Euclidean(V_d, V_s) \tag{1}$$

As in summarization task, main human values within the source document are what we concerned for, we determine the l dimension human values by setting a threshold θ. The dimension of human values above threshold θ of the source document is regarded as the l dimension human values.

4   Proposed Method

Figure 1 illustrates our two-stage extract-generate framework designed to achieve human values alignment between the document and the generated summary. In the first stage, known as the extractive stage, important sentences containing both semantic information and human values are extracted from documents. Then in the generation stage, an overall summary is generated via LLMs based on the extracted sentences. Both stages are described in Sect. 4.1 and Sect. 4.2.

4.1 Extractive Stage to Capture Human Values

In the extractive stage, given document D, our goal is to extract two types of sentences: E_h representing human values and E_s capturing semantic information. A diffusion model is trained to identify E_h and E_s. Since no existing annotated datasets are available to train a diffusion model for human values sentences, we present the training procedure of the diffusion model, along with a description of our greedy-based approach for human values dataset annotation.

Training Diffusion Extractive Model. Diffusion model uses a parameterized Markov chain to gradually generate the sample $x_0 \sim p(x_0)$ from a latent variable x_T in simple distribution, which can be formulated as :

$$p_\theta(x_0) := \int p_\theta(x_{0:T}) \, dx_{1:T} \tag{2}$$

where $p_\theta(x_{0:T})$ denotes the joint distribution of x_0 and latent variables $x_1, ..., x_T$, θ denote the model parameters.

In order to use discrete texts as inputs for the continuous diffusion model, we first encode D with E_s and E_h using Sentence-BERT [57] into contextualized sentence embeddings of R_d, R_s and R_h, respectively. Note that $R_d \in \mathbb{R}^{m \times k}$, $R_s \in \mathbb{R}^{n \times k}$ and $R_h \in \mathbb{R}^{g \times k}$, where k denotes the dimension of the encoded sentence embeddings. Then we concatenate R_d, R_s and R_h together to get $R_w \in \mathbb{R}^{(m+n+g)k}$ and use a one-step Markov transition to obtain the initial continuous input x_0, which can be formulated as:

$$q(x_0 | R_w) = N(R_w, \beta_0 I) \tag{3}$$

Having x_0, we denote x_0^d, x_0^s and x_0^h as the transition of R_d, R_s and R_h, respectively, where $x_0^d = x_0[: m]$, $x_0^s = x_0[m : m+n]$, and $x_0^h = x_0[m+n :]$.

In the forward process of diffusion model, to facilitate the generation of diffusion model conditionally on different documents, we maintain x_0^d while only adding noise into $x_0^s \oplus x_0^h$, which is formulated as :

$$
\begin{aligned}
q(x_t^s \oplus x_t^h | x_{t-1}^s \oplus x_{t-1}^h) = \\
N(x_t^s \oplus x_t^h; \sqrt{1 - \beta_t}(x_{t-1}^s \oplus x_{t-1}^h), \beta_t I)
\end{aligned}
\tag{4}
$$

where $\oplus$ denotes the concatenation operation.

Hence, x_t at t-th step in the forward process is determined as:

$$x_t = x_0^d \oplus q(x_t^s \oplus x_t^h | x_{t-1}^s \oplus x_{t-1}^h) \tag{5}$$

Following T steps of the forward process, a reverse process was applied to sample random noise to reconstruct x_0. The conditional distribution $p_\theta(x_{t-1}^s \oplus x_{t-1}^h | x_t^s \oplus x_t^h)$ is modeled as a Gaussian distribution:

$$\mathcal{N}(x_{t-1}^s \oplus x_{t-1}^h; \mu_\theta(x_t^s \oplus x_t^h, t), \sigma_\theta(x_t^s \oplus x_t^h, t)) \tag{6}$$

where $\mu_\theta(\cdot)$ and $\sigma_\theta(\cdot)$, which predict the mean and standard variation at each step, are computed by a Transformer network.

During the training process, the objective of the diffusion model is to minimize the mean square error (MSE) loss between the Kullback-Leibler (KL) divergence of p_θ and q. For detailed optimization techniques of the diffusion loss, we refer to Ho et al. [58]. The diffusion loss is defined as:

$$\mathcal{L}_{\text{diffusion}} = \sum_{t=2}^{T} \|x_0 - f_\theta(x_t, t)\|^2 +$$
$$\|R_w - f_\theta(x_1, 1)\|^2 + \mathcal{R}(x_0) \tag{7}$$

where $f_\theta(x_t, t)$ represents the predicted x_t and $\mathcal{R}(x_0)$ is a L_2 regularization term.

To enable the diffusion model to generate R_s and R_h jointly conditioned on R_d, we adopted the approach by Zhang et al. [4], defining both contrastive loss and matching loss. While both losses are computed to update R_s, only the matching loss is calculated to update R_h, as R_s and R_h may contain same sentences.

The contrastive loss aims to minimize the distance between R_s and R_d while maximizing the distance within R_s, the contrastive labels y_p^c are defined as follows:

$$y_p^c = \begin{cases} q, & \text{if } p \leq n \text{ and } d_p = s_q \\ q, & \text{if } p = n + q \\ 0, & \text{otherwise} \end{cases} \tag{8}$$

$$\mathcal{L}_{\text{contra}} = \frac{-1}{2N_{y_p^c} - 1} f(r, y^c) \tag{9}$$

$$f(r, y^c) = \sum_{p=1}^{n+m} \sum_{\substack{q=1; \\ q \neq p; \\ y_q^c = y_p^c}}^{n+m} \log \frac{\exp\left(r_p \cdot r_q^T / \tau\right)}{\sum_{\substack{k=1; \\ p \neq k}}^{n+m} \exp\left(r_p \cdot r_k^T / \tau\right)} \tag{10}$$

where $N_{y_p^c}$ is the total number of sentences in the document that have the same label y_p^c and τ is a temperature hyperparameter.

The matching loss aims to optimize the cross-entropy loss between predicted labels $\hat{y}_j$ and ground-truth labels y_j. During training process, $\hat{y}_j$ can be computed as follows:

$$\hat{y}_j = \text{softmax}\left(r_j \cdot R_d^T\right) \tag{11}$$

where r_j denotes the generated sentence representation. For the summary loss, r_j belongs to R_s and for the value sentence loss, r_j belongs to R_h.

Hence, the matching loss is defined as:

$$\mathcal{L}_{\text{match}} = \sum_{i=1}^{m} \text{CrossEntropy}(\mathbf{y}_i, \hat{\mathbf{y}}_j) \tag{12}$$

The value sentence loss $\mathcal{L}_{\text{value}}$ equals to the matching loss only for value sentences, while the summary loss $\mathcal{L}_{\text{summary}}$ contains both matching loss and contrastive loss, expressed as:

$$\mathcal{L}_{\text{summary}} = \mathcal{L}_{\text{match}} + \lambda \mathcal{L}_{\text{contra}} \tag{13}$$

where λ adjusts the contribution of the contrastive loss.

The overall loss can be defined as:

$$\mathcal{L} = \alpha \mathcal{L}_{\text{diffusion}} + \beta \mathcal{L}_{\text{summary}} + \gamma \mathcal{L}_{\text{value}} \tag{14}$$

where α, β and γ are hyperparameters that balance the loss components.

Dataset Annotation. In this section, we present dataset annotation procedure built for training the diffusion model to extract human values sentences. The greedy optimization algorithm, widely used for generating extractive summary oracles [54,59,60], maximizes the ROUGE between extracted sentences and the document. In our approach, we propose a greedy extraction algorithm to extract value sentences E_h that minimize the human values distance with document. This distance is calculated by Euclidean Distance as:

$$\text{Distance}_{V_d, V_s} = \sqrt{\sum_{i=1}^{l} (V_{di} - V_{si})^2} \tag{15}$$

The proposed greedy extraction algorithm starts with a set containing only the first sentence d_1 of D. Then we iteratively select sentences from the document. If adding a selected sentence to the set reduces the human values distance from the document, we include it in the set. We determine whether to include the first sentence in the final set by evaluating the human values of the set without it. We annotate human values labels on extractive summarization dataset. Hence, we can finally obtain a dataset containing both labeled value sentences E_h and summary E_s.

4.2 Summary Generation Stage

With a trained diffusion model, we can obtain R_w by denoising $x_T^s \oplus x_T^h$, where $x_T^s \oplus x_T^h$ denotes random noise. Subsequently, we derive extracted sentences Top_k by selecting the k sentences in R_w with the highest similarity with R_d. This process can be described as below:

$$Top_k = \text{argsort}(\underset{r_j \in R_w}{\text{softmax}} (r_j \cdot R_d^T))[1:k] \tag{16}$$

Based on Top_k, the generation stage leverages the text summarization generation capacity of LLMs to generate overall summaries S. Previous studies showed that LLMs exhibit better performance on question-answering tasks through in-context learning (ICL) settings [27]. This insight motivates us to design prompts either with or without ICL to explore the impact of ICL on human values alignment. The prompts are present in Table 2.

5 Experiment

5.1 Experiment Settings

Datasets. As human values alignment has more significance in news and social media domains, we perform experiments on CNN/DM [30] and Reddit [31] datasets covering both domains. CNN/DM [30] is the most commonly used summarization dataset containing news articles and associated human-written highlights as summaries. In this work, we use the non-anonymized version. The CNN/DM is split into three sets: 287,084 for training, 13,367 for validation and 11,489 for testing. Reddit [31] contains posts collected from social media platform Reddit and their highlights. Here, we use the TIFU-long version of Reddit. The Reddit is also split into three sets: 41,675 for training, 645 for validation and 645 for testing.

Baselines. In this work, to compare with our methods, we designed four LLM-based text summarization settings either with or without ICL. As LLMs outperform the previous fine-tuned models in text summarization task [14], these fine-tuned models are not considered as the baselines. The baselines are described below:

- **Abs** : Abs setting is the most commonly used LLM-based summarization setting. This setting leverages LLMs to generate summaries directly on source documents.
- **Abs+ICL**: Abs+ICL setting first introduces human values definition to LLMs, and then asks LLMs to generate summaries preserving the human values within source documents.
- **Ext-Abs**: Ext-Abs setting follows the two-stage pattern of our method by leveraging LLMs for extractive stage. LLMs first extracts sentences from the source document, and then generate overall summary based on extracted sentences.
- **Ext-Abs+ICL**: Ext-Abs+ICL setting requires LLMs to extract sentences reflecting the human values within the source document by introducing the human values definition to LLMs. The overall summary then can be generated based on extracted sentences via LLMs.

For each setting, we designed respective prompt as detailed in Table 2.

Evaluation Metrics. We conducted text summarization evaluation on 1000 examples from each testing set of the datasets. To evaluate the quality of generated summaries, we utilize ROUGE [61] for calculating the semantic similarity between generated summaries and references. To investigate the hallucination behind generated summaries, we employ FactCC [48] for factuality detection. Besides the above mainstream summarization evaluation methods, we also explore human values similiarity between summaries and the source documents by computing the Euclidean distance.

Table 2. Prompts used for our methods and baselines. Ext_q denotes sentences extracted by ChatGPT. q denotes the length of sentences extracted. n denotes the length of overall summary. Human values definition is given in Table 1.

Settings	Prompt
Diff-Abs+ICL	**System:**You are an abstractive summarizer that follows the output pattern **User:**{human values definition} Please summarize the document. The summary should contain {n} sentences and be aligned with document's values according to the given human values definition. Document:{Top_k}
Diff-Abs	**System:**You are an abstractive summarizer that follows the output pattern **User:** Please summarize the document. The summary should contain {n} sentences. Document:{Top_k}
Abs	**System:**You are an abstractive summarizer that follows the output pattern **User:** Please summarize the document. The summary should contain {n} sentences. Document:{D_m}
Abs+ICL	**System:**You are an abstractive summarizer that follows the output pattern **User:** {human values definition} Please summarize the document. The summary should contain {n} sentences and be aligned with document's values according to the given human values definition. Document:{D_m}
Ext-Abs	**System:**You are an extractive summarizer that follows the output pattern **User:** Please extract {q} sentences from the document as the summary. Document:{D_m} **System:**You are an abstractive summarizer that follows the output pattern **User:** Please summarize the document. The summary should contain {n} sentences. Document:{Ext_q}
Ext-Abs+ICL	**System:**You are an extractive summarizer that follows the output pattern **User:** {human values definition} Please extract {q} sentences from the document as the summary. The summary should be aligned with document's values according to the given Value Definition. Document:{D_m} **System:**You are an abstractive summarizer that follows the output pattern **User:** Please summarize the document. The summary should contain {n} sentences. Document:{Ext_q}

Implementation Details. In the first extractive stage, the Transformer with 8 layers and 12 attention heads is used for sentence encoding. For diffusion model, the diffusion steps T is set to 500, and the temperature? is set to 0.07. The channel number is set to 128 and the hidden size is set to 768. An AdamW optimizer is used for optimizing. The scaling factors α, β, γ and λ are set to 100, 0.7, 0.3 and 0.001 respectively. The diffusion model is trained on the PC equipped with NVIDIA RTX A5000.

In the second generation stage, we leverage GPT-3.5-turbo to generate overall summaries. The temperature for GPT-3.5-turbo is 1. The length of extracted sentences is typically 70% of the length of the source document. The length of overall summary refers to the average length of ground-truth. For CNN/DM, the length of overall summary is 3. For Reddit, the length of overall summary is 2. The threshold θ for determining the human values dimension is set to 0.3.

5.2   Experiment Results

The overall results are shown in Table 3. We can observe Diff-Abs setting leads to significant improvements on both human values similiarity and FactCC while preserving the performance on ROUGE. To compare with the strongest baseline, for human values similiarity, Diff-Abs setting outperforms Abs setting by

Table 3. Experiment results of baselines and our methods on Reddit and CNN/DM dataset. Dis(hv.): human values distance between summaries and documents. The best score is highlighted in **bold**.

Dataset	Setting	Dis(hv.)↓	ROUGE-1	ROUGE-2	ROUGE-L	FactCC
Reddit	Abs	0.42	**20.64**	**4.03**	**16.93**	37.46
	Abs+ICL	0.48	19.70	3.66	16.40	42.68
	Ext-Abs	0.43	19.97	3.68	16.38	39.13
	Ext-Abs+ICL	0.47	13.99	1.38	11.50	43.74
	Diff-Abs	**0.41**	18.59	2.87	15.30	**46.96**
	Diff-Abs+ICL	0.50	14.97	1.64	12.41	43.73
CNN/DM	Abs	0.27	**27.30**	**7.41**	**25.05**	59.34
	Abs+ICL	0.30	25.79	6.38	23.48	60.40
	Ext-Abs	0.27	25.41	6.13	23.16	54.96
	Ext-Abs+ICL	0.29	23.73	5.40	21.77	59.78
	Diff-Abs	**0.25**	25.05	6.05	22.85	61.66
	Diff-Abs+ICL	0.31	23.96	6.62	23.82	**63.06**

1% and 2% in Reddit, and CNN/DM correspondingly. For FactCC, Diff-Abs setting outperforms Ext-Abs+ICL setting by 3% in Reddit and outperforms Abs+ICL setting by 1% in CNN/DM. Diff-Abs+ICL setting, however, exhibited poor performance on human values similiarity, demonstrating the failure of LLMs in understanding human values definition.

It's worth noticing that in LLM-only settings, human values similiarity and FactCC seemed to show an inverse relationship, indicating LLM-only settings are not able to preserve the human values similiarity while remaining factual consistency. It's shown that extractive setting and ICL can improve FactCC while harming the performance on human values alignment. For instance, in Reddit, Ext-Abs+ICL setting outperformed Abs setting by 9.54% on FactCC score but human values similarity is 5% lower. The striking discordance between human values similiarity and FactCC reveals the significance of Diff-Abs setting, which improves both human values similiarity and FactCC.

Although the ROUGE for Diff-Abs setting is lower than LLM-only settings by about 2%, we stress our improvement on both human values similiarity and FactCC score is substantially larger than the decrease in ROUGE. Previous study has shown the negative relationship between factuality metrics and ROUGE [62,63]. In our work, results suggest that ROUGE can not reflect human alignment between summaries and source documents too.

The overall results demonstrate that our method is competitive in both human values similiarity and FactCC, compared to LLM-only methods. Although some settings of LLM-only methods showed satisfying performance on specific evaluation metrics, they can not balance the performance on different evaluation metrics. Hence, our method outperforms LLM-only methods in terms of balancing the performance on both human values alignment and factual consistency, while preserving the semantic information.

5.3   Ablation Study

Table 4. Ablation study results on Reddit and CNN/DM dataset. The best score is highlighted in **bold**.

Dataset	Setting	Dis(hv.)↓	ROUGE-1	ROUGE-2	ROUGE-L	FactCC
Reddit	Ext-Abs	0.43	**19.97**	**3.68**	**16.38**	39.13
	Ran-Abs	0.42	19.18	3.30	16.11	44.48
	Diff-Abs(ours)	**0.41**	18.59	2.87	15.30	**45.73**
CNN/DM	Ext-Abs	0.27	25.41	6.13	23.16	54.96
	Ran-Abs	0.28	**26.01**	**6.61**	**23.82**	57.38
	Diff-Abs(ours)	**0.25**	25.05	6.05	22.85	**61.66**

To understand the necessity of diffusion model in the extractive stage, we perform an ablation study by selecting random sentences for generating overall summaries to remove the diffusion model. The setting is named as Ran-Abs setting. We conduct the ablation study on both Reddit and CNN/DM. The results of both Diff-Abs setting and Ran-Abs setting are shown in Table 4. We also include Ext-Abs setting for comparison.

The results demonstrate that performance on human values similiarity and FactCC drops when removing the diffusion model in the extractive stage. In Reddit, the performance drops about 2% and 1% on human values similiarity and FactCC respectively. In CNN/DM, the performance drops about 3% on human values similiarity and 4% on FactCC. What's more, Ext-Abs setting exhibited an even worse performance compared to the Ran-Abs setting in terms of FactCC and human values similarity, indicating the failure of LLMs in extracting essential sentences. The overall results revealed the significance of diffusion model in extracting essential sentences from source documents. These extracted sentences played an important role in both human values alignment and factual consistency.

6   Conclusion

In this paper, we propose an extract-generate summarization framework to preserve human values from documents, aiming for less hallucination and more factuality in overall summaries. Our method utilizes a diffusion model to extract sentences containing both semantic information and human values within source documents. These sentences are then fed into LLMs to generate overall abstractive summaries. The experiment results show that our hybrid method can significantly improve the factuality of overall summaries while preserving human values compared to LLM-only methods. We hope that our findings will inspire future research in the field of summarization tasks to consider both human values alignment and factual consistency between summaries and documents.

Disclosure of Interests. The authors have no competing interests to declare that are relevant to the content of this article.

References

1. Kumar, Y., Kaur, K., Kaur, S.: Study of automatic text summarization approaches in different languages. Artif. Intell. Rev. **54**(8), 5897–5929 (2021)
2. Subbalakshmi, C., Pareek, P.K., Narayana, M.: A gravitational search algorithm study on text summarization using nlp. In: International Conference on Artificial Intelligence and Data Science, pp. 144–159. Springer, Heidelberg (2021)
3. Liu, Y.: Fine-tune bert for extractive summarization. arXiv preprint arXiv:1903.10318 (2019)
4. Zhang, H., Liu, X., Zhang, J.: Diffusum: generation enhanced extractive summarization with diffusion. arXiv preprint arXiv:2305.01735 (2023)
5. Nallapati, R., Zhou, B., Gulcehre, C., Xiang, B., et al.: Abstractive text summarization using sequence-to-sequence rnns and beyond. arXiv preprint arXiv:1602.06023 (2016)
6. Shi, T., Keneshloo, Y., Ramakrishnan, N., Reddy, C.K.: Neural abstractive text summarization with sequence-to-sequence models. ACM Trans. Data Sci. **2**(1), 1–37 (2021)
7. Gupta, V., Lehal, G.S.: A survey of text summarization extractive techniques. J. Emerg. Technol. Web Intell. **2**(3), 258–268 (2010)
8. Lin, H., Ng, V.: Abstractive summarization: a survey of the state of the art. In: Proceedings of the AAAI Conference on Artificial Intelligence, vol. 33, pp. 9815–9822 (2019)
9. Moratanch, N., Chitrakala, S.: A survey on extractive text summarization. In: 2017 International Conference on Computer, Communication and Signal Processing (ICCCSP), pp. 1–6. IEEE (2017)
10. Zhang, T., Ladhak, F., Durmus, E., Liang, P., McKeown, K., Hashimoto, T.B.: Benchmarking large language models for news summarization. arXiv preprint arXiv:2301.13848 (2023)
11. Keswani, G., Bisen, W., Padwad, H., Wankhedkar, Y., Pandey, S., Soni, A.: Abstractive long text summarization using large language models. Int. J. Intell. Syst. Appl. Eng. **12**(12s), 160–168 (2024)
12. Bhaskar, A., Fabbri, A., Durrett, G.: Prompted opinion summarization with gpt-3.5. In: Findings of the Association for Computational Linguistics: ACL 2023, pp. 9282–9300 (2023)
13. Van Veen, D., et al.: Clinical text summarization: adapting large language models can outperform human experts. Res. Square (2023)
14. Pu, X., Gao, M., Wan, X.: Summarization is (almost) dead. arXiv preprint arXiv:2309.09558 (2023)

15. Goyal, T., Li, J.J., Durrett, G.: News summarization and evaluation in the era of gpt-3. arXiv preprint arXiv:2209.12356 (2022)

16. Jones, E., et al.: Teaching language models to hallucinate less with synthetic tasks (2023)

17. Xu, Z., Jain, S., Kankanhalli, M.: Hallucination is inevitable: an innate limitation of large language models. arXiv preprint arXiv:2401.11817 (2024)

18. Liu, Z., Chen, Z., Zhang, M., Ren, Z., Ren, P., Chen, Z.: Zero-shot position debiasing for large language models (2024)

19. Chen, G.H., Chen, S., Liu, Z., Jiang, F., Wang, B.: Humans or llms as the judge? a study on judgement biases. arXiv preprint arXiv:2402.10669 (2024)

20. Chhabra, A., Askari, H., Mohapatra, P.: Revisiting zero-shot abstractive summarization in the era of large language models from the perspective of position bias. arXiv preprint arXiv:2401.01989 (2024)

21. Zunke, S., et al.: Research paper summarization using extractive summarizer. In: 2024 IEEE International Students' Conference on Electrical, Electronics and Computer Science (SCEECS), pp. 1–6. IEEE (2024)

22. Mishra, N., Khan, F., Mishra, A.: Revolutionizing text summarization: a breakthrough in content compression. Int. J. Performabil. Eng. **20**(1) (2024)

23. Jing, L., Zuo, J., Zhang, Y.: Fine-grained and explainable factuality evaluation for multimodal summarization. arXiv preprint arXiv:2402.11414 (2024)

24. Ramprasad, S., Krishna, K., Lipton, Z.C., Wallace, B.C.: Evaluating the factuality of zero-shot summarizers across varied domains. arXiv preprint arXiv:2402.03509 (2024)

25. Schwartz, S.H.: Universals in the content and structure of values: theoretical advances and empirical tests in 20 countries. In: Advances in Experimental Social Psychology, vol. 25, pp. 1–65. Elsevier (1992)

26. Durmus, E., et al.: Towards measuring the representation of subjective global opinions in language models. arXiv preprint arXiv:2306.16388 (2023)

27. Zhang, Z., Bai, F., Gao, J., Yang, Y.: Measuring value understanding in language models through discriminator-critique gap (2023)

28. Gehman, S., Gururangan, S., Sap, M., Choi, Y., Smith, N.A.: Realtoxicityprompts: evaluating neural toxic degeneration in language models. arXiv preprint arXiv:2009.11462 (2020)

29. Weidinger, L., et al.: Ethical and social risks of harm from language models. arXiv preprint arXiv:2112.04359 (2021)

30. Hermann, K.M., et al.: Teaching machines to read and comprehend. Adv. Neural Inf. Process. Syst. **28** (2015)

31. Kim, B., Kim, H., Kim, G.: Abstractive summarization of reddit posts with multi-level memory networks. In: Proceedings of NAACL-HLT, pp. 2519–2531 (2019)

32. Shaw, D., Grehan, E., Shiu, E., Hassan, L., Thomson, J.: An exploration of values in ethical consumer decision making. J. Cons. Behav. Int. Res. Rev. **4**(3), 185–200 (2005)

33. Schwartz, S.H.: An overview of the schwartz theory of basic values. Online Read. Psychol. Cult. **2**(1), 11 (2012)

34. Chen, J., Hsieh, G., Mahmud, J.U., Nichols, J.: Understanding individuals' personal values from social media word use. In: Proceedings of the 17th ACM Conference on Computer Supported Cooperative Work & Social Computing, CSCW '14, pp. 405–414. Association for Computing Machinery, New York (2014). https://doi.org/10.1145/2531602.2531608

35. Boyd, R., Wilson, S., Pennebaker, J., Kosinski, M., Stillwell, D., Mihalcea, R.: Values in words: using language to evaluate and understand personal values. In: Proceedings of the Ninth International AAAI Conference on Web and Social Media (2015). https://doi.org/10.1609/icwsm.v9i1.14589

36. Kiesel, J., Alshomary, M., Handke, N., Cai, X., Wachsmuth, H., Stein, B.: Identifying the human values behind arguments. In: Proceedings of the 60th Annual Meeting of the Association for Computational Linguistics, vol. 1: Long Papers, pp. 4459–4471 (2022)

37. Kiesel, J., et al.: SemEval-2023 task 4: ValueEval: identification of human values behind arguments. In: Ojha, A.K., Doğruöz, A.S., Da San Martino, G., Tayyar Madabushi, H., Kumar, R., Sartori, E. (eds.) Proceedings of the 17th International Workshop on Semantic Evaluation (SemEval-2023), pp. 2287–2303. Association for Computational Linguistics, Toronto (2023). https://doi.org/10.18653/v1/2023.semeval-1.313. https://aclanthology.org/2023.semeval-1.313

38. Schroter, D., Dementieva, D., Groh, G.: Adam-smith at SemEval-2023 task 4: discovering human values in arguments with ensembles of transformer-based models. In: Ojha, A.K., Doğruöz, A.S., Da San Martino, G., Tayyar Madabushi, H., Kumar, R., Sartori, E. (eds.) Proceedings of the 17th International Workshop on Semantic Evaluation (SemEval-2023), pp. 532–541. Association for Computational Linguistics, Toronto (2023). https://doi.org/10.18653/v1/2023.semeval-1.74. https://aclanthology.org/2023.semeval-1.74

39. Kang, D., Park, J., Jo, Y., Bak, J.: From values to opinions: predicting human behaviors and stances using value-injected large language models (2023)

40. Yi, X., Yao, J., Wang, X., Xie, X.: Unpacking the ethical value alignment in big models. arXiv preprint arXiv:2310.17551 (2023)

41. Liu, R., Zhang, G., Feng, X., Vosoughi, S.: Aligning generative language models with human values. In: Carpuat, M., de Marneffe, M.C., Meza Ruiz, I.V. (eds.) Findings of the Association for Computational Linguistics: NAACL 2022, pp. 241–252. Association for Computational Linguistics, Seattle (2022). https://doi.org/10.18653/v1/2022.findings-naacl.18. https://aclanthology.org/2022.findings-naacl.18

42. Luhn, H.P.: The automatic creation of literature abstracts. IBM J. Res. Dev. **2**(2), 159–165 (1958)

43. Kanitha, D., Mubarak, D.M.N.: An overview of extractive based automatic text summarization systems. AIRCC's Int. J. Comput. Sci. Inf. Technol., 33–44 (2016)

44. Raffel, C., et al.: Exploring the limits of transfer learning with a unified text-to-text transformer (2023)

45. Lewis, M., et al.: Bart: denoising sequence-to-sequence pre-training for natural language generation, translation, and comprehension (2019)

46. Zhang, J., Zhao, Y., Saleh, M., Liu, P.J.: Pegasus: pre-training with extracted gap-sentences for abstractive summarization (2020)

47. Tay, Y., et al.: Long range arena: a benchmark for efficient transformers (2020)

48. Kryściński, W., McCann, B., Xiong, C., Socher, R.: Evaluating the factual consistency of abstractive text summarization. arXiv preprint arXiv:1910.12840 (2019)

49. Chen, Y.C., Bansal, M.: Fast abstractive summarization with reinforce-selected sentence rewriting. arXiv preprint arXiv:1805.11080 (2018)

50. Bajaj, A., et al.: Long document summarization in a low resource setting using pretrained language models. arXiv preprint arXiv:2103.00751 (2021)

51. Pilault, J., Li, R., Subramanian, S., Pal, C.: On extractive and abstractive neural document summarization with transformer language models. In: Proceedings of the 2020 Conference on Empirical Methods in Natural Language Processing (EMNLP), pp. 9308–9319 (2020)

52. Zhang, Y., et al.: An exploratory study on long dialogue summarization: what works and what's next. arXiv preprint arXiv:2109.04609 (2021)
53. Hsu, W.T., Lin, C.K., Lee, M.Y., Min, K., Tang, J., Sun, M.: A unified model for extractive and abstractive summarization using inconsistency loss. arXiv preprint arXiv:1805.06266 (2018)
54. Mao, Z., et al.: Dyle: dynamic latent extraction for abstractive long-input summarization. arXiv preprint arXiv:2110.08168 (2021)
55. Zhang, H., Liu, X., Zhang, J.: Extractive summarization via chatgpt for faithful summary generation. arXiv preprint arXiv:2304.04193 (2023)
56. Danielsson, P.E.: Euclidean distance mapping. Comput. Graphics Image Process. **14**(3), 227–248 (1980)
57. Reimers, N., Gurevych, I.: Sentence-bert: sentence embeddings using siamese bert-networks. arXiv preprint arXiv:1908.10084 (2019)
58. Ho, J., Jain, A., Abbeel, P.: Denoising diffusion probabilistic models. arXiv preprint arxiv:2006.11239 (2020)
59. Nallapati, R., Zhai, F., Zhou, B.: Summarunner: a recurrent neural network based sequence model for extractive summarization of documents. In: Proceedings of the Thirty-First AAAI Conference on Artificial Intelligence, AAAI 2017, pp. 3075–3081. AAAI Press (2017)
60. Akhmetov, I., Gelbukh, A., Mussabayev, R.: Greedy optimization method for extractive summarization of scientific articles. IEEE Access **9**, 168141–168153 (2021)
61. Lin, C.Y.: Rouge: a package for automatic evaluation of summaries. In: Text Summarization Branches Out, pp. 74–81 (2004)
62. Gunel, B., Zhu, C., Zeng, M., Huang, X.: Mind the facts: knowledge-boosted coherent abstractive text summarization (2020)
63. Kryściński, W., Keskar, N.S., McCann, B., Xiong, C., Socher, R.: Neural text summarization: a critical evaluation. arXiv preprint arXiv:1908.08960 (2019)

ChineseKorean Cross-Language Transfer Method That is Based on Language Features

Jia Meng, Guozhe Jin$^{(\boxtimes)}$, Yahui Zhao, and Rongyi Cui

Yanbian University, Yanji, China
jinguozhe@ybu.edu.cn

Abstract. In view of the current problem of insufficient utilization of language characteristics in the field of cross-language transfer between China and South Korea, a method of cross-language transfer between China and South Korea based on linguistic features of Chinese characters is proposed. This method first uses Hanjaro to process the Chinese characters in the Korean sentence into the traditional Chinese form. Second, the original, unprocessed Korean sentence is constructed into a positive example sentence via random dropout. These two sentences are used as a positive example pair for contrastive learning training. Moreover, the preprocessed Korean sentences and the processed Korean sentences are used as a set of parallel sentences for confrontation training. The training goal is to enable the pretrained model encoder to generate parallel sentences. It can confuse the sentence embedding of the discriminator, thereby bringing the sentence vector closer. Finally, we use our model and the original multilingual pre-trained language model to train on the Chinese dataset and conduct experiments on the Korean dataset in a zero-shot manner to obtain the transfer results. This paper uses the public dataset PAWS-X to conduct the experiments. The experimental findings demonstrate that our model enhances the cross-language transfer capability from Chinese to Korean in comparison with the original multilingual pre-trained model.

Keywords: Cross-language · Transfer learning · Chinese words · Zero-shot · Language characteristics

1 Introduction

Recent advancements in deep learning have significantly enhanced the performance of various natural language processing (NLP) models. The presence of extensive annotated datasets has significantly contributed to the enhancement of model performance. However, not all languages have sufficient large-scale annotated datasets for our use. For many languages, large-scale annotated datasets are lacking for various reasons, such as a relatively small number of speakers or difficulty with the language [1]. Cross-language transfer learning (CLTL)

H. Yu et al. (Eds.): CCF NCCA 2024, CCIS 2274, pp. 231–240, 2024.
https://doi.org/10.1007/978-981-97-9671-7_15

provides a method for learning a target language model via annotated data in other languages (source language) [2], which cleverly solves this problem.

Owing to the excellent performance of multilingual pre-trained models, academia and industry have begun to pay attention to the use of multilingual pre-trained models. For example, mBERT (Devlin et al., 2019) [3], XLM (Conneau and Lample, 2019) [4], XLM-R (Conneau et al., 2020) [5], achieve zero-shot cross-language. The purpose of transfer learning is to use annotated data of high-resource languages to fine-tune multilingual models to achieve good performance in low-resource language tasks. However, we have overlooked the inextricable relationship between languages. When two languages are spoken relatively close to each other, they are affected by foreign cultures, and their language habits and pronunciation are similar.

In this study, we use the "bridge" between Chinese and Korean, that is, Chinese characters, to improve the cross-language transfer effect between Chinese and Korean languages from the perspective of language characteristics. Chinese words in Korean account for a large part of the Korean vocabulary. They can be converted into well-known Chinese through plug-ins, so they are of great help to us in studying migration from Chinese to Korean. Although there are large linguistic differences between the two languages, this shared vocabulary can make the two languages closer, thereby improving zero-shot cross-language transfer from Chinese to Korean [6].

2   Related Work

2.1   Cross-Language Transfer Methods

"Multilingual translation" based on multilingual language models or machine translation models has amazing results. These representations facilitate a variety of different tasks (Hu et al., 2020) [7]. However, they require extensive datasets for training and are therefore not practical for low-resource languages; in this case, Korean is a relatively low-resource language.

Artetxe et al. enable unsupervised transfer learning between multiple languages by using a sentence embedding method for zero-shot cross-language transfer [8]. Nooralahzadeh et al. used meta-learning methods to achieve cross-language transfer under zero-sample conditions, effectively transferring knowledge from one language to another [9]. Huang et al. (2021) [10] also created robust regions within the embedding space to withstand noise in contextual embeddings. However, these methods are not optimal for closely related languages, as words that appear similar in form may still be significantly distant from each other within the embedding space.

2.2   Sentence Embedding

Sentence embeddings are a representation that maps text sentences into a high-dimensional vector space. This vector representation can capture the semantic information of sentences, making sentences with similar semantics closer in the vector space, and can be used in tasks such as text similarity calculations, sentence classification, and semantic matching. As a result, it has become extensively utilized in natural language processing and holds significant importance. Recently, considerable research has been conducted on this topic. Yan et al. [11] devised an effective natural language enhancement method by integrating four data augmentation strategies, including adversarial attack, cutoff, and discard, to achieve improved sentence embeddings. Gao et al. [12] proposed SimCSE, a straightforward contrastive sentence embedding model that can produce improved sentence embeddings from either unlabeled or labeled data. To develop sentence embeddings that can capture the differences between an original sentence and its edited version, Chuang et al. [13] introduced the DiffCSE algorithm. In this approach, the edited sentence is generated by randomly masking parts of the original sentence and then sampling replacements for the masked sections via a language model.

3   Method

To better utilize the linguistic features of Chinese characters in Korean, we design a model using contrastive learning methods and adversarial training methods for differential prediction targets.

Our experiment consists of three main steps. First, we train the model to adjust the parameters of the pretrained model mBERT. Next, we use the finetuned pretrained model to train the Chinese language model. Finally, in the third step, we test the Chinese language model, which was trained on the Chinese dataset, on Korean in a zero-shot manner.

In the first step of training, since we do not have a directly available Korean corpus with Chinese characters annotated, we need to use Hanjaro to convert the Korean and Chinese words in the Korean corpus into traditional Chinese. Next, we convert traditional Chinese to simplified Chinese. A converted Korean corpus with Chinese characters can be obtained. We use the original, nonconverted Korean corpus to randomly dropout to generate corresponding positive examples and form positive example pairs with them, and other Korean sentences are negative examples to conduct comparative learning training. The converted Korean sentences with Chinese characters and the original Korean sentences form parallel data for adversarial learning training, and the encoder mBERT is the generator in the adversarial learning framework. The ultimate training purpose of the model is to improve the correspondence between Chinese character words and their corresponding original Korean words at the sentence level. Our training goal is to make the sentence embedding of the original Korean sentence and the sentence embedding of the converted Korean sentence as close as possible. The specific content of the model is introduced below.

3.1   Contrastive Learning

The core idea of contrastive learning is to learn feature representations by learning the differences between samples in the data. In contrastive learning, the model is asked to place samples from the same category closer to each other and samples from different categories further apart from each other. Assume a set of paired examples: x and x^+. In contrastive learning, x and x^+ are pairs of sentences with very similar semantics. Using the comparison framework of Chen et al. [14], h_i and h_i^+ are used to represent the sentence embeddings of x and x^+. Where $sim(h_i, h_i^+)$ is the cosine similarity calculation of the two sentence embedding representations, which is the temperature hyperparameter. An important issue in contrastive learning is how to construct the (x, x^+) pair mentioned above. In the field of computer vision, methods such as flipping and cropping of images are generally adopted. Recently, Wu et al. [15] and Meng et al. [16] used operations such as replacement or deletion of words in the field of natural language processing to achieve a data enhancement effect. In this work, we follow the SimCSE method and use dropout to construct positive example pairs.

In our experiments, we utilize mBERT, a pre-trained multilingual model, to encode the input sentences. The input sentences here are the Korean training set in the PAWS-X dataset, which is the original Korean corpus that has not been converted by Hanjaro. Then, the contrastive learning target (Formula 1) is used to fine-tune all the parameters, and finally, the loss of the contrastive learning part is Loss1.

$$l_i = -log \frac{e^{sim(h_i,h_i^+)/\tau}}{\sum_{j=1}^{N} e^{sim(h_i,h_i^+)/\tau}} \tag{1}$$

3.2   Difference-Based Contrastive Learning

In recent years, DiffCSE, proposed by Chuang et al. [13], has achieved excellent results in the generation of sentence embeddings. The so-called DiffCSE can be divided into two parts. One part is the standard contrastive learning goal of SimCSE, and the other part is the difference prediction goal conditioned on sentence embedding.

For the SimCSE part, we need to be given an unprocessed input sentence x. Here, we use the Korean training set in PAWS-X. SimCSE constructs a positive example x^+ by applying different dropout masks. Then, by using the mbert encoder, that is, f, we can obtain the sentence embedding of x as $h = f(x)$.

The other part is the difference prediction target part used in the ELECTRA framework proposed by Clark et al. [17], which contains a generator and a discriminator. The generator part is an mBERT encoder, which we call G; the discriminator is another mBERT encoder, which we call D. A new pretraining task is used in ELECTRA called Replacement Label Detection (RTD). Inspired by generative adversarial networks (GANs), ELECTRA trains a model to distinguish between "real" and "fake" input data. Through adversarial training, the generator continuously learns to generate more realistic samples, the discriminator continuously learns to improve the accuracy of real samples, and the discriminative ability of the generated samples finally reaches a balanced state. In our experiments, we use it to distinguish between Korean data and data processed by the Chinese character word plug-in. The training goal is to enable the generator mBERT to continuously approach the original Korean sentence embedding of the Chinese character word processed sentence embedding. The cross-entropy loss for a single sentence is as follows:

$$L_{RTD}^{X} = \sum_{t=1}^{T}(-1(x_{(t)}^{''} = x_{(t)})logD(x^{''}, h, t) - 1(x_{(t)}^{''} \neq x_{(t)})log(1 - D(x^{''}, h, t)) \tag{2}$$

The cross-loss entropy of batch data is as follows:

$$\sum_{i=1}^{N} L_{RTD}^{x_i} \tag{3}$$

In the experiment, we record the cross-loss entropy of batch data as Loss2. The overall model diagram is shown in Fig. 1 below. In the model, our inputs are the original Korean sentence x_1 and the Korean sentence x_2 containing Chinese characters. The sentences are encoded separately through the same mBERT; to obtain better sentence embedding, x_1 is trained via contrastive learning, and the output is the loss of contrastive learning, which we record as Loss1. The cls in the output of mBERT is used as the sentence embedding of x_1 and is input into the discriminator together with the result of the word embedding of x_2, and the discriminator can obtain two different results. When it is 0, it means that it is the same as the word in the original sentence, and when it is 1, it means it is not the same as the word in the original sentence, and it is a converted Chinese word. We record the output of this part as Loss2.

The above model diagram uses the language characteristics of Chinese characters to fine-tune the parameters of the multilingual pre-trained model mBERT. After we obtain a fine-tuned mBERT after multiple rounds of training, we can use the mBERT before and after fine-tuning as pretraining. The model was trained as a Chinese model. For the Chinese dataset, we selected the Chinese training set in PAWS-X and transferred the two different trained Chinese models directly to the PAWS-X Korean dataset and used the Korean test set to test and compare them.

4 Experiment

4.1 Experimental Setup

In our experiments, we followed the settings of DiffCSE and built our model on the basis of the PyTorch implementation. We also use mBERT's checkpoints as the initialization of the sentence encoder f. We take the [CLS] representation obtained from the encoded output of the mBERT encoder as the sentence embedding and add an MLP layer based on [CLS].

The dataset uses the public multilingual dataset PAWS-X. The parts used in this experiment are the Chinese dataset and the Korean dataset in PAWS-X.

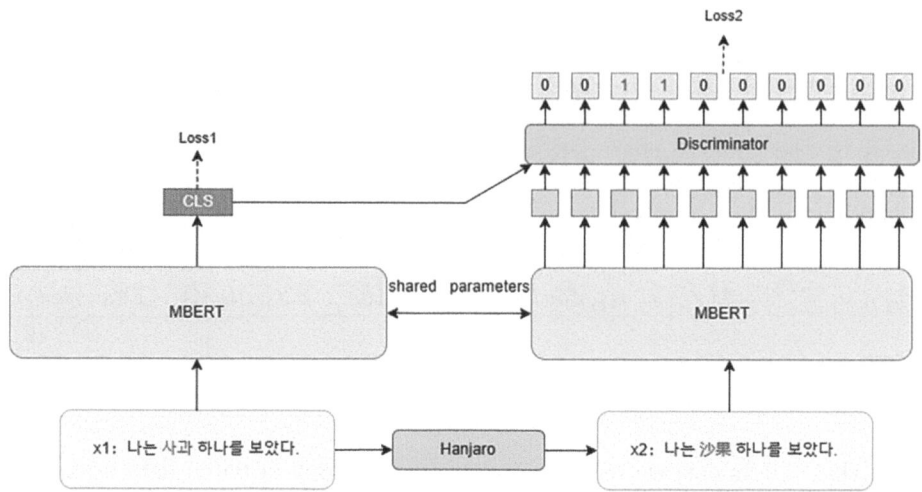

Fig. 1. Overall model diagram

Table 1. Experimental data format

	Example Sentences
Original Korean sentence	1560 년 10 월 파리에서 그는 비밀리에 영국 대사 인 니콜라스 트록 모튼을 만났고 스코틀랜드를 통해 영국으로 돌아갈 여권을 요구했다.
(After conversion of Chinese characters)	1560 年 10 月 파리에서 그는 密裏에 英大使 人 니콜라스 트錄 모튼을 만났고 스코틀랜드를 통해英으로 돌아갈 圈을 要求했다.
(The Chinese characters will be converted into simplified Chinese characters after conversion)	1560 年 10 月 파리에서 그는 秘密里에 英大使 人 니콜라스 트 모튼을 만났고 스코틀랜드를 통해英으로 돌아갈 圈을 要求했다.

Due to the needs of the experiment, the Korean dataset needs to be converted into Chinese characters so that all the Chinese words in the Korean sentence are converted through the processing of the plug-in, and the final Korean sentence is a mixture of Korean and Chinese words. Since the converted result contains much traditional Chinese, the next step is to convert traditional Chinese to simplified Chinese. The format of the data is illustrated in Table 1 below:

Finally, we have the following Korean sentence: 1560 년 10 월 파리에 서 그 는 비밀리에 영국 대사 인 니콜라스 트록 모튼을 만났고 스코틀랜드 를 통해 영 "국으로 돌아갈 여권을 요구했다.   "

And the result after conversion to simplified Chinese: 1560 年 10 月 파리에서 그는 秘 密里에 英大使 人 니콜라스 트 모튼 "을 만났고 스코틀랜드를 통해英 으로 돌아갈 圈을 要求했다 .   "

4.2   Parameter Settings

This experiment was conducted on a 10 GB 2080Ti GPU. In the fine-tuning stage, 3 epochs were set for the experiment, and the batch size was set to 8. The BERT used is bert-base-multilingual-cased, the dimension is set to 768 dimensions, and the learning rate is set to 5e-7. Subsequently, bert-base-multilingual-cased and fine-tuned bert were used as multilingual pre-trained models to train the classifier of Chinese data in the PAWS-X dataset. At this time, the training stage of the classifier is set to 15 epochs, and the batch size is set to 8. The Korean test set in PAWS-X is input into two trained classifiers to compare the transfer effects of zero-shot. When evaluating the performance of the model, the accuracy value is used as the evaluation index of the dataset PAWS-X to evaluate the performance of the model. Each experiment was tested multiple times, and the average was taken as the final result. The parameters are shown in Table 2 below:

Table 2. Parameter settings

Fine-tuning Parameter Stage		Downstream Task Stage	
Device	10 GB GPU 2080Ti	Device	10 GB GPU 1080Ti
Epoch	3	Epoch	15
Batch size	8	Batch size	8
Learning rate	5e-7	Learning rate	1e-3

4.3   Experimental Results

Experimental Results on the PAWS-X Dataset. To unify the evaluation indicators, we choose to use accuracy and the F1 value as evaluation indicators, compare our model with related methods in recent years, and present the relevant results in Table 3 below. Since the process of fine-tuning mBERT is divided into two parts of loss, we add the loss1 of the first part of comparative learning and the loss2 of adversarial training in proportion during the training process. Our final loss is as follows:

$$Loss = Loss1 + weight * Loss2 \qquad (4)$$

Different weight values have different migration effects on downstream tasks. From our experiments, we know that when the weight value is 2, the migration effect on downstream tasks is the best. Table 3 shows the corresponding results when the weight value is 2.

The final comparison of the results of testing our method on the Korean test set in PAWS-X with the results of other paper models on the Korean dataset of PAWS-X is shown in Table 3:

Table 3. Transfer results of the PAWS-X dataset

Model	Accuracy	F1
mBERT [10]	62.4	Null
mBERT-RS-RP [10]	69.9	Null
mBERT-RS-DA [10]	71.8	Null
mBERT†	75.1	75.0
Our model	**78.5**	**78.1**

The model names with "†" are the migration results of our original mBERT pretraining parameters on the PAWS-X Korean dataset.

The results presented in Table 3 clearly indicate that our method performs well in the cross-language transfer task from Chinese to Korean on the PAWS-X dataset. Therefore, the use of Chinese characters, one of the language characteristics of Korean, in our model can improve the transfer ability from Chinese to Korean.

To assess how varying the weight parameter impacts the downstream Korean test set, we conducted a series of experiments with different weight values. The results of these experiments are presented in Table 4 below:

Table 4. Transfer results with different weight values

Weight	Accuracy	F1
0.02	75.6	74.7
0.2	76.6	76.6
2	**78.5**	**78.1**
20	76.2	75.5

Table 4 shows that if the weight value is too large or too small, it affects the migration effect. After experimental verification, a weight value of 2 is more appropriate.

Experimental Results on the Sentiment Analysis Dataset To validate the model's effectiveness, we also conducted relevant experimental verifications in analyzing the sentiment of various comments. Among them, the Chinese dataset we chose was the sentiment classification dataset of car platform reviews, and the Korean dataset for testing was the sentiment classification dataset of movie reviews. Experiments were also conducted on the original mBERT and our model. The outcomes are presented in Table 5 below:

The model names with "†" are the migration results of our original mBERT pretraining parameters on the sentiment analysis dataset.

Table 5. Transfer results of sentiment analysis datasets

Model	Accuracy	F1
mBERT†	59.5	59.0
Our model	**61.2**	**61.1**

The outcomes demonstrate that our model achieves a certain enhancement in the transfer capability of the original mBERT pretrained parameters on sentiment analysis datasets, which proves that our model also has certain improvements on sentiment analysis datasets in different fields.

5   Conclusion and Outlook

To address the issue of insufficient annotated data for Korean, a low-resource language, this paper introduces a novel approach for cross-language transfer from Chinese to Korean. It aims to use the language feature between Chinese and Korean-Chinese character words, which is an important tool, to fine-tune the distance between Korean sentence embeddings and Korean sentence embeddings containing Chinese character words closer in the semantic space; thus, the parameters of the multilanguage pre-trained model are better tuned, and finally, the transfer effect from Chinese to Korean is improved.

Although Chinese words have very similar characteristics to modern Chinese translated from Korean itself, they are not the same. Some Chinese words converted through plug-in processing are far from the actual Chinese meaning. Therefore, future work will focus on how to optimize the processing of Chinese characters and make better use of them to improve their ability to transfer from Chinese to Korean.

Acknowledgements. Jin Guozhe (PhD Startup Fund) - Research on cross-language transfer learning methods between Chinese and Korean. Project number: 602024041

References

1. Chen, X., Awadallah, A.H., Hassan, H., Wang, W., Cardie, C.: Multi-source cross-lingual model transfer: Learning what to share. arXiv preprint arXiv:1810.03552 (2018)
2. David, Y., Grace, N., Richard, W., et al.: Inducing multilingual text analysis tools via robust projection across aligned corpora. In: Proceedings of the First International Conference on Human Language Technology Research, pp. 1–8 (2001)
3. Devlin, J.: Bert: Pre-training of deep bidirectional transformers for language understanding. arXiv preprint arXiv:1810.04805 (2018)
4. Lample, G., Conneau, A.: Cross-lingual language model pretraining. arXiv preprint arXiv:1901.07291 (2019)
5. Conneau, A., et al.: Unsupervised cross-lingual representation learning at scale. arXiv preprint arXiv:1911.02116 (2019)

6. Park, J., Zhao, H.: Korean-to-chinese machine translation using chinese character as pivot clue. arXiv preprint arXiv:1911.11008 (2019)

7. Aepli, N., Sennrich, R.: Improving zero-shot cross-lingual transfer between closely related languages by injecting character-level noise. arXiv preprint arXiv:2109.06772 (2021)

8. Artetxe, M., Schwenk, H.: Massively multilingual sentence embeddings for zero-shot cross-lingual transfer and beyond. Trans. Assoc. Comput. Linguist. **7**, 597–610 (2019)

9. Nooralahzadeh, F., Bekoulis, G., Bjerva, J., Augenstein, I.: Zero-shot cross-lingual transfer with meta learning. arXiv preprint arXiv:2003.02739 (2020)

10. Huang, K.H., Ahmad, W.U., Peng, N., Chang, K.W.: Improving zero-shot cross-lingual transfer learning via robust training. arXiv preprint arXiv:2104.08645 (2021)

11. Sun, M., Huang, D.: Conisi: A contrastive framework with inter-sentence interaction for self-supervised sentence representation. In: China National Conference on Chinese Computational Linguistics, pp. 31–47. Springer (2022)

12. Gao, T., Yao, X., Chen, D.: Simcse: Simple contrastive learning of sentence embeddings. arXiv preprint arXiv:2104.08821 (2021)

13. Chuang, Y.S., et al.: Diffcse: Difference-based contrastive learning for sentence embeddings. arXiv preprint arXiv:2204.10298 (2022)

14. Chen, T., Kornblith, S., Norouzi, M., Hinton, G.: A simple framework for contrastive learning of visual representations. In: International Conference on Machine Learning, pp. 1597–1607. PMLR (2020)

15. Wu, Z., Wang, S., Gu, J., Khabsa, M., Sun, F., Ma, H.: Clear: Contrastive learning for sentence representation. arXiv preprint arXiv:2012.15466 (2020)

16. Meng, Y., Xiong, C., Bajaj, P., Bennett, P., Han, J., Song, X., et al.: Coco-lm: Correcting and contrasting text sequences for language model pretraining. Adv. Neural. Inf. Process. Syst. **34**, 23102–23114 (2021)

17. Clark, K., Luong, M.T., Le, Q.V., Manning, C.D.: Electra: Pre-training text encoders as discriminators rather than generators. arXiv preprint arXiv:2003.10555 (2020)

A Knowledge–Enhanced Text Clustering Based Adversarial Learning for Text Generation

Wenbin Zhao[1]($\boxtimes$), Luhan Liu[1], Feng Wu[2], ShuoKai Pan[1], Di Wu[1], and Haonan Wang[1]

[1] Shijiazhuang Tiedao University, Shijiazhuang 050043, Hebei, China
zhaowb2013@stdu.edu.cn
[2] Hebei Institute of Science and Technology Information, Shijiazhuang 050021, Hebei, China

Abstract. Aiming at the problem that users find it difficult to quickly filter out effective information when faced with massive amounts of data. The advantage of using a generative adversarial network mod el is that it does not require supervision and data labeling. Traditional generative adversarial network text generation models are prone to mode collapse. A knowledge-enhanced cluster text generation adversarial network model is proposed. This model improves the model structure, designs an auxiliary training module, combines the clustering algorithm of knowledge learning, integrates the generative adversarial network and the clustering algorithm to generate text, and adjusts the generative network by training the results generated by the discriminator. And parameters in clustering algorithms to generate more diverse, high-quality text. In this article, we verified the text under two different data sets. Compared with the evaluation indicators of other models in the same environment, the BLUE index of this model increased by 0.9% ~ 19.8%. The evaluation index is better than the baseline model, the results show that the model has high feasibility and effectiveness.

Keywords: Text Generation · Clustering · Generative Adversarial Network · Knowledge Enhancement

1 Introduction

In the digital age, information is growing explosively. How to quickly and accurately extract the information required by users from massive amounts of data has become an important direction in technology research and application. Knowledge graph, as a structured semantic knowledge base, represents the relationship between entities in the form of a graph, providing a rich information base and reasoning basis for intelligent question answering systems. As a bridge between users and information interaction, the question and answer system's accuracy and efficiency directly determine the user's information acquisition experience.

With the continuous development of technologies such as natural language generation [1], information retrieval, and artificial intelligence, text generation technology based on knowledge graphs is also constantly improving. Based on previous developments, there is reason to believe that with the in-depth research and application of this technology, it will bring users a better information acquisition experience.

H. Yu et al. (Eds.): CCF NCCA 2024, CCIS 2274, pp. 241–255, 2024.
https://doi.org/10.1007/978-981-97-9671-7_16

Goodfellow [2] and others have become popular data synthesis models. The training process of GAN is essentially a process of continuous confrontation between the generation network and the discriminant network. As an unsupervised data generation model, the core idea of GAN is to capture the real data distribution by promoting competition between the generator and the discriminator and then jointly generate reasonable data. When GAN first appeared, it mainly focused on image synthesis, and after continuous improvement and exploration, it finally achieved high-quality synthetic data results (Arjovsky et al. [3], 2017; Radford et al. [4], 2015; Karras et al. [5, 6], 2018, 2021).

In addition to this, GANs also play an important role in the field of computer vision, especially in data enhancement through image synthesis (Sandfort et al. [7], 2019; Bowles et al. [8], 2018; Antoniou et al. [9], 2018; Tran et al. [10], 2021). The generator of GAN can learn the implicit density based on the loss of the discriminator without directly referring to the training data. Therefore, GAN has the ability to memorize data in advance; that is, the model can reproduce the training data. What's more worth mentioning is that GAN can use random noise rather than a specific starting point (such as a preset starting mark) to synthesize a variety of data. GAN has demonstrated excellent performance in many fields, and its advantages of no need for supervision and data labeling make it have broad application prospects in fields such as text generation.

The most common text generation method is a language model based on auto regression. For example, people have studied the Recurrent Neural Network (RNN) using Long Short Term Memory (LSTM) [11, 12] cells. The model has been extensively studied. By using LSTM, Graves et al. [13] successfully predicted sequences and generated handwritten text, while Wen et al. [14] synthesized sentences under specific conditions. Bowman et al. [15] further combined the autoregressive LSTM model and the Variational Auto Encoder (VAE) architecture to generate text after learning the text embedding space. They use Policy Gradient with BLEU (PG-BLEU) to compute the Bilingual Evaluation Understudy (BLEU) of synthetic sentences and use it as a reward when updating the generator using policy gradient. In addition, there are many research attempts to use GAN for text synthesis. Yu. et al. [16] proposed the SeqGAN model, trying to solve the backpropagation problem through reinforcement learning and reward systems based on gradient strategies. However, SeqGAN faces the problem of reward sparsity, which prompted Lin et al. [17] to introduce RankGAN, which replaces the regression-based discriminator with a new ranker. RankGAN trains the discriminator to assign higher scores to more realistic sentences. MaskGAN [18], on the other hand, utilizes an LSTM-based generator to fill in the masked parts of sentences during training. Since MaskGAN uses discrete data, the generator cannot perform gradient backpropagation. To overcome this challenge, they adopt an actor-critic approach, leveraging the discriminator's probabilistic ability to reward candidate tokens during training. Che et al. [19] proposed the maximum likelihood augmented discrete adversarial model (Mali GAN), which synthesizes text by minimizing Kullback-Leibler divergence. Leak GAN alleviates problems related to reward sparsity and lack of intermediate information by providing information leaked by the discriminator.

2   Related Work

Generative adversarial networks include a generative network (generator G) and a discriminator network (discriminator D). These networks are neural networks. The two networks compete with each other to generate text. The function of the generation network is to generate false text that can deceive the discriminator network, while the function of the discriminator network is to distinguish the authenticity of the generated text. After many games between these two networks, the discriminator network finally thinks The text that generates online students is real.

The generator G is a random variable z, and its output is a generated sample G(z) sampled from a certain distribution. The goal of the generative network is to master the data distribution characteristics of real samples through continuous learning and generate generated samples from random variables with the same distribution as the real samples. The input of the discriminator D consists of real samples and generated samples. Its output is a probability value, which is used to characterize the probability that the input comes from a real sample, and the result is fed back to the generation network to help it train more effectively. The goal of the discriminant network is to accurately determine whether the input is a real sample through iterative training. Adversarial training is to train two network models based on a game framework. The training process is a binary mini-max (mini-max) adversarial process. The goal is to minimize the difference between the generated sample distribution and the real sample distribution, that is, KL divergence (Kullback-Leibler divergence). Therefore, the objective function is defined as:

$$\min_{G} \max_{D} V(D, G) = E_{x \sim P_{data(x)}} \left[log D(x) \right] + E_{z \sim P_{z(x)}} [log(1 - D(G(z)))] \tag{1}$$

It is the largest relative to the discriminator D and the smallest relative to the generator Q. It can be found that for the fixed distribution generator Q, the optimal discriminator D is:

$$D_{(X)} = \frac{P(x)}{P(x) + Q(x)} \tag{2}$$

Among them is the binary cross-entropy function, which represents the real data distribution and the generated data distribution. The generation network calculates the gradient of the parameters based on the feedback of the discriminant network; the discriminant network calculates the gradient of its parameters based on the labels of the real samples and the generated samples. Gradient information backpropagation uses the chain rule to transfer the gradient of the loss function from the output layer of the network to the input layer of the network, thereby updating the network parameters. The generator is trained to minimize the objective function, while the discriminator is trained to maximize it. Through iterative training, the two networks gradually optimize and improve and finally reach a stable state until the data generated by the generator makes the discriminator D unable to distinguish the data. The training will stop only when the authenticity of the training is achieved.

Basic GAN will encounter some problems in the process of generating text. After the generator G receives the data of the random variable z, it will generate false sample data G(z), pass it to the discriminator D, and the discriminator D will The received real sample data x is discriminatively processed, and the generated results are output. The generated result is the probability of whether the input of the discriminator D is a real distribution. If it is 1, the data is real data, and if it is 0, it is false data. At the same time, the discriminator D will return the output results to the generator G for its training. When the output probability value of D is about 0.5, indicating that the data source cannot be distinguished, that is, the model has reached the optimal condition, then the training is stopped. If the stopping condition causes the discriminator D to be misadjusted, it will cause problems such as non-convergence, mode collapse, gradient weakening, or even disappearance. When the gradient disappears, the trained generator G is equivalent to the state without training. This is also the reason why GAN models are difficult to train.

Therefore, Yu et al. [16] proposed the sequence generation model SeqGAN. The core idea is to regard GAN as a reinforcement learning system, use the policy gradient algorithm to update the parameters of the generated network, and use the Monte Carlo search algorithm to evaluate the incomplete sequence at any time step. Despite this, the SeqGAN model still has flaws; that is, when longer texts are to be generated, the sparsity of the guidance signal of the discriminant network causes a lack of intermediate information related to the text structure during the generation process, resulting in less than ideal results. Subsequently, Guo et al. [20] proposed the LeakGAN model based on the encoder-decoder structure. The discriminant network passes the extracted features through leakage to the generation network, and the generation network uses it as a stepwise guidance signal to generate text with key information. At the same time, reference hierarchical reinforcement learning provides information and features to the generative network. The discriminant network in GAN trains the MAGER module with the feature vector obtained by feature extraction through CNN. The purpose is to provide more feature information, and based on the hierarchical structure, the sparsity problem can be further solved by dividing subtasks. However, using this kind of hierarchical reinforcement learning will cause the generative adversarial network to encounter many difficulties in training, and the generated text will lack diversity. SidiLu et al. [21] proposed cooperative training (CoT), a new algorithm for training likelihood-based discrete data generation models by directly optimizing the estimated Jensen-Shannon divergence. The CoT coordinates the training of the generation module G and the auxiliary prediction module M, called the intermediary, to guide G in a collaborative manner. The algorithm improves in terms of generation performance, generalization ability, and computational performance in both synthetic and real-life scenes. Ehsan Montahaei et al. [22] proposed a new discrete data generation model adversarial training framework (DGSAN). In this framework, the gradient transfer problem is solved by considering the explicit distribution of generators (due to the advantage of finite discrete domains) and finding a closed relationship between the next generator, the current generator, and the current discriminator. In this method, the generator and discriminator are unified in one network. The network provides a probability distribution of the data and also prepares the (conditional) probability of assigning the input to the real data class. The result of this integration is that the gradient transfer problem is bypassed, and training stability is achieved. Zhang et al. [23] proposed a new text GAN model called text feature GAN (TFGAN). In order to solve this problem, reinforcement learning (RL) or continuous relaxation is often used to calculate the gradient in the learning process, resulting in high variance. or biased estimates. Furthermore, existing text GANs often suffer from mode collapse (i.e., they have limited generative diversity). In this model, adversarial learning is performed in a continuous text feature space.Lee et al. [24] proposed the Text Embedding Space Generative Adversarial Network (TESGAN), which generates a continuous text embedding space instead of discrete labels to solve the gradient backpropagation problem. In addition, TESGAN performs unsupervised learning and does not directly refer to the text of the training data to overcome the data memory problem. Furthermore, TESGAN enables unconditional text synthesis in the inference stage by using random noise instead of markers or cues for text synthesis.

In short, these methods are trying to improve the accuracy and diversity of text generation in different ways to meet the needs of different fields and applications, but there are some drawbacks and problems that have not yet been solved and improved. Based on the above, this paper proposes a new cooperative clustering discrete data generation adversarial network framework and named it CDGSAN, as shown in Fig. 1.

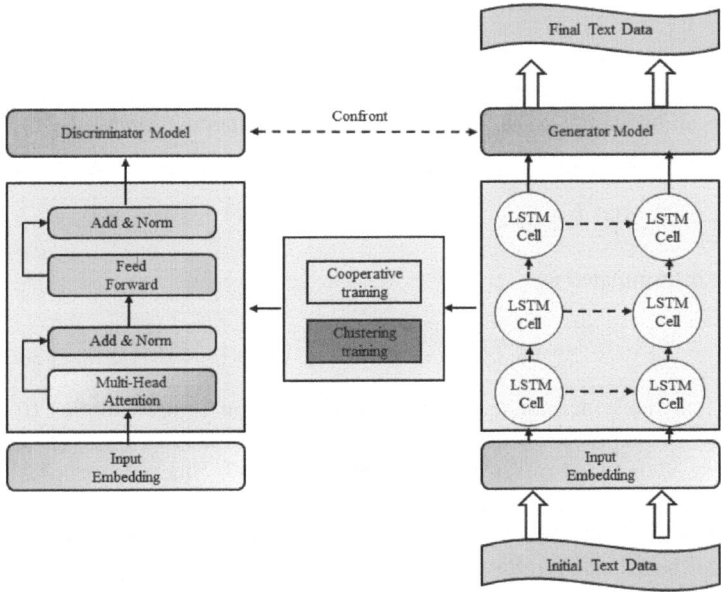

Fig. 1. Cooperative clustering discrete data generation adversarial network model

3 Methodology

3.1 Model of CDGSAN

In this framework, by considering adding a knowledge training module after the generator G, which is called cooperative training, the text generated by it is put into a pre-trained model that has undergone knowledge learning. In addition, a clustering algorithm is added after cooperative training, allowing the generated text to be keyword clustered, which not only greatly reduces the training time but also makes the generated text more authentic and diverse, further empirically and theoretically proving the effectiveness of the algorithm proposed in this article in synthetic and real-life scenarios. Outperforms many strong baselines in terms of generation performance, generalization ability, and computational performance.

3.2 Generator G and Discriminator D

First, the initial generative adversarial network model is optimized, and an explicit generative model, Q, needs to be rebuilt. Therefore, we start with an initial model and iteratively improve it. The original generator is named Qold, from which a new generator, Qnew, is found. In order to form the objective function that optimizes Q, a discriminator between the real distribution and Qold is required.

The discriminator is modeled as:

$$D_{(X)} = \frac{Q^{new}(x)}{Q^{new}(x) + Q^{old}(x)} \tag{3}$$

Using the above discriminator is the optimal discriminator based on GAN. In fact, if we want Qnew to be optimal (i.e. equal to the real distribution P), the optimal discriminator can be expressed as. To optimize the new generator Qnew, use the following goals:

$$\max_D E_{x \sim P_{data(x)}}[lnD(x)] + E_{x \sim Q^{old}}[ln(1 - D(G(z)))] \tag{4}$$

Can be reformulated as:

$$\min_G E_{x \sim P_{data(x)}}\left[ln\frac{Q^{\theta}(x)}{Q^{\theta}(x) + Q^{old}(x)}\right] + E_{z \sim P_{z(x)}}\left[ln\frac{Q^{old}(x)}{Q^{\theta}(x) + Q^{old}(x)}\right] \tag{5}$$

Then, find the generator using the true distribution P and samples from the old generator Q^{θ}. This process is done iteratively. In each model iteration, Q^{θ} a new Qnew is learned based on the results of the previous iteration. Q^{θ} Since the samples come from Qold, the optimization is based on θ, thus bypassing the difficulty of gradient propagation through the discreteness of the Q^{θ} output. During the implementation phase, we use the equality goal based on Softplus [25]:

$$\min_G E_{x \sim P_{data(x)}}\left[Softplus\left(lnQ^{old}(x) - lnQ^{\theta}(x)\right)\right] + E_{x \sim Q^{old}}\left[Softplus\left(lnQ^{\theta}(x) - lnQ^{old}(x)\right)\right] \tag{6}$$

GAN essentially works by optimizing the approximate estimate between the current learning distribution and the target distribution, namely the Jensen-Shannon divergence (JSD). GANs have shown promising results in a variety of unsupervised and semi-supervised learning tasks. The success of GAN marks a new paradigm of deep generative models (i.e., adversarial networks). For sequential discrete data, since the data distribution is decomposed into the sequential product of finite-dimensional multinomial distributions and this decomposition is always based on the form of softmax, the situation of Arjovsky et al. [3] will not occur, that is, when generating When the data distribution is far from the actual data distribution, JSD cannot be effectively optimized. Therefore, optimizing JSD is feasible in this situation. However, according to previous research, no algorithm has yet been able to provide a direct and low-variance JSD optimization method.

3.3 Cooperative and Clustering Training

In this paper, such a model is trained using the method proposed by SidiLu et al. [21] to directly optimize a well-estimated JSD.The text generated by the above generator needs to be re-entered into the pre-trained model that has undergone knowledge training. There is a wealth of knowledge in the pre-trained language model. Making full use of the knowledge inside the pre-trained language model will be helpful for downstream tasks as well as for generation. Whether the semantics of the statement is correct plays an important role. The process is shown in Fig. 2.

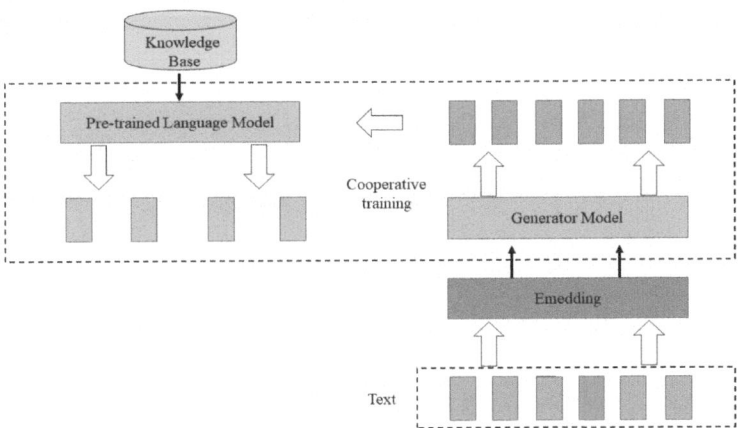

Fig. 2. Cooperative Training

In text generation, in order to speed up text generation and make the generated text more authentic, a clustering algorithm is added to the above basis. After the text is generated, clustering is regenerated through keyword learning.

Clustering is one of the most popular data mining algorithms and has been extensively studied in textual environments. It has a wide range of applications in classification, visualization, document organization, etc. Clustering is the task of finding groups of similar documents in a collection of documents. Calculate the similarity using the similarity function. Text clustering can be at different levels of granularity, where clusters can be documents, paragraphs, sentences or terms. Clustering is one of the main techniques used to organize documents to enhance retrieval and support browsing.

K-means clustering [26] is the most widely used partitioning algorithm in data mining. k-means clustering, divides n documents in the context of text data into k clusters. The representation around which clusters are built. The basic form of the k-means algorithm is: Finding the optimal solution for k-means clustering is computationally difficult

(NP-hard), however, there are some effective heuristics to quickly converge to a local optimum. The main disadvantage of k-means clustering is that it is indeed very sensitive to the initial choice of k. Therefore, there are some techniques for determining the initial k, such as using another lightweight clustering algorithm such as agglomerative clustering algorithm. In this experiment, the generated text is taken as input, the similarity measure S, the number of clusters k is output, and the k cluster initialization sets randomly select k data points as the starting centroid. In case of non-convergence, documents are assigned to centroids based on closest similarity. Calculate the cluster centroids of all clusters. At the end, k clusters are returned, and the regenerated text is output directly. In this paper, the clustering algorithm is applied to the text clustering after the generator generates the text for the first time to select the optimal and logical text in the generated text, and then re-enters it into the discriminator for multiple training of adversarial generation. The process is shown in Fig. 3.

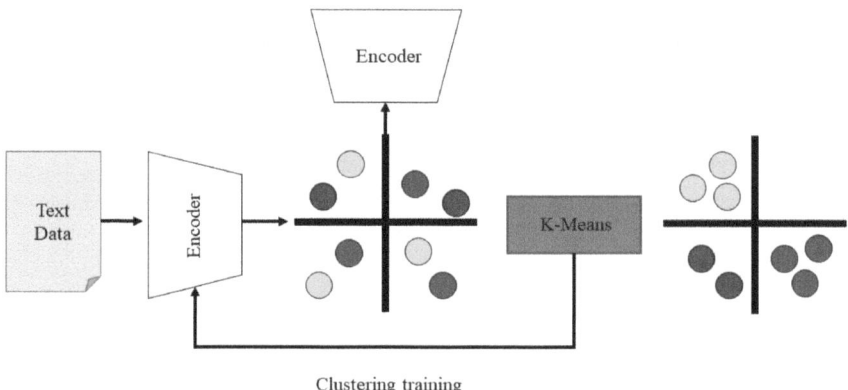

Clustering training

Fig. 3. Clustering training process

4 Experiment

In this section, we conduct experiments to test the proposed approach for question prompt text generation and compare it with other baseline approaches.

4.1 Evaluation Indicators and Data Information

Firstly, the negative log-likelihood (NLL) is introduced as a common metric for evaluating the generative model, and then the BLEU [27] (BL) metric is introduced.

Negative log-likelihood (NLL) represents the negative log-likelihood of the real data in the model. Gan-based methods often have poor NLL scores on training and test data because this metric is very sensitive to modal collapse. It should be noted that NLL is closely related to another well-known metric called perplexity. Therefore, the use of NLL can be used as a reference. NLL_gen represents the negative log-likelihood of the generated model, and the lower the value, the more accurate the generated text. NLL_div represents a measure of diversity, and the lower the value, the lower the diversity of the generated answers, its calculation formula is as follows:

$$NLL_{gen} = -E_{r_{1:T \sim P_r}} log P_q(r_1, \ldots, r_T) \tag{7}$$

$$NLL_{div} = -E_{y_{1:T \sim P_q}} log P_r(y_1, \ldots, y_T) \tag{8}$$

BLEU is an indicator used to evaluate machine translation models. It measures the n-gram similarity between the test sentence and the reference set and then takes the geometric mean of the n-gram similarity to generate a score for the test sentence. As mentioned, the model that repeatedly generates a high-quality sentence can obtain a high BLEU score without completely losing diversity. Therefore, BLEU evaluates the quality of the sample and is not sensitive to its diversity. BLEU is obtained by the weighted geometric mean, and the calculation formula is as follows:

$$BLEU_N(C, S) = b(C, S) exp(\sum_{n=1}^{N} w_n log(CP_N(C, S))) \tag{9}$$

BLEU- measures the accuracy of the word level, and higher-order BLEU can measure the fluency of the sentence. In this paper, n-gram takes 2, 3, 4, and w_n generally takes a constant value for all n, that is, 1/n.

Self-BLEU(SBL) is a self-assessment index, which is usually used to evaluate the diversity and quality of the generated model when generating a set of text samples. It measures the similarity between the generated text and other samples in the generated set. Specifically, self-BLEU calculates the average BLEU score of each sample in the generated set by comparing it with other generated samples. Self-BLEU can help researchers and developers understand the diversity and uniqueness of the generated model so as to optimize and improve it.

This paper uses two real-world datasets: The ImageCOCO, and MR15 data to cover a wide range of linguistic datasets. The data in the above dataset have been pre-processed with word splitting.

The Image COCO dataset is extracted from Amazon Mechanical Turk, which belongs to the annotation type of image captions; and the training and test sets are composed of 10,000 sentences, respectively, with the sample size set to 10,000, the maximum length of a single sentence to be 37, and the vocabulary size to be 4658.

The MR15 dataset contains 1,000 English phrases or sentences on 15 different topics. These themes include books, cars, movies, food, music, news, politics, products, technology, sports, travel, TV, videos, weather, and family. The training and test sets consist of 10,000 sentences each, with a set sample size of 10,000, a maximum individual sentence length of 19, and a glossary size of 6224. The following Table 1 shows the statistical information of the dataset.

4.2 Compare and Contrast Models

In order to validate the performance of the proposed CDGSAN model, experiments are conducted to compare the generated adversarial network models: the DGSAN, COT, SeqGAN, and RelGAN, as well as the baseline model MLE(MaliGAN [19]).

Table 1. Dataset information table.

Dataset	Train set size	Validation set size	Test set size	Vocabulary size	Maximum length	MeanLength
ImageCOCO	10000	10000	10000	4658	37	11.62
MR15	10000	10000	10000	6224	19	6.85

(1) MaliGAN [19]: MLE achieves better results in generative adversarial networks and is a simple generative adversarial network model trained using LSTM.

(2) SeqGAN [16]: Yu et al. used reinforcement learning as a framework to consider individual words using a policy gradient algorithm and Monte Carlo search, respectively.

(3) COT [21]: SidiLu et al. proposed Cooperative Training (CoT), a new algorithm for training likelihood-based generative models for discrete data by directly optimising the estimated Jensen Shannon scatter. CoT coordinates the training generator module G and the auxiliary predictor module M, called the mediator, to collaboratively guide G. The mediator is called the mediator.

(4) DGSAN [22]: A new framework for adversarial training of discrete data generation models, DGSAN, has been proposed by EhsanMontahaei et al. In this framework, the gradient transfer problem is solved by considering the explicit distribution of the generators (due to the advantage of the finite discrete domains) and by finding the closure relation between the next generator, the current generator and the current discriminator. In this method, the generators and discriminators are unified in a network. The network provides the probability distribution of the data and also prepares the (conditional) probability of assigning the input to the real data class. The result of this integration is that the gradient transfer problem is bypassed and training stability is achieved.

In order to verify the performance of CDGSAN model in text generation, comparative experiments are carried out on the ImageCOCO and MR15 datasets. The results of different model processing on the ImageCOCO dataset are shown in Table 2, and the results of different model processing on the MR15 dataset are shown in Table 3, On ImageCOCO, it can be concluded that the CDGSAN model proposed in this chapter has improved performance compared with other models. The CDGSAN model has improved the evaluation index (BLEU-2,3,4,5) of other models in the same environment by 0.9%, 2.1%, 1.7%, and 1.7%, in (Self-BLEU -2,3,4) are also improved. Respectively. Moreover, the lowest NLL_gen index indicates that the generated text is more accurate, and the relatively highest NLL_div indicates the diversity of the generated text. On MR15, it can be concluded that the CDGSAN model proposed in this paper has improved the performance compared with other models. The CDGSAN model has improved the evaluation indicators (BLEU-2,3,4,5) of related models in the same environment by 6.7%,14.5%, 19.8%, and 18.6%, in (Self-BLEU-2,3,4) are also improved. Respectively. Moreover, the lowest NLL_gen index indicates that the generated text is more accurate, and the highest NLL_div index indicates that the generated text is diverse.

In the small-scale English data set CDGSAN model, the clustering mechanism is introduced to enhance the text generation speed of the model through keywords. Through collaborative training before text generation, in an environment of limited resources, the feature extraction ability of the CDGSAN model is improved to a certain extent after parameter adjustment.

Table 2. The results of different model processing on the ImageCOCO dataset table.

Model	BL2	BL3	BL4	BL5	SBL2	SBL3	SBL4	NLL_{gen}	NLL_{div}
CDGSAN	**0.901**	**0.731**	**0.525**	**0.352**	**0.902**	**0.729**	**0.537**	**0.7601**	**1.1521**
DGSAN	0.892	0.71	0.508	0.335	0.892	0.715	0.498	0.7723	0.8199
COT	0.876	0.684	0.471	0.300	0.885	0.707	0.488	1.0942	0.9861
SeqGAN	0.85	0.645	0.434	0.269	0.88	0.675	0.451	0.9955	0.7884
MaliGAN	0.87	0.668	0.446	0.289	0.871	0.689	0.487	1.0089	1.0594

Table 3. The results of different model processing on the MR15 dataset table.

Model	BL2	BL3	BL4	BL5	SBL2	SBL3	SBL4	NLL_{gen}	NLL_{div}
CDGSAN	**0.884**	**0.67**	**0.505**	**0.388**	**0.905**	**0.753**	**0.597**	**0.7755**	**1.8306**
DGSAN	0.801	0.499	0.302	0.202	0.848	0.578	0.352	1.2183	1.5875
COT	0.817	0.468	0.251	0.161	0.859	0.599	0.333	2.1726	1.7811
SeqGAN	0.815	0.514	0.286	0.186	0.829	0.542	0.315	1.2983	1.8156
MaliGAN	0.811	0.525	0.307	0.201	0.826	0.537	0.302	1.2746	0.9331

According to the data results, the GEN line chart is given. The results of ImageCOCO dataset on different models are shown in Fig. 4 and the results of MR15 dataset on different models are shown in Fig. 5.

In order to observe the quality of the text data generated by the model more intuitively, this paper gives the text sentences generated by the two data sets after training under different models. By comparison, it can be found that the sentences given by other models will have grammatical errors and inaccurate expressions. It is difficult to generate smooth and natural sentences with clear structure and accurate expression. Under the training of the CDGSAN model proposed in this paper, both ImageCOCO and MR15 datasets can generate naturally smooth and logically structured high-quality statements, such as 'an airplane jet flying through the clear sky ', and generate text samples as shown in Table 4.

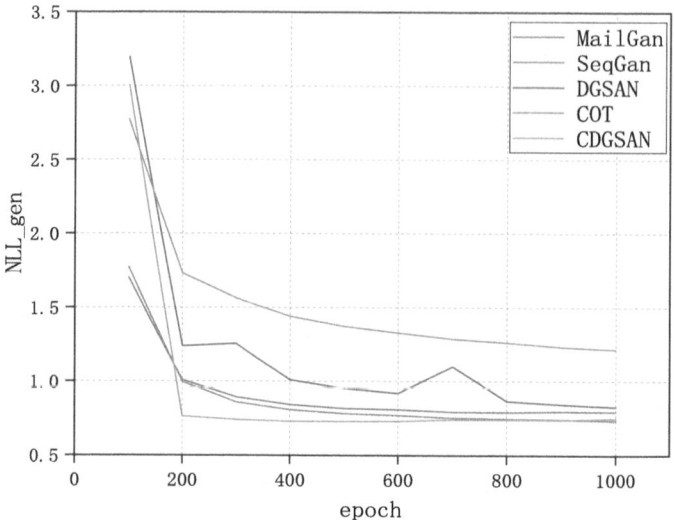

Fig. 4. ImageCOCO on different models NLL_gen graph

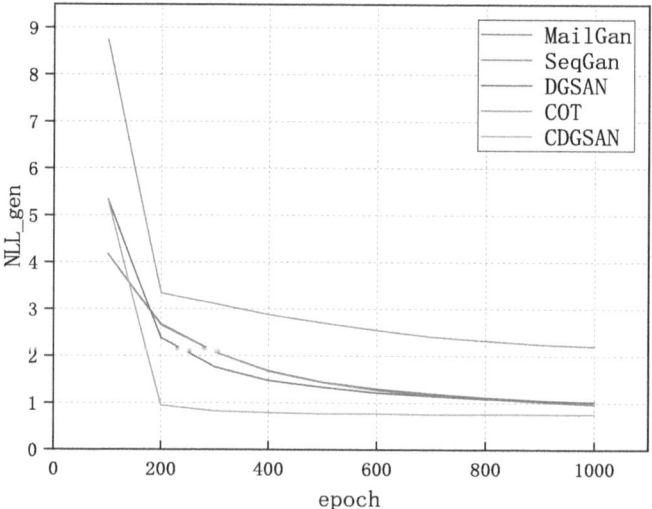

Fig. 5. MR15 on different models NLL_gen graph

Table 4. Comparison of two data set generation statements under different models.

ImageCOCO	Model	MR15
(1)a person on a motorcycle line up by traffic lighting (2)an airplane jet flying through the clear sky	CDGSAN	(1)a film is a blend of exciting elements that unfortunately lack depth or taste (2)a tasty appetizer that leaves you wanting about this year
(1)an eifel plane has islands in the distance in trees on right (2)a picture of a group of car around from a green pasture	DGSAN	(1)soderbergh skims the fat mamet through the ideas, heart-felt drama (2)smith 's best - violent moments
(1)a skier is driving past a parking lot on a trail (2)a toilet is in an airplane is parked on the wall	COT	(1)fails to be a rainy characters (2)nothing just does n't have been a made-for-home-video and it is what it is a symptom
(1)there is a skateboard that has a barn (2)the sink of a very tiny bathroom with all neon sinks	SeqGAN	(1)a plate on a table with chairs, white stove and a stove (2)a few people are sitting on the bathroom windows near a sink with a white tile car next to a window
(1)a woman in a restroom holding chocolate bars (2)a motor scooter parked next to an airport tarmac	MaliGAN	(1)two men in a parking area being instructed by something (2)the living room is includes a washer, a sink and sink

5   Conclusion

This paper proposes a generative adversarial network model. The experimental results show that on the two real datasets of ImageCOCO and MR15, the CDGSAN model performs better than other models, which shows that the feature vector with global semantics can be obtained through the model in this paper so as to enhance the keyword feature extraction ability of the text and generate higher quality text. It can improve the ability of the discriminator to better distinguish between true and false data, and the output feedback value can also effectively guide the optimization direction of the generator so that it can generate high-quality text content. In addition, this paper also uses five different generation models for comparative experiments to verify the effectiveness of the proposed model. By analyzing the experimental results, the values of the model proposed in this paper on the two evaluation indicators of NLL and BLEU are significantly better than other comparison models, so it can be concluded that the model proposed in this paper is effective. In the future, more detailed text feature extraction methods can be considered to counter the generated text.

Acknowledgments. This study was supported by the National Natural Science Foundation of China (61373160), Hebei Province Central Guidance Local Science and Technology Development Fund Project (No. 236Z0306G), the Natural Science Foundation of Hebei Province (F2021210003), Hebei Provincial Science and Technology Plan Project (22567636H).

Disclosure of Interests. The authors declare no conflicts of interest.

References

1. LeCun, Y., Bengio, Y., Hinton, G.: Deep learning. Nature **521**(7553), 436–444 (2015)
2. Goodfellow, I., et al.: Generative adversarial nets. In: Ghahramani, Z. (eds.) NIPS 2014, Advances in Neural Information Processing Systems 27, pp. 2672–2680. Curran Associates, Inc. (2014)
3. Arjovsky, M., Chintala, S., Bottou, L.: Wasserstein generative adversarial networks. In: Proceedings of the 34th International Conference on Machine Learning, vol. 70 of Proceedings of Machine Learning Research, pp. 214–223. PMLR (2017)
4. Radford, A., Metz, L., Chintala, S.: Unsupervise drepresentation learning with deep con volutional generative adversarial networks (2015)
5. Karras, T., et al.: Alias-free generative adversarial networks (2021)
6. Karras, T., Laine, S., Aila, T.: Astyle based generator architectureforgenerativ eadversar ial networks (2018)
7. Sandfort, V., Yan, K., Pickhardt, P., Summers, R.: Data augmentation using generative adversarial networks (CycleGAN) to improve general izability in CT segmentation tasks. Scientific Reports 9 (2019)
8. Bowles, C., et al.: GAN augmentation: Augmenting training data using gen erative adversarial networks (2018)
9. Antoniou, A., Storkey, A., Ed wards, H.: Data augmentation generative adver sarial networks (2018)
10. Tran, N.-T., Tran, V.-H., Nguyen, N.-B., Nguyen, T.-K., Cheung, N.-M.: On data augmentation for GAN training. Trans. Img. Proc. **30**, 1882–1897 (2021)
11. Williams, R.J., Zipser, D.: A learning algorithm for continually running fully recurrent neural networks. Neural Comput. **1**(2), 270–280 (1989)
12. Hochreiter, S., Schmidhuber, J.: Long short-term memory. Neural Comput. **9**(8), 1735–1780 (1997)
13. Graves, A.: Generating Sequences with Recurrent Neural Networks. CoRR abs/1308.0850 (2013)
14. Wen, T.-H., Gašić, M., Mrkšić, N., Su, P.H., Vandyke, D., Young, S.: Semantically conditioned LSTM-based natural language generation for spoken dialogue systems. In: Proceedings of the 2015 Conference on Empirical Methods in Natural Language Processing, pp. 1711–1721. Lisbon, Portugal: Association for Computational Linguistics (2015)
15. Bowman, S.R., Vilnis, L., Vinyals, O., Dai, A., Jozefowicz, R., Bengio, S.: Generating sentences from a continuous space. In: Proceedings of The 20th SIGNLL Conference on Computational Natural Language Learning, pp. 10–21. Berlin, Germany: Association for Computational Linguistics (2016)
16. Yu, L., Zhang, W., Wang, J., Yu, Y.: SeqGAN: sequence Generative Adversarial Nets with Policy Gradient. Proceedings of the AAAI Conference on Artificial Intelligence 31(1) 2017

17. Lin, K., Li, D., He, X., Zhang, Z., Sun, M.-T.: Adversarial ranking for language generation. In: Proceedings of the 31st International Conference on Neural Information Processing Systems, NIPS'17, pp. 3158–3168. Red Hook, NY, USA: Curran Associates Inc. (2017)
18. Fedus, W., Goodfellow, I.J., Dai, A.M.: MaskGAN: better text generation via filling in the _. In: 6th International Conference on Learning Representations, ICLR 2018, Vancouver, BC, Canada, April 30-May 3, 2018, Conference Track Proceedings. OpenReview.net (2018)
19. Che, T., et al.: Maximum-likelihood augmented discrete generative adversarial networks. arXiv preprint arXiv:1702.07983 (2017)
20. Guo, J., et al.: Long text generation via adversarial training with leaked information. Proceedings of the AAAI Conference on Artificial Intelligence, vol. 32, no. 1 (2018)
21. Lu, S., et al.: COT: cooperative training for generative modeling of discrete data. In: Chaudhuri, K., Salakhutdinov, R. (eds.) International Conference on Machine Learning. PMLR (2019)
22. Montahaei, E., Alihosseini, D., Baghshah, M.S.: DGSAN: discrete generative self-adversarial network. Neurocomputing **448**, 364–379 (2021)
23. Zhang, H., Cong, Y., Wang, Z., et al.: Text feature adversarial learning for text generation with knowledge transfer from GPT2. IEEE Transactions on Neural Networks and Learning Systems, 2022
24. Lee, J.-M., Ha, T.-B.: Unsupervised Text Embedding Space Generation Using Generative Adversarial Networks for Text Synthesis. ArXiv Preprint arXiv:2306.17181 (2023)
25. Dugas, C., Bengio, Y., Bélisle, F., Nadeau, C., Garcia, R.: In: Leen, T.K., Dietterich, T.G., Tresp, V. (eds.) Ad-vances in Neural Information Processing Systems 13, Papersfrom Neural Information Processing Systems (NIPS) 2000, pp. 472–478. CO, USA, MIT Press, Denver (2000)
26. Likas, A., Vlassis, N., Verbeek, J.J.: The global k-means clustering algorithm. Pattern Recogn. **36**(2), 451–461 (2003)
27. Papineni, K., Roukos, S., Ward, T., Zhu, W.J.: BLEU: a method for automatic evaluation of machine translation. In: Proceedings of the 40th Annual Meeting of the Association for Computational Linguistics, pp. 311–318. Association for Computational Linguistics, Philadelphia, Pennsylvania, USA (2002)

Dialogue Understanding and Generation of Sequence Template and Path Retrieval Based on Knowledge Enhancement

Wenbin Zhao[1]([⊠]), Keqiang Liu[1], Yan Ren[2], Chaocheng Zhang[1], Shuokai Pan[1], and Zixuan Zheng[1]

[1] Shijiazhuang Tiedao University, Shiiiazhuang 050043, Hebei, China
zhaowb2013@stdu.edu.cn
[2] Hebei Institute of Science and Technology Information, Shiiiazhuang 050021, Hebei, China

Abstract. With the rapid development of deep learning technology, pre-trained language models have made breakthrough achievements in the field of dialogue systems, and these models have demonstrated their excellent ability to capture complex contextual relationships and semantic features. However, they still face challenges in understanding entities in cross-domain data; The sentences generated by the current dialogue model have some fluency, but the generated statements contain the problem of dialogue illusion. To solve these problems, this paper proposes a dialogue understanding and generation method of sequence template and path retrieval based on knowledge enhancement. The dialogue understanding part first selects the knowledge related to the text from the external knowledge, and constructs the knowledge into the graph and the text content as the input of the model. The attention mechanism is also used to evaluate knowledge. Finally, by using knowledge, lexeme and its lexeme category to build a generative template, the model can predict the lexeme entity on the template. The dialogue generation part mainly constructs the dialogue as a graph structure, learns the relationship information between nodes, enhances the use of knowledge, injects knowledge and dialogue to generate dialogue, and re-identifies and retrieves the knowledge in the generated dialogue. The implementation of this method enables the model to show better adaptability and superior performance in processing natural language processing tasks.

Keywords: attention mechanism · generating template · dialogue mapping · knowledge retrieval

1 Introduction

A dialog system is an intelligent platform that allows users to engage in multiple rounds of conversation with a computer through natural language to achieve a specific goal. [1] Such systems provide users with an intuitive and convenient way to interact, greatly improving the efficiency of daily communication and reducing communication costs. The core purpose of the dialogue system is to meet the interactive needs of users, and

© The Author(s), under exclusive license to Springer Nature Singapore Pte Ltd. 2024
H. Yu et al. (Eds.): CCF NCCA 2024, CCIS 2274, pp. 256–273, 2024.
https://doi.org/10.1007/978-981-97-9671-7_17

its typical applications such as smart speakers and mobile phone assistants are gradually gaining the favor of a wide range of user groups. Due to its great potential and commercial value, the dialogue system is also receiving increasing attention. The dialog system can be divided into two categories: task-based and open domain. The task-based dialog system breaks down user requirements into a series of specific tasks in specific vertical fields (such as buying air tickets, booking hotels, etc.), and develops specialized dialog systems for these specific scenarios, ultimately providing one-stop services by integrating multiple task-based systems [2]. Although the research of task-based dialogue system has made remarkable achievements, its dialogue ability is not enough universal. On the one hand, the specific dialogue scene tasks are numerous and difficult to exhaust; On the other hand, existing task-based dialog systems based on specific scenarios are difficult to flexibly migrate to other scenarios. In order to achieve more general dialogue capabilities, the research focus in recent years has shifted to the development of open domain dialogue systems [3]. The characteristic of open domain dialogue system is that its dialogue field and topic are highly open. Currently, researchers are mainly taking a data-driven approach to building such systems in order to enable more flexible and extensive conversational interactions.

In many areas of life, including the economy, government and public life, we can collect a large amount of conversational system data. This data often exists in an unstructured and unordered state, which requires fine processing of the raw data. With the help of knowledge injection technology, the dialogue system can recognize and understand the entity concepts in different fields, and then apply them more effectively.

As the heart of the dialogue system, the technology of dialogue generation has made significant progress to create a dialogue experience that is close to human level. However, pre-trained language models are designed to overemphasize user-centered interactions and strive for consistency with user instructions, which sometimes leads to a tendency to generate dialogue that is inconsistent with actual facts or knowledge, or even wrong. This problem usually stems from the neglect of the context of dialogue and the use of knowledge in the dialogue model. Therefore, through effective use of knowledge base, the accuracy and reliability of language model in dialogue generation can be significantly improved [4]. With the rapid progress of dialogue system technology, this study aims to integrate rich knowledge into dialogue models to overcome the lack of knowledge faced by dialogue models in understanding dialogue entities and generating responses.

Our contributions are summarized as follows:

(1) The external knowledge is embedded into the pre-trained language model to enhance knowledge, the attention mechanism is introduced to screen more appropriate knowledge, the generation template containing knowledge, lexeme and lexeme category information is constructed, and the pre-trained language model is used to make generative prediction with the help of these templates, so as to identify the entity in the text and the corresponding entity category. It complements the model's lack of knowledge in dealing with complex sequence labeling tasks.

(2) By modeling the whole conversation as a one-way global graph. In this way, the model can learn knowledge node information and complex interdependence information from the dialogue background, build an enhancement graph for nodes with

bidirectional relationship in the global graph. The global graph and the enhance-ment graph enhance knowledge, and input knowledge and dialogue into the model for dialogue generation. After dialogue generation, a discriminator is first used to retrieve the problem entity node in the generated statement. Centered on the entity, the appropriate knowledge entity is retrieved to correct the generated statement.

2 Related Work

2.1 Dialogue Understanding

Named Entity Recognition (NER) constitutes the core of the function of the dialog system, which provides key data identification and classification capabilities for the dialog system [5]. Through accurate named entity recognition processing, the data is optimized to improve the efficiency and application value of the dialog system. In the field of natural language processing, named entity recognition is often structured as a sequential labeling problem designed to predict the entity boundaries mentioned in conversation data and their corresponding categories. Recent studies have shown that pre-trained language models have achieved remarkable results in a wide range of natural language processing tasks, including text classification, question answering, text generation, and machine translation. These models are highly regarded for their ability to capture rich semantic information and contextual dependencies. Although pre-trained language models show strong universality and performance in multiple tasks, they encounter specific challenges when applied to named entity recognition [6]. The main challenge is that named entity recognition essentially focuses on the details of sequence annotation rather than just text generation. The pre-trained language model is designed for smooth text generation tasks, so it may not achieve optimal processing results when facing sequence annotation problems that require accurate boundary identification and classification.

In order to bridge the gap between pre-trained language models and named entity recognition (NER) tasks, Wang et al. [7] proposed a methodology to transform the sequence labeling problem into a text generation problem. Specifically, they converted sentences containing named entities, such as "Paris is a city," into sequences of text with a special tag: "@ @ Paris ## is a city." Here, "@ @" and "##" are used as special markers to identify the start and end boundaries of the entity, respectively. This transformation method can solve the problem that pre-trained language models are originally designed for text generation but not directly applicable to sequence annotation, thus enhancing the applicability and efficiency of these models in NER tasks. Laskar et al. [8] conducted a comprehensive evaluation of an advanced language model, ChatGPT[9], to test its efficacy on multiple academic datasets. These datasets cover a range of tasks including question answering, text summarization, code generation, common-sense reasoning, mathematical problem solving, machine translation, bias detection, and ethical considerations. By in-depth analysis of the 255,000 responses generated by the model in these datasets, the study reveals the importance of choosing the right language model for different tasks. Kim et al. [10] explored how GPT-2 and BERT could be used to resolve entity ambiguity in dialogue systems, emphasizing the utility of a hybrid approach. During the training phase, GPT-2 is used as a generator and combined with input sentences is used with the target entity. In the reasoning phase, the input sentence guides an evaluation process to determine whether the target entity can accurately match the context. Although the GPT-2-based approach is slightly less accurate than BERT, the study shows that there is significant potential for combining pre-trained language models to apply to specific conversational entity understanding tasks.

The application of pre-trained language models to named entity recognition (NER) in specific fields is also increasing. For example, De Toni et al. [11] conducted a study on historical texts where they tested models that had not been specifically trained in entity recognition or evaluated in the field. Experiments on multilingual historical datasets show that the accuracy of model predictions is affected by the language of the document and the historical period to which it belongs. Although this model does not perform well compared to the most advanced models of historical documents in dialogue understanding, it demonstrates a unique ability to identify the language and date of publication of documents. The study by Kammer et al. [12] focused on the processing of compound nominal phrases in German medical texts, which constitute a unique linguistic structure in which certain components or coordinating parts of the sentence are omitted in order to avoid repetition. For example, "Vitamins C, E, and A" may be abbreviated as "vitamins C, E, and A," a linguistic phenomenon that increases the difficulty of entity extraction and disambiguation. To address this challenge, the researchers created a dataset of more than 4,000 manually annotated compound noun phrases and used models to parse such phrases in German medical texts in an end-to-end manner. The results show that the model achieves high accuracy in the task of understanding the data set of compound noun phrase, which proves the effectiveness of this method in dealing with certain types of text.

2.2 Session Generation

Natural Language Generation (NLG) plays an important role in the field of Natural Language Processing (NLP) [13]. It transforms diverse data inputs from text, images, tables and knowledge bases into human-like, trusted and easy-to-read text outputs. After decades of development and innovation, text generation technology has been widely used in various fields, among which dialogue generation has become the focus of research in recent years. The dialogue system uses advanced language models and deep knowledge to generate fluent and informative responses. However, as language models increasingly emphasize user-centered interactive experiences and a high degree of consistency with user instructions, these models often produce the so-called "dialogue illusion" in the process of statement generation [14]. This phenomenon refers to the fact that the model, in pursuit of alignment with user input, may generate responses that are inconsistent with the content of previous conversations or known facts, raising questions about the authenticity of the generated text. A striking problem with the current practice of large-scale language models is that they sometimes produce text that is inconsistent with objective facts or tends to be misleading. For example, advanced models such as GPT-4 have the potential to produce imprecise or even made-up statements. [15] In addition, statements generated during conversation generation may inadvertently violate the sensitive boundaries of privacy, raising potential privacy security issues.

Given the central role of dialogue generation in the dialogue task, evaluating the quality of dialogue generation has become a critical issue. Qiu et al. [16] propose a method to identify problems with conversational statements in summaries and extend this to English as well as other languages. Mundler et al. [17] proposed a method of differentiation based on the self-contradiction of conversational statements. A language model contradicts itself when it produces two contradictory sentences in the same conversational context, revealing the absence of some fact-based dialogue generation. The more capable dialogue is of generating models, the less self-contradiction it creates. For example, ChatGPT had a low self-contradiction rate, with only 35.8% of conversations generated being unverifiable on Wikipedia.

Knowledge-based dialogue generation improves the accuracy of dialogue output by leveraging an external authoritative knowledge base rather than relying on internal knowledge of potentially outdated training data or models [18]. Cao et al. [19] propose a decomposition and query framework that guides models to use external knowledge when reasoning with reliable information. In the prediction phase, the framework uses external tools to query a reliable question-and-answer database, allowing backtracking and launching new searches when needed, thus improving the robustness of the conversation generation task.

Optimizing the architecture of the dialog system can also significantly improve the quality of dialog generation. Shi et al. [20] introduced context-aware decoding technology, which enhances the difference between outputs by comparing the output probability distribution of models with and without context. This decoding method can effectively override the innate knowledge of the model when it is inconsistent with the provided context, and significantly improve the dialog generation performance. Yoon et al. [21] proposed a text coordination mitigation framework, which adopts the strategy of copying input text to generate responses when the dialogue problem cannot be understood. The framework takes advantage of the regularization loss of improved dialogue information between the language model and the proposed enhanced language model. By minimizing this loss, it helps to reduce meaningless text duplication, thereby further improving the performance of dialogue generation. These studies provide a new perspective and methodology for improving the performance of dialogue systems and promote the technical progress in the field of natural language processing.

3 Method

3.1 Dialogue Understanding of Sequence Templates

In the traditional named entity recognition (NER) approach, the problem is modeled as a sequence labeling task. Each word is assigned a label, which consists of two parts: one is a sequential segmentation component, such as "B" representing the starting word of the entity, "I" representing the middle word within the entity, and "O" representing the non-entity word; The other part is the entity type label, such as "person" or "location." For example, the label "b-person" means that the word is the first word of a "person" type entity, while the label "i-location" indicates that the word is part of the name of a place. Named entity recognition is the prediction of entity and entity type label by the model.

Define a sentence in the data set as $X^H = \{x_1^H, x_2^H, \ldots, x_n^H\}$, and $L^H = \{l_1^H, l_2^H, \ldots, l_n^H\}$ is its corresponding label sequence. The input text is represented as $X = \{x_1, x_2, \ldots, x_n\}$, the entity type of the input text segment $x_{i:j}$ is represented as y_k.

In this study, we propose a novel model framework to enhance the performance of conversational entity understanding tasks. The framework consists of three core components, each of which is optimized for knowledge information processing processes (Fig. 1).

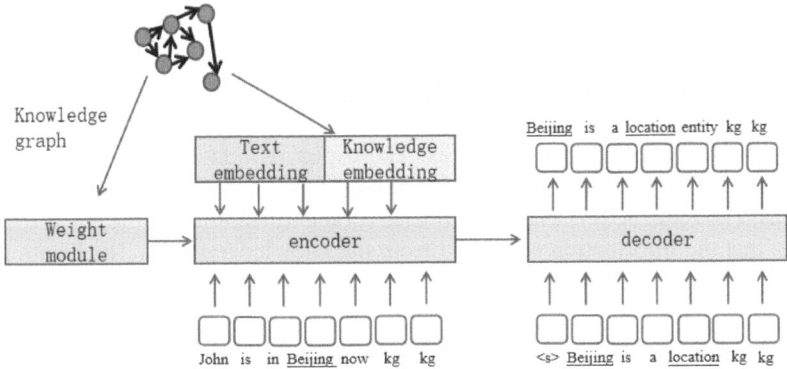

Fig. 1. Named entities identify the overall structure.

(1) Knowledge enhancement module: This module is responsible for co-embedding the current text with relevant knowledge in order to capture the structural information of knowledge. In order to achieve this goal, we adopt a multi-dimensional embedding method, including lexical embedding, entity type embedding, triplet type embedding, sequence type embedding and position embedding. This comprehensive embedding strategy enables the model to understand and represent knowledge information at multiple levels.

(2) Graph weight calculation: In the process of constructing the knowledge graph, the relevant knowledge triples are first retrieved from the knowledge base. In order to accurately screen out the knowledge most relevant to the current text, we introduce the graph weight calculation module. The module independently calculates attention scores for both entity and relational parameters and selects the top knowledge based on these scores, thus ensuring that only the most relevant knowledge is integrated into subsequent processing.

(3) Generate templates: This module builds a generate template by combining knowledge, entities, and categories. Using this template, the pre-trained language model can transform the task of understanding dialogue entities into a language model ordering problem. The output of the model is in the form of a labeled sentence, and the model accomplishes the task of dialog entity understanding by predicting the empty entities and entity types in the template. In addition, the model can learn the correlation between labels and types in different domains.

The knowledge sequence contains the details associated with the knowledge graph (i.e., subject entities, relations, and object entities). In order to integrate the structural information of knowledge, this method adopts a comprehensive embedding technology, which embedding the entity, relationship, and category information of the entity relationship. Under this framework, triples in the knowledge graph are converted into linear sequences for easy use as inputs.

For example, suppose that the knowledge graph to be embedded consists of two triples associated with a subject entity that together shape the knowledge view of that entity. At the Token level, the input of the sequence begins with a specific token and is filled in accordance with the order of the triple type and the arrangement of knowledge, in the form of: where the identification of the types inside the triplet represents the three different types of the subject, the relation and the object in the triplet. The entity type layer starts with all knowledge and type tags associated with entities being categorized as entity types, and parts are labeled as. The triplet type layer is labeled so that the first triplet associated with it is labeled as, and the second triplet associated with it is labeled as. The sequence type tag layer is used for identification, and the rest is recorded, meaning that the information is derived from the knowledge graph. The location label layer marks the location information of knowledge elements in order (Fig. 2).

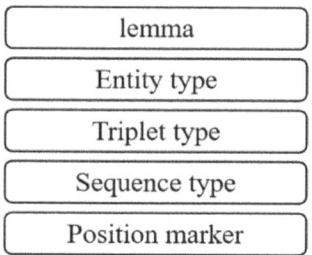

Fig. 2. Knowledge enhancement structure diagram.

In the input sequence, the text part is distinguished from the knowledge part by specific tags. The lexical embedding layer is responsible for storing text-related information. Whether it is knowledge embedding or text embedding, positional embedding encodes the position of each word marker in the sequence and the entire input sequence to preserve the overall structural information of the input sequence. Location coding ensures that the original semantic information of the text is not disturbed, while also avoiding affecting the meaning of other words in the text.

In this study, the Knowledge attention module first retrieves knowledge triples related to a given text using an external knowledge base. Then, a weighted knowledge graph is constructed, which synthesizes the relative importance of different knowledge by calculating the weights of entities and relationships independently.

The entity weights were calculated using the pre-trained language model RoBERTa[22]. The process involves associating each entity in the knowledge graph with the input text and using the pre-trained model to calculate a probability score. The pre-trained language model calculates the probability that an entity is associated with the input text. This probability value serves as the weight of the entity and reflects the correlation between the entity and the input text. The calculation formula is as follows:

$$E_{iw} = LM_{head} (LM_{enc} ([X ; E_i])) \tag{1}$$

where $LM_{head} \circ LM_{enc}$ represents the probability of entity Ei calculated by the language model.

Graph convolutional neural network (GCN) [23] was used to calculate the relational weights. The filtered knowledge is constructed as a graph structure, the entities of the knowledge triplet are regarded as nodes, and the relations in the knowledge are regarded as edges of the graph. The weight of each edge is calculated using a graph neural network. The feature vector is obtained by calculating the cosine similarity between the embeddings of the knowledge graph (containing entities and relationships) and the text embeddings, and no additional learning parameters are required to capture the relational information. By setting the corresponding positions of relation and entity to 1 and 0, and multiplying the entity value with the relation value, the final relation weight is obtained. The graph weight calculation module is used to obtain the top n entities and relationships most relevant to the input text from the knowledge graph.

Through the construction of generating templates, we can realize the effective use of structured knowledge, and thus enhance the understanding and reasoning ability of natural language processing models. Specifically, we embed the information extracted from the knowledge graph in order to inject this rich structural information into deep learning models.

In the implementation process, a series of generation templates are defined, which are built based on lexical elements, lexical categories, and the knowledge associated with them. Each generated template is a string of predictors and lexical categories that are used to be replaced by a specific lexical when predicting. The positive template of a lexeme expresses what entity class an entity belongs to, and accordingly, we design negative templates to express non-existent relationships between certain entities and specific knowledge (Tables 1, 2, 3, 4, 5 and 6).

For example, the type of entity $x_{i:j}$ is y_k, and kg represents knowledge. To illustrate this approach more intuitively, the following table lists some examples of generating templates based on entities, entity categories, and related knowledge :

Table 1. Sample table of sequence templates.

Forward template	Negative template
$< x_{i:j} >$ is type $< y_k >$ $< kg >$	$< x_{i:j} >$ is not type $< y_k >$ $< kg >$
The type of $< x_{i:j} >$ is $< y_k >$ $< kg >$	The type of $< x_{i:j} >$ is not $< y_k >$ $< kg >$
Type $< y_k >$ includes $< x_{i:j} >$ $< kg >$	Type $< y_k >$ does not include $< x_{i:j} >$ $< kg >$

This strategy can help the model learn to distinguish between existing and non-existing knowledge relationships, and improve its identification accuracy of entities. We bring together all possible positive and negative templates into a general set of

Table 2. Data sample.

U.N	NNP	I-NP	I-ORG
official	NN	I-NP	O
Ekeus	NNP	I-NP	I-PER
heads	VBZ	I-VP	O
for	IN	I-PP	O
Baghdad	NNP	I-NP	I-LOC

Table 3. Comparison table.

Model	CoNll03			GENIA		
	P	R	F1	P	R	F1
Biaffine	93.7	93.3	93.5	81.8	79.3	80.5
DiffusionNER	92.99	92.56	92.78	82.1	80.97	81.53
PIQN	93.29	92.46	92.87	**83.24**	80.35	81.77
SBWNER	92.61	93.87	93.24	78.57	79.3	78.93
DENER	92.78	93.57	93.17	81.04	77.21	79.08
Our	**93.91**	**93.96**	**93.93**	82.33	**81.32**	**81.82**

Table 4. Ablation experiment data sheet.

Ablation	CoNLL03		
	P	R	F1
BART	89.61	91.64	90.61
+ Knowledge enhancement	91.72(2.11↑)	93.40(1.76↑)	91.90(1.29↑)
+ Generate templates	93.91(2.19↑)	93.96(0.56↑)	93.93(2.03↑)

Table 5. Data field distribution table.

Domain	Movies	Books	Sports	Music	ALL
#of dialogs	6429	5891	2495	858	15673
#of turns	37838	34035	14344	4992	91209

Table 6. Comparison table.

Model	BLEU4↑	ROUGE-L↑	FEQA↑	QuestEval↑		EntityCoverage(%)↑		
				RD	RF	P	R	F1
EARL	7.97	23.61	39.93	37.88	35.59	86.61	45.17	64.44
GPT2	10.27	29.59	39.60	46.86	42.07	91.62	33.26	52.30
GPT2NPH	10.41	29.93	40.83	**47.45**	42.45	95.61	33.39	53.96
BART	14.45	33.33	39.00	46.97	42.75	96.99	44.96	62.87
KGBART	13.92	33.31	41.87	45.55	42.86	97.68	45.63	64.58
Our	**14.79**	**34.42**	**42.35**	47.23	**43.25**	**97.99**	**46.79**	**66.37**

templates to cover different types of entities and knowledge relationships. This collection is manually created and can be extended and adjusted as needed to fit the statements in the data set.

In this way, we can translate complex structured knowledge into a form that the model can understand, and further improve the recognition and reasoning ability of the model for complex relationships through machine learning techniques. This method not only enhances the model's understanding of the relationship between entities, but also improves the model's adaptability and generalization ability in different scenarios. In this study, we take an innovative approach to encode and utilize structured knowledge using a pre-trained language model. The process involves embedding all possible segments of text in a sentence into our carefully constructed template to achieve efficient capture of structural information about these segments. To ensure the efficiency and feasibility of the process, we limit the maximum number of text segments that can be filled per template to 10. This limitation reduces computational complexity while ensuring information richness.

We then use the pre-trained generation language model to score each of the filled templates based on the fit between each specific fill template and the language understanding ability of the pre-trained model. The scoring function of the template is shown in equation:

$$f(\boldsymbol{T}) = \sum_{c=1}^{m} \log p(t_c|t_{1:c-1}, \boldsymbol{X}) \tag{2}$$

where X represents the input text and T represents the build template.

By comparing the scores of the positive and negative templates for each entity type, we are able to determine how well each entity type matches a given text. Finally, the entity type with the highest score is assigned to the corresponding text fragment, giving that text fragment an explicit entity type. In the training phase, we extract all the entities, entity types, and related knowledge from the training set and fill them into the generation template to create informative generation template sentences. If the text segment is of a non-entity type, it is also filled into the template to construct the target sentence.

In order to further optimize the training process of the model, in addition to using all the entities in the training set to construct positive samples, we also create negative samples by random sampling. The number of negative samples is set to 1.5 times the number of positive samples to provide a more diverse training signal and enhance the model's ability to learn the degree of differentiation between different entity types. The loss function is:

$$\mathcal{L} = -\sum_{c=1}^{m} \log p(t_c | t_{1:c-1}, X) \tag{3}$$

With this approach, our model can not only accurately identify and classify entities and their types in text, but also effectively leverage external knowledge sources to enhance its understanding of complex linguistic phenomena, significantly improving the performance of natural language processing tasks.

3.2 Dialog Generation for Path Retrieval

We propose an innovative dialogue generation method based on global graph and path retrieval. We represent the entire conversation as a global graph that is unidirectional and incorporates syntactic and discourse structure information. To enhance semantic learning, we construct an enhancement graph between nodes with bidirectional dependencies. Our model uses this conversation global graph and enhancement graph to reinforce the knowledge base and input this knowledge along with the conversation context into the pre-trained language model GPT-2 to generate the conversation.

After the conversation is generated, we introduce a knowledge discriminator that first identifies the problematic knowledge entity in the conversation statement. Then, starting from these identified problem entities, through the path retrieval method, the entity set associated with the dialogue statement is found in the knowledge contained in the dataset, and scored. Finally, we replace the problem entity node with the highest-scoring entity, improving the accuracy and quality of conversation generation. Figure 3 shows the structure details of the model.

Firstly, the dialogue is encoded globally, and a unidirectional global graph is constructed from the answer part of the dialogue, and an enhanced graph is constructed by using nodes with bidirectional relationship in the global graph.

The answer part of dialogue can be regarded as a kind of semantic structure between interactive discourses to ensure the semantic consistency of information transmission. The utterances of each dialogue will be connected to subsequent utterances, where either the response is correct (i.e. the case of the speaker's reciprocal reply), or both come from the same speaker. In the case of cross-speakers, the current utterance is linked to the previous utterance and an "ans" label is assigned to the edge. In addition, these utterances are associated with the same speaker with a "smsp" (same speaker) label of the same speaker (Fig. 4).

Enhanced image: The dialog global graph is a one-way graph structure, and there is a bidirectional dependency relationship between some parameter entities in the dialog global graph, so a bidirectional enhanced graph is created between these parameter pairs to carry out sufficient relational reasoning (such as the interdependence between

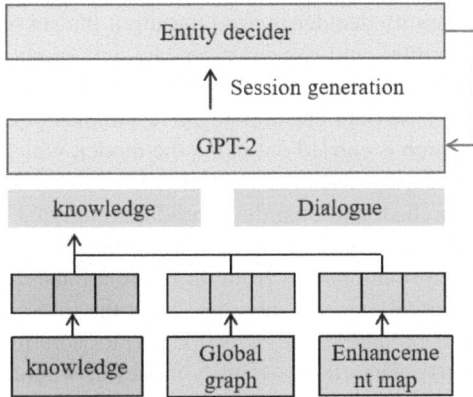

Fig. 3. General structure of conversational illusion.

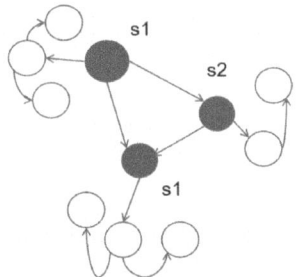

Fig. 4. Session global graph.

entities). The enhanced graph can learn more global information about the dialog. Some parameters are mentioned many times in the dialog global graph. And are scattered throughout the dialog global, so these parameters are merged from the dialog global graph.

Each edge of the dialogue global graph is labeled, and the graph convolutional network is used to encode the entire conversation. In the diagram $G = (V, E)$, for nodes $\pi_{i,j} \in E$, from nodes v_i to v_j, when there is an edge in the middle, convert to 1, If not, it is 0. A "self" tag is added to each node as a self-loop edge to aggregate information. The graph convolutional network representation of nodes v_i is calculated as shown in equation:

$$e_i^d = \text{ReLU}\left(\sum_j \gamma_{i,j}\left(W_1 \cdot h_j + W_2 \cdot x_{i,j}^\pi + b\right)\right) \tag{4}$$

In order to improve the quality of dialogue generation, it is necessary to first find the exact source that affects the quality of dialogue in the generated discourse. Dialogue generation errors are usually related to knowledge, which is mainly in the form of knowledge entities in the dialogue data, and there are many kinds of entities mentioned, such as the name of the region, the name of the music, etc. To label the entities used in the conversation data, an entity decider is used to output the set of entities by entering the conversation history, triples, and conversation data, i.e., predicting a binary label at each word position. To label the training data, a synthetic dataset is created, consisting of positive samples of the base dialogue and negative samples of errors.

The node-centered search is carried out using the model, which is divided into relational search and entity search. The first is relational search: the set of all relations related to the entity is searched, and then the candidate relational set and possible path set are obtained and used for scoring and screening. For example, in the first search, the model selects the top 3 relationships from all the relationships associated with the head entity Beijing as {capital, country, region}. Then the inference path of the entity node Beijing's top three candidate relations is updated to {(Beijing, capital), (Beijing, country), (Beijing, region)}. Then there is entity retrieval, where the model filters out the best set of entities by calculating the query score of the set of relationships, such as: The score of both entity nodes of China and China Capital Region is 1, and since the relation capital, country, and region are all connected to the tail entity with only one relation, the current inference path is updated to {(Beijing, Capital, China), (Beijing, country, China), (Beijing, region, China Capital Region)}.

$$\mathcal{L}_{\text{NCE}} = -\log(s(t)) - \log\left(s(t) + \sum_{j=1}^{n} s(t')\right) \tag{5}$$

In the training, the entity determiner is used to find the problem entities, obtain their related information representation, retrieve the related knowledge set through the way of path retrieval, and then use the scoring function to calculate the correlation of knowledge.

$$\mathcal{L} = \mathcal{L}_{\text{MLE}} + \mathcal{L}_{\text{NCE}} \tag{6}$$

4 Experiments and Results

4.1 Dialogue Understanding Experiment

4.1.1 Datasets and Evaluation Indicators

In this study, we used the CoNLL-2003 dataset [24], which includes data in both English and German. However, this study only used the English portion of the data, which was made up primarily of news reports from Reuters in August.

In the CoNLL-2003 dataset, data is divided into four different types of entities: person (PER), organization (ORG), place (LOC) and MISC (MISC). In the CoNLL-2003 dataset, each line of each data file contains one word. Empty lines are used to represent sentence boundaries. At the end of each line, there is a tag that indicates whether the current word is in a named entity, and that also encodes the type of the named entity. Table 1 lists the data samples.

In Table 1, each row contains four fields: the word, its part-of-speech tag, the block tag, and its named entity tag. Tagged words are outside the named entity, tagged words are inside the named entity of the type. When two entities of a type are next to each other, the first word of the second entity will be labeled B.

The measurement values of precision P, recall rate R and F1 were used as evaluation indexes.

$$P = \frac{Nm}{Np}, N = \frac{Nm}{Nr}, F1 = \frac{2 \times P \times R}{P + R} \qquad (4)$$

where N_m,N_p, and Nr represent the total number of matching entities, predicted entities, and real entities respectively. In the field of named entity recognition, Precision, Recall, and F1 scores are commonly used to evaluate model performance. These indicators help to understand the model's ability to predict data and provide a quantitative measure of the quality of the model's predicted results.

When one of the accuracy and recall rates is low, F1 scores are also low, reflecting the model's poor performance in accurately identifying relevant entities. Thus, F1 scores provide a balance point designed to optimize both accuracy and recall. By combining P, R and F1 scores, the performance of the model on the named entity recognition task can be evaluated.

4.1.2 Experimental Analysis

The visual bar chart of the comparison experiment is shown in the Figs. 5, 6 below:

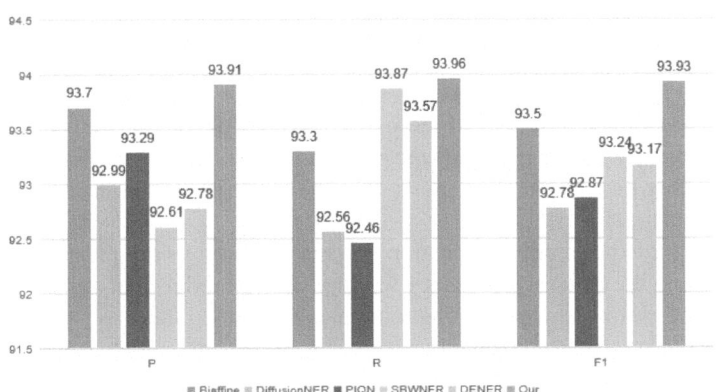

Fig. 5. Visual histogram of CoNll03 data set.

In this study, the proposed method based on knowledge embedding and template generation shows significant performance improvement in experiments. Compared with the sub-optimal data, the proposed method performs particularly well on the CoNLL-2003 dataset, where Precision (P), Recall (R) and F1 scores are increased by 0.21, 0.09 and 0.43, respectively. This significant improvement demonstrates the high efficiency of the method in processing named entity recognition tasks. This result shows that the recognition ability of entities in different fields of text can be effectively improved by optimizing the model structure.

The knowledge embedding design adopted in this study provides an efficient way for pre-trained language models to deal with knowledge triples. This design not only

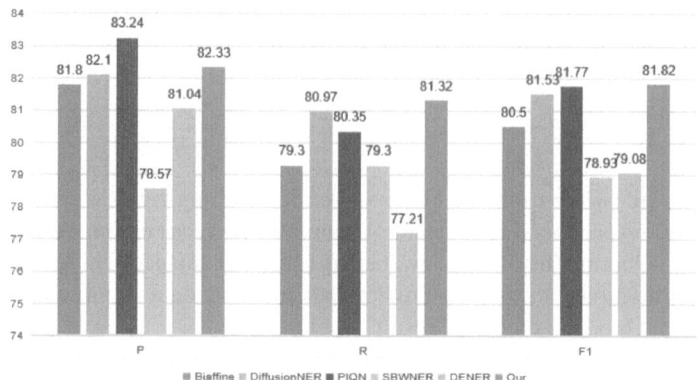

Fig. 6. GENIA data set visual bar graph.

enhances the application flexibility of the model in the knowledge graph, but also selects the knowledge triplet through the attention mechanism, so that the model can better learn the graph triplet representation of knowledge. More importantly, knowledge enhancement, combined with the structure that generates the template, gives the model the ability to retain knowledge structure information, and the model is also able to effectively capture and use this structure information. This is crucial when the model is dealing with complex knowledge extraction tasks, as it ensures the effectiveness and accuracy of the model's use of knowledge.

In summary, the method proposed in this chapter not only achieves significant performance improvement in statistical sense, but also provides a new perspective and powerful tool for processing natural language processing tasks such as knowledge extraction and named entity recognition through its unique knowledge enhancement and template generation techniques.

After careful experimental evaluation, the research results are clearly shown in the table. Analyzing the data revealed a key finding: BART models that did not incorporate knowledge embedding and generating templates performed the least well. In contrast, when knowledge embedding technology was introduced, the model accuracy (P), recall (R) and F1 scores improved significantly by 2.11 percentage points, 1.76 percentage points and 1.29 percentage points, respectively. The application of template generation also had a positive impact on model performance, increasing P, R and F1 scores by 2.19 percentage points, 0.56 percentage points and 2.03 percentage points, respectively.

It is worth noting that although knowledge embedding and generated templates have similar contributions in improving accuracy, knowledge embedding has a more significant effect in improving recall rate. In the F1 score improvement, the generation of templates has played a greater role. This result not only highlights the respective areas of strength of the two technologies, but also shows how complementary they are when they work together.

The results of ablation experiments provide a strong validation for the techniques proposed in this chapter, and confirm the significant effect of knowledge enhancement and template generation methods in improving the comprehensive performance of the model. These findings highlight the value of structured information in natural language processing tasks and the potential to leverage these techniques in building more accurate, higher-recall models.

4.2 Session Generation Experiment

4.2.1 Datasets and Evaluation Indicators

OpenDialKG is an open domain conversation dataset launched by Facebook that contains approximately 91,000 rounds of conversations[25]. These dialogues deeply integrate entities and relationships in the knowledge graph, and each dialogue is linked to a corresponding knowledge path through which the mentioned knowledge points and relationships are connected. The corpus and its corresponding knowledge graph path contain the logical reasoning process of human in the dialogue, and support the distributed operation of the information in the knowledge graph.

This dataset contains conversations that span multiple domains and cover a wealth of knowledge, such as movies (including titles, actors, directors), books (including titles, authors), sports (including athletes, teams), and music (including singers). In order to reduce the interference of irrelevant knowledge, the marginal entities were screened according to the importance of the entities, resulting in a dataset containing 1,190,658 knowledge triples, 100,813 entities, and 1,358 relationships.

In the experimental study, we divided the dataset sample containing knowledge and randomly assigned it to the training set (80%), verification set and test set (10% each) for in-depth analysis and evaluation.

4.2.2 Experimental Analysis

The experimental results reveal the significant improvement of the proposed method compared to the suboptimal performance. Specifically, The method has good performance in BLEU4, ROUGE-L, FEQA, QuestEval-RF, Precision (P), Recall (Recall, R) and F1 scores were improved by 0.34, 1.09, 0.48, 0.39, 0.31, 1.16 and 1.79, respectively. These results fully prove the validity of the dialog generation model which integrates the dialog global graph and path retrieval. By integrating the conversation global graph and the bidirectional gain graph into the pre-trained language model, we can capture the global context characteristics of the conversation level and the complex dependencies between nodes. The path-dependent searcher effectively identifies the use of wrong entities in dialogue generation, and corrects the knowledge misuse in sentences by taking the problem node as the core and gradually retrieving and correcting it in the knowledge base. In addition, visual path retrieval not only deepens the depth of knowledge retrieval, but also makes the knowledge reasoning process more transparent and explainable, thus improving the accuracy of knowledge use. These improvements together verify the practical benefits of this research method in enhancing the ability of dialogue generation.

5   Conclusion

In this paper, we propose a knowledge-enhanced dialogue understanding and generation approach. First, the external knowledge is filtered, and the related knowledge is constructed as a graph structure. Through knowledge embedding, the model successfully captures the rich structural information contained in the knowledge graph, thus providing the model with a deep insight into the relationship between the knowledge triples. At the same time, the integration of knowledge, entities, and entity categories into generating templates improves the ability to perform the task of named entity recognition. In the dialogue illusion part, the dialogue text is first modeled as a dialogation-level global graph to facilitate more accurate feature learning. The global graph makes the sentences generated by the dialogue more in line with the dialogue background. The error types of dialogue statements are identified and refined through knowledge retrieval, which improves the reasoning ability of knowledge depth and reduces the dialogue illusion in sentences. This method improves the adaptability and superiority of the pre-trained language model in different data domains. These findings highlight the potential of structured information in natural language processing and provide a valuable reference for future research.

Acknowledgments. This study was supported by the National Natural Science Foundation of China (61373160), Hebei Province Central Guidance Local Science and Technology Development Fund Project (No. 236Z0306G), the Natural Science Foundation of Hebei Province (F2021210003), Hebei Provincial Science and Technology Plan Project (22567636H).

Disclosure of Interests.   The authors declare no conflicts of interest.

References

1. Zhai, C., Wibowo, S.: A systematic review on artificial intelligence dialogue systems for enhancing English as foreign language students' interactional competence in the university Comput. Educ.: Artif. Intell. **4**, 100134 (2023)
2. Addlesee, A., Sieinska, W., Gunson, N., et al.: Data collection for multi-party task-based dialogue in social robotics. In: International Workshop on Spoken Dialogue Systems Technology 2023 (2023)
3. Rodríguez-Cantelar, M., Zhang, C., Tang, C., et al.: Overview of robust and multilingual automatic evaluation metrics for open-domain dialogue systems at dstc 11 track 4[J]. arXiv preprint arXiv:2306.12794, 2023
4. Wölfel, M., Shirzad, M.B., Reich, A., et al.: Knowledge-based and generative-ai-driven pedagogical conversational agents: a comparative study of grice's cooperative principles and trust[J]. Big Data Cogn. Comput. **8**(1), 2 (2023)
5. Qu, X., Gu, Y., Xia, Q., et al.: A survey on arabic named entity recognition: Past, recent advances, and future trends. IEEE Trans. Knowl. Data Eng. (2023)
6. Yu, J., Li, Z., Wang, J., et al.: Grounded multimodal named entity recognition on social media. In: Proceedings of the 61st Annual Meeting of the Association for Computational Linguistics (Volume 1: Long Papers). 2023: 9141–9154 (2023)
7. Wang, S., Sun, X., Li, X., et al.: GPT-NER: Named Entity Recognition via Large Language Models[J]. arXiv e-prints, 2023: arXiv: 2304.10428

8. Laskar, M.T.R., Bari, M. S., Rahman, M., et al.: A systematic study and comprehensive evaluation of chatgpt on benchmark datasets[J]. arXiv preprint arXiv:2305.18486, 2023

9. Wu, T., He, S., Liu, J., et al.: A brief overview of ChatGPT: The history, status quo and potential future development. IEEE/CAA J. Automat. Sinica **10**(5), 1122–1136 (2023)

10. Kim, Y., Hwang, Y., Shin, J., et al.: Injecting comparison skills in task-oriented dialogue systems for database search results disambiguation. In: Findings of the Association for Computational Linguistics: ACL 2023, pp. 4047–4065 (2023)

11. De Toni, F., Akiki, C., de la Rosa, J., et al.: Entities, Dates, and Languages: Zero-Shot on Historical Texts with T0[J]. arXiv e-prints, 2022: arXiv: 2204.05211 (2022)

12. Kammer, N., Borchert, F., Winkler, S., et al.: Resolving elliptical compounds in german medical text. In: The 22nd Workshop on Biomedical Natural Language Processing and BioNLP Shared Tasks, pp. 292–305 (2023)

13. Wong, M.F., Guo, S., Hang, C.N., et al.: Natural language generation and understanding of big code for ai-assisted programming: a review[J]. Entropy **25**(6), 888 (2023)

14. Kuhn, L., Gal, Y., Farquhar, S.: Semantic uncertainty: Linguistic invariances for uncertainty estimation in natural language generation[J]. arXiv preprint arXiv:2302.09664 (2023)

15. Rawte, V., CHAKRABORTY, S., Pathak, A., et al.: The troubling emergence of hallucination in large language models-an extensive definition, quantification, and prescriptive remediations. In: The 2023 Conference on Empirical Methods in Natural Language Processing (2023)

16. Qiu, Y., Ziser, Y., Korhonen, A., et al.: Detecting and mitigating hallucinations in multilingual summarisation. Proceedings of the 2023 Conference on Empirical Methods in Natural Language Processing, pp. 8914–8932 (2023)

17. Mündler, N., He, J., Jenko, S., et al.: Self-contradictory hallucinations of large language models: evaluation, detection and mitigation. In: The Twelfth International Conference on Learning Representations (2023)

18. Zhong, S., Huang, Z., Wen, W., et al.: Sur-adapter: Enhancing text-to-image pre-trained diffusion models with large language models. In: Proceedings of the 31st ACM International Conference on Multimedia, pp. 567–578 (2023)

19. Cao, H., An, Z., Feng, J., et al.: A Step Closer to Comprehensive Answers: Constrained Multi-Stage Question Decomposition with Large Language Models[J]. arXiv e-prints, 2023: arXiv: 2311.07491

20. Shi, W., Han, X., Lewis, M., et al.: Trusting your evidence: Hallucinate less with context-aware decoding[arXiv:2305.14739J]. arXiv preprint (2023)

21. Yoon, S., Yoon, E., Yoon, H.S., et al.: Information-theoretic text hallucination reduction for video-grounded dialogue[J]. arXiv preprint arXiv:2212.05765 (2022)

22. Liu, Y., Ott, M., Goyal, N., et al.: Roberta: A robustly optimized bert pretraining approach[J]. arXiv preprint arXiv:1907.11692 (2019)

23. Vashishth, S., Yadav, P., Bhandari, M., et al.: Confidence-based graph convolutional networks for semi-supervised learning. In: The 22nd International Conference on Artificial Intelligence and Statistics. PMLR, pp. 1792–1801 (2019)

24. Sang, E.F., De Meulder, F.: Introduction to the CoNLL-2003 shared task: Language-independent named entity recognition. arXiv preprint cs/0306050 (2003)

25. Moon, S., Shah, P., Kumar, A., et al.: Opendialkg: explainable conversational reasoning with attention-based walks over knowledge graphs. In: Proceedings of the 57th Annual Meeting of the Association for Computational Linguistics, pp. 845–854 (2019)

DGCBA: A Novel Medical Point Cloud Segmentation Network Based on Dilated Graph Convolution and Boundary Awareness

Wenbin Zhao[1]($\boxtimes$), Longbiao Jia[1], Haoyang Zhao[1], Jianming Wang[1], and Pingsheng Dai[2]

[1] Shijiazhuang Tiedao University, Shijiazhuang Hebei, Shijiazhuang 050043, China
zhaowb2013@stdu.edu.cn
[2] Beijing Jingxing Huidong Technology Co., LTD, BeiJing 100000, China

Abstract. Medical point cloud analysis is a task that applies point cloud data for analysis and processing in the medical field. However, there is little research based on medical point cloud analysis. With the continuous development of graph neural networks and attention mechanisms, significant research progress has been made in the field of point cloud analysis. However, current methods focus on designing powerful feature extractors, ignoring the segmentation of geometric edge features, which in turn reduces segmentation performance. In response to the above issues, We propose a medical point cloud based segmentation network based on dilated graph convolution and boundary awareness, DGCBA. This network includes an extended geometric convolution DGConv (Dilated Graph Convolution), a multi-graph boundary-aware module (MGBA), and a learnable difference pooling module (LDP) for medical point cloud segmentation. Thus improving the accuracy of boundary segmentation. By using DGConv to construct feature relationship graphs, complex geometric edge structure information can be learned. MGBA is introduced to check global knowledge propagation on channel graphs, enrich feature representations, and LDP is introduced to obtain different global feature vectors, reducing the limitations of single pooling and fully capturing local and global feature information of point clouds. The extensive experiments of the network proposed in this article on the medical point cloud dataset IntrA and the common point cloud dataset ShapeNetPart show that our method outperforms advanced methods in segmentation tasks and also has good generalization performance.

Keywords: Medical point cloud segmentation · Feature extractor · Dilated graph convolution · Multi-Graph boundary-aware

1 Introduction

In recent years, with the rapid development of 2D image processing tasks in the field of computer vision, traditional 2D images can no longer meet many social needs. People have gradually shifted their research direction to 3D data, and the processing tasks

H. Yu et al. (Eds.): CCF NCCA 2024, CCIS 2274, pp. 274–292, 2024.
https://doi.org/10.1007/978-981-97-9671-7_18

related to 3D data have received widespread attention. Point cloud data represents complex 3D objects with low memory requirements, enabling real-time 3D visual applications on resource constrained devices. Therefore, point cloud data is widely used in fields such as autonomous driving [1], 3D reconstruction [2], forest monitoring [3], medical care, scene reconstruction [4], digital cities, etc. Point cloud has significant significance in promoting the development of digital society. This article mainly introduces the widespread application of point cloud analysis in the medical field, such as segmentation when processing 3D scanning data of dental models, which helps with related operations such as tooth extraction, removal, and correction [5]; On the other hand, when performing intracranial aneurysm resection surgery, the accuracy of aneurysm segmentation becomes a crucial factor. Compared to 2D images, the three-dimensional form of cerebral blood vessels provides richer boundary information, thus providing important support for high-precision navigation and operation of the surgery.

Although current models perform well on public 3D point cloud datasets, their performance may sometimes be affected in medical data due to domain differences. For precise medical pathological fragments, they are crucial for the accurate diagnosis and treatment of diseases. However, the particularity of medical point clouds lies in the possibility of containing incomplete pathological structures, which makes accurate differentiation from healthy parts within an object challenging [6]. The limited availability of medical data samples poses a challenge for learning unique shape descriptors. Therefore, it is necessary to propose an effective deep learning method that can demonstrate excellent performance on medical datasets and achieve good generalization ability on non medical datasets. Solving this problem is crucial for deep learning research in the medical field, as it not only helps to overcome the problem of insufficient medical data, but also ensures that the model has strong shape representation ability in different fields, thereby more comprehensively dealing with diverse data features.

PointNet [7] used multi-layer perceptrons to process point cloud data, pioneering the use of deep learning to directly process point cloud tasks. However, it lacked sufficient exploration of local contextual features. To solve this problem, PointNet + + [8] constructed local layer modules, each consisting of sampling layers, grouping layers, and feature extraction layers, ultimately solving the problem of local point cloud sampling and local feature extraction. However, due to the use of Max Pooling alone for single pooling, other feature information was ignored, and the geometric relationships between points were not fully utilized. DGCNN [9] introduced a dynamic graph convolutional network, which achieves more powerful feature learning of point clouds by modeling the dynamic weights of edges on the local neighborhood of the point cloud. However, in the process of constructing a dynamic graph, KNN is used to display and partition the local area, which results in being very sensitive to the density distribution of the point cloud and ignoring the directional structural information between points. Inspired by the knowledge of Transformer [10] and graph convolution theory, this paper proposes a medical point cloud based segmentation model DGFNet based on graph convolution and attention mechanism. By iteratively expanding the operation of geometric convolution DGConv, the model effectively learns the complex local geometric structure of medical point clouds. By introducing a learnable difference pooling (LDP) module, the model can fully utilize the information of all point features, including non maximum point features, to avoid information loss, and utilize the multi-graph boundary-aware module to enrich the representation of global features, thereby improving the accuracy of edge segmentation.

Our contributions are summarized as follows:

(1) We propose a medical point cloud based segmentation network DGCBA based on dilated graph convolution and boundary awareness. By dilated graph convolution DGConv, we fully explore the topological relationships of point clouds to supplement the edge information of local point clouds. We also introduce learnable difference pooling to aggregate global features and improve the network's feature extraction ability.

(2) To compensate for the shortcomings of single pooling, a learnable differential pooling module is proposed to obtain feature information of all available points.

(3) In order to solve the problem of insufficient medical point cloud data leading to poor training performance, a multi-graph boundary-aware module based on graph convolution and graph inference is proposed to enrich the representation of global features, which helps to improve the model's understanding and representation of medical point clouds.

(4) Our model has achieved advanced segmentation performance on medical datasets through Intra, while also demonstrating good generalization performance on the common point cloud dataset ShapeNetPart.

2 Related Work

2.1 Traditional Point Cloud Segmentation Algorithms.

Traditional point cloud segmentation algorithms are based on edge based and region growing methods, which have played a key role in the segmentation of ultrasound and CT images. Gupta [11] proposed a hybrid edge based segmentation method specifically applied to ultrasound medical images. The core idea of this method is to combine edge information and put forward high requirements for the extraction of edge information. Dong et al. [12] proposed an improved supervised voxel 3D region growth method aimed at precise segmentation and reconstruction of pulmonary nodular ground glass nodules. Chen et al. [13] modeled the risk of occlusal caries guided by dental plaque based on three-dimensional surface morphology analysis. Through curvature analysis and region growth method, they extracted pits and fissure areas from the three-dimensional tooth model, and preliminarily completed the segmentation task of dental pits and fissures.

Although existing segmentation methods have achieved certain results in lesion segmentation when processing point cloud data, complex image preprocessing steps are required before automatic segmentation to obtain high-quality point cloud datasets. This type of method typically requires manual labeling of point clouds and extraction of relevant features, followed by the selection of appropriate classifiers for semantic label partitioning. However, there are also a series of problems: the expression ability of manually extracting features is relatively weak, and it requires professional personnel to execute, with strong subjectivity; The differences in constraint conditions in different application scenarios result in relatively weak generalization performance of the model; The selection of the optimal classifier is time-consuming and cumbersome, requiring the support of professional experience.

2.2   Point Cloud Segmentation Algorithm Based on Deep Learning.

In 2017, QI et al. [7]. From Stanford University proposed using deep learning models to directly process raw point clouds - PointNet, pioneering the use of deep learning to directly process point cloud tasks. The PointNet network can directly input the original point cloud, and then use the T-Net module to perform affine transformation on the input point cloud matrix. The multi-layer perceptron is used to extract point cloud features and the maximum pooling layer is used to aggregate global features, effectively solving problems such as point cloud disorder, displacement invariance, and rotation invariance. However, PointNet ignores the exploration of local geometric information. Based on this problem, QI et al. [8] proposed PointNet + +, which builds a local layer module on the basis of PointNet. Each layer consists of sampling layers, grouping layers, and feature extraction layers, ultimately solving the problem of local point cloud sampling and local feature extraction, while also improving classification accuracy. Later, some research work was improved based on PointNet + + [14–17]. In order to solve the problem of high computational complexity in extracting local features, Ma Et al. [17] proposed a pure residual network called PointMLP, which abandons computationally complex local networks. The feature extractor and the introduction of a lightweight local geometric affine module have significantly improved inference speed. With the development of graph convolution and attention mechanisms, many scholars have begun to apply graph convolution [18] and attention mechanisms [19] and related knowledge to point cloud segmentation tasks.

Among the classic methods of graph convolution, Wang et al. [9] proposed DGCNN (Dynamic GCNN) based on Graph Convolutional Neural Network (GCNN), which designed edge convolution operators to extract edge vectors of center points and K-nearest neighbors. Through continuous stacking and cycling, global features were finally obtained, which improved performance in point cloud classification and segmentation tasks. Chen et al. [20] were inspired by research on graph convolution and proposed GAPNet by combining graph convolution with attention mechanisms. By embedding graph attention mechanisms in stacked multi-layer perceptron MLP layers to learn local geometric features, GAPLayer is used to extract attention features of points. To enhance the connection between local features, a multi head mechanism is used to aggregate attention features from different GAPLayers. Finally, attention pooling layers are used to capture important features. This network has achieved advanced performance in point cloud segmentation and component segmentation tasks. Wang et al. [21] proposed a new Graph Attention Convolution (GAC) and constructed GACNet. The convolutional kernel can be dynamically sculpted into feature structure shapes, and the shape of the convolutional kernel is determined by the learned attention weight distribution to adapt to different object structures.

With the rise of point cloud analysis technology and deep learning, research on classification and segmentation of medical images has become relatively popular, and research based on medical point clouds has also attracted widespread attention and exploration.

Yang et al. [6] proposed a publicly available three-dimensional dataset for intracranial aneurysms, IntrA, to replace two-dimensional medical images for diagnosing intracranial aneurysms. They compared the performance of classification and segmentation on classic deep learning point cloud classification and segmentation networks, making direct point based deep learning classification and segmentation models possible. This helps researchers to gain a deeper understanding of the morphology, size, location, and other characteristics of this disease, and promotes progress in the field of medical point cloud and cerebrovascular disease diagnosis. Ghazvinian [22]et al. designed a method for semantic segmentation of teeth based on the PointCNN [23] to capture local and global feature information, which is used for semantic segmentation of individual teeth and gums from point clouds scanned within the oral cavity of teeth. This method has potential important implications for clinical dentistry, medical research, and dental treatment.

3 Method

3.1 Reviewing Local Point Feature Learning

Given a point cloud consisting of N points $P = \{p_i \mid i = 1, 2,..., N\} \in \mathbb{R}^{N \times 3}$, and a data matrix $X = [x_1, x_2, x_3,..., x_n] \in \mathbb{R}^{N \times d}$ containing their feature vectors, for each point pi, local feature extraction can be expressed as:

$$gi = M\left(h(pi, pj, xi, xj) \mid pj \in N(pi)\right) \tag{1}$$

Among them, gi represents the learned local feature vector, $N(p_i)$ represents the set of neighbors of pi, $h(\cdot)$ encodes the coordinates p_i of the i-th point, the coordinates pj of its neighboring points, as well as their feature vectors xi and xj. In addition, M represents the aggregation function, such as maximum and average.

In previous research, a significant amount of work has focused on designing $h(\cdot)$ encoding functions., The EdgeConv method is widely popular due to its application in Dynamic Graph Convolutional Neural Networks (DGCNN). In EdgeConv, the neighborhood of a point is defined by its K-nearest neighbor (KNN) in the feature space. The feature difference $xi - xj$ is used to represent the pairwise local relationship between a point and its adjacent points. These relationships are connected to the original point features. The connected feature pairs pass through multi-layer perceptron θ Process and summarize using the maximum (MAX) aggregation function. The entire process can be described as:

$$gi = MAX\left(h\theta(xi - xj) \| xi\right) \tag{2}$$

where $\|$ represents the connection operation, $h\theta$ A nonlinear function that represents learnable parameters in a multi-layer perceptron.

3.2 Overall Architecture

We have designed a separate network architecture for point cloud segmentation tasks using the proposed DGconv layer. The network architecture is shown in Fig. 1. The overall network structure consists of fourDilated Graph Convolutions (DGConv) modules, one Graph Interactive Feature Fusion (GIFF) module, and five learnable Difference

Pooling modules. DGConv is mainly used to extract local features of point cloud geometric edge position information, the Graph Interactive Feature Fusion module is used to enhance global correlation in feature representation, and the learnable Difference Pooling is used to pool the output point features of each DGConv through learnable Difference Pooling (LDP) and concatenate them to form the feature information of superimposed points. Finally, through the learnable Difference Pooling LDP module, dynamic weight adjustment is used to synthesize the feature information of each point, generating representative features. The information of multi-resolution features is used to further improve segmentation accuracy.

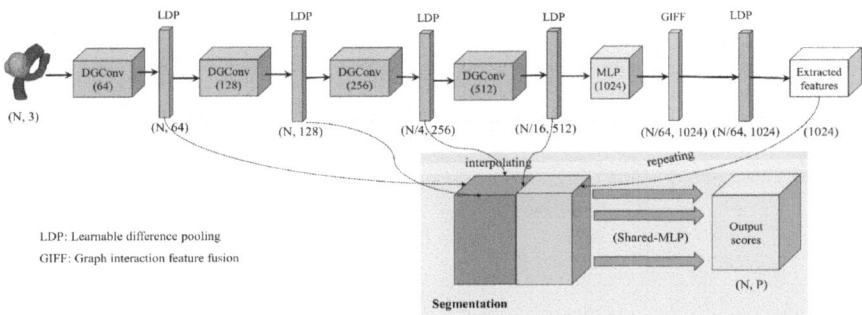

Fig. 1. Overall Architecture.

3.3 Dilated Graph Convolution

In order to achieve precise segmentation around boundaries, it is necessary to fully explore and utilize semantic features and edge geometric feature information. This is crucial in certain medical applications, such as in surgical procedures such as aneurysm resection or tooth extraction, where precise boundary segmentation directly affects the success of the surgery. To meet the needs of these applications, we propose the Dilated Geometric Boundary Convolution (DGConv). DGConv is a local feature extraction method proposed for point cloud data, which improves segmentation accuracy by utilizing the edge geometric structure information of points. When EdgeConv down-samples points in the middle layer, especially when a large acceptance domain is set, edge geometric information may be lost,

DGConv explicitly uses relative point coordinates and edge features as point features when processing point cloud data, enabling better preservation of edge geometric information during convolution operations. DGConv can be represented as:

$$g_{new} = CA\left(h_{\theta 3}\left(h_{\theta 2}\left(f_i, f_j - f_i\right), h_{\theta 1}\left(x_j - x_i, \left\|x_j - x_i\right\|_2\right)\right)\right) \tag{3}$$

Among them, CA represents channel attention, $x_j - x_i$ and $\|x_j - x_i\|_2$ respectively represent relative position and Euclidean distance. By combining the two relative position features, local geometric shapes can be preserved on the convolutional layer regardless of the size of the receptive domain. This advantage enables the network to learn the fine geometric structure of complex structures, significantly improving segmentation accuracy.

Firstly, the extended K-nearest neighbor (Dilated-KNN) method is used to construct a local graph, which expands the receiving field without increasing computational costs. Then, each local point and relative position feature is input into a shared multi-layer perceptron, which elevates the features to the same number of channels to maintain equal importance. Next, through the channel attention mechanism, the features are recalibrated to suppress unnecessary features and enhance the weights of important features. As shown in Fig. 1.

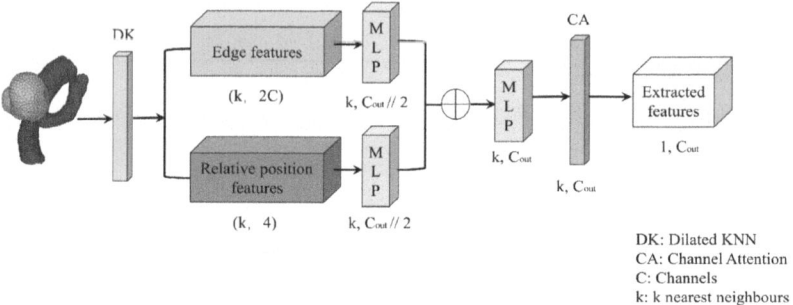

Fig. 2. Proposed local feature extraction methods: DGConv.

3.4 Learnable Difference Pooling

In point cloud data analysis networks, the role of pooling layers is to pool the extracted features to generate global feature vectors. Given that the representation of three-dimensional shapes can be any arrangement of a set of points, the processing network of point cloud data needs to overcome the challenge of unordered input data arrangement Therefore, when selecting a pooling function, it is necessary to ensure that it maintains invariance in the arrangement of input data. Existing networks typically choose symmetric pooling functions such as Max Pooling and Mean Pooling Fig. 2.

The maximum pooling function extracts the maximum value from a set of local feature vectors as the global feature vector. For the input point cloud X, the global feature vector extracted by maximum pooling can be represented as:

$$M\{N\} = \max\{g(x1), g(x2), ..., g(xn)\} \tag{4}$$

Among them, N represents the length of the feature vector, max is the maximum value function, and g represents the feature extraction function for the input point cloud.

$$A\{N\} = avg\{g(x1), g(x2), ..., g(xn)\} \tag{5}$$

Among them, N represents the length of the feature vector, avg is the average function, and g represents the feature extraction function for the input point cloud.

learnable Difference Pooling.

Although max pooling and average pooling have to some extent solved the problem of point cloud disorder and achieved certain results in point cloud classification and segmentation tasks, max pooling only retains the features of the maximum value and ignores the information of non maximum points, resulting in the loss of non maximum point information. Similarly, choosing only to use average pooling often results in the loss of key information for the maximum value.

To compensate for the shortcomings of single pooling, this paper proposes a learnable differential pooling module by introducing a multi head attention mechanism [24], as shown in the Fig. 3. By calculating the correlation between point features and learnable parameters, all point features are weighted and aggregated, which is essentially an information retrieval process that uses a series of learnable queries to extract key related information from values. The key and value both originate from the same point feature tensor, while the query is a learnable parameter. This module enables the network to dynamically adjust weight parameters based on input data, adjust channel size of input point features through linear transformation, match with expected output channel size and multi head attention mechanism, learn appropriate query parameters, so that the retrieved feature information is related to the learning target. Meanwhile, LDP is also a symmetric pooling function, so this function is permutation invariant for points in the point cloud. Compared with traditional max pooling and average pooling, the proposed learnable differential pooling function can more comprehensively preserve point feature information, thereby generating more representative global features. The LDP formula can be defined as:

$$\Psi(Q, F) = \delta\left(\frac{QW_q(FW_k)^T}{\sqrt{d_k}}\right)FWv \tag{6}$$

where F is the input point cloud feature matrix, Q, K, and V are the query matrix, key matrix, and value matrix in the attention mechanism, respectively. Therefore, W_q, W_k, and Wv can be used for the learnable linear transformation weight matrix of queries, keys, and values. In addition, δ is the Softmax function, where d_k is a scaling factor proportional to the feature dimension.

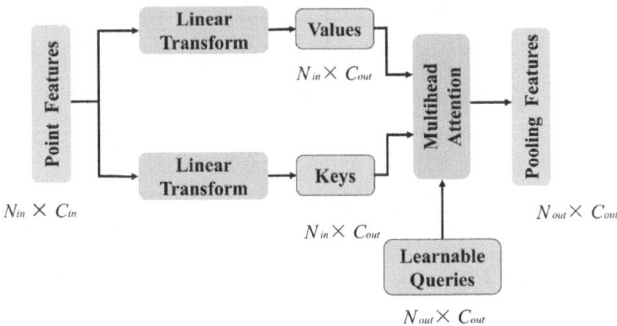

Fig. 3. The structure of LDP.

3.5 Multi-graph Boundary-Aware

Due to the complexity of medical point cloud production and the limited number of medical point cloud datasets available for training, feature extraction is crucial for model performance. In this article, we propose a multi-graph boundary-aware module based on graph inference [25] and graph convolution [26], which fully mines and enhances global correlations in feature representation. Compared with modeling channel information solely through a single graph structure, GIFF can use a learnable adjacency matrix to initialize multiple graphs simultaneously, enhancing the diversity of node features through different graph states, thereby improving the model's representation ability. Therefore, the output Fout of the GIFF module can be represented as:

$$Fout = [\text{ReLU}((V + I)ai)]i \in Ck, ai \in \mathbb{R}^{Cv \times Cv} \tag{7}$$

Among them, I is the identity matrix, used to introduce self loops, that is, each node itself to its own edge; a_i is a learnable adjacency matrix that represents the weights of edges between nodes; By simultaneously computing the C_k graph and corresponding adjacency matrix through C_v channel nodes, contextual interactions between nodes can be modeled, and structural relationships can be captured for feature learning.

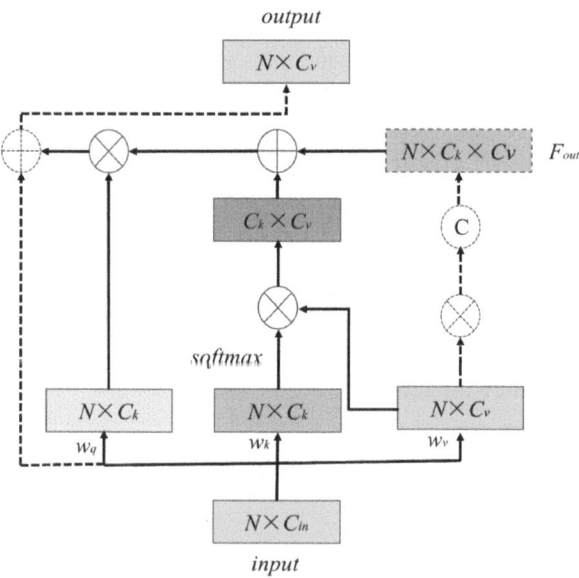

Fig. 4. The structure of multi-graph boundary-aware.

4 Experiments and Results

In order to verify the effectiveness of the method in this chapter, segmentation experiments were first conducted on the common point cloud dataset ShapeNetPart and the medical point cloud dataset IntrA. This chapter's experiment is developed based on the deep learning framework Pytorch, using Linux Ubuntu 20.04 as the operating system, with a memory size of 32GB, a CPU model of Intel (R) Xeon (R) Platinum 8352V, and a GPU model of NVIDIA RTX A4090 (24GB). The basic parameter settings for each experiment are shown in Table 1.

Table 1. Experimental environment settings.

Datasets	Optimizer	Learning Rate	Weight Decay	Batch Size	Epoch
ShapeNetPart[27]	AdamW	3×10^{-3}	2×10^{-4}	32	350
IntrA[6]	AdamW	1×10^{-3}	1×10^{-4}	32	300

4.1 Datasets and Evaluation Indicators

ShapeNet is a three-dimensional point cloud dataset developed jointly by research teams from Stanford University, Princeton University, and Toyota Institute of Technology in Chicago. This chapter uses a partial segmentation subset of the dataset, ShapeNetPart, to evaluate the effectiveness of the point cloud segmentation method in this chapter. The ShapeNetPart dataset consists of 16880 three-dimensional samples, including 14006 samples for training and 2874 samples for testing, covering 16 different categories and a total of 50 different partial labels. These categories cover various common objects, such as airplanes, chairs, and guitars. The visualization of the ShapeNetPart dataset is shown in Fig. 5.

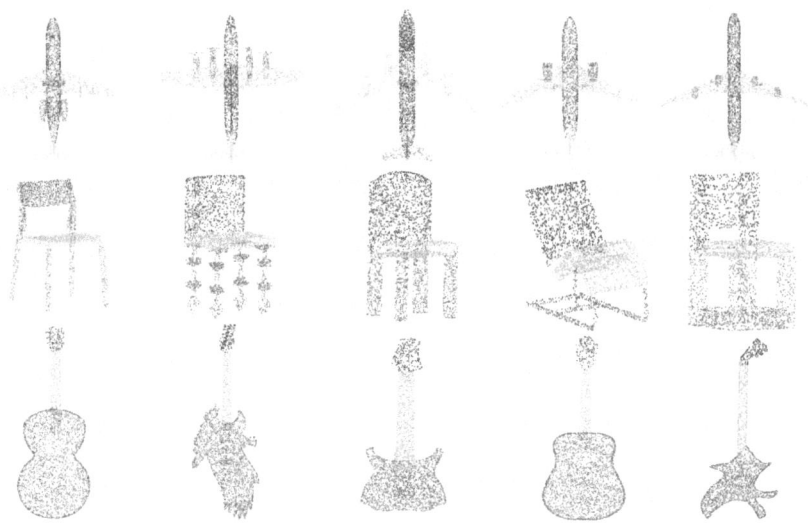

Fig. 5. The visualization of the ShapeNetPart dataset.

IntrA is an open-source 3D intracranial aneurysm dataset used to support the application of point cloud and grid based classification and segmentation models for segmentation tasks. It consists of 1909 vascular point cloud segments extracted from real patient 3D surface models, including 1694 healthy vascular segments and 215 aneurysm segments for diagnosis. Among them, 116 aneurysm segments were manually divided and labeled by medical experts. The size of each aneurysm segment was determined based on the needs of preoperative examination, which is a binary segmentation task. Red represents healthy vascular segments, and blue represents aneurysm segments. The visualization of the IntrA dataset is shown in Fig. 6.

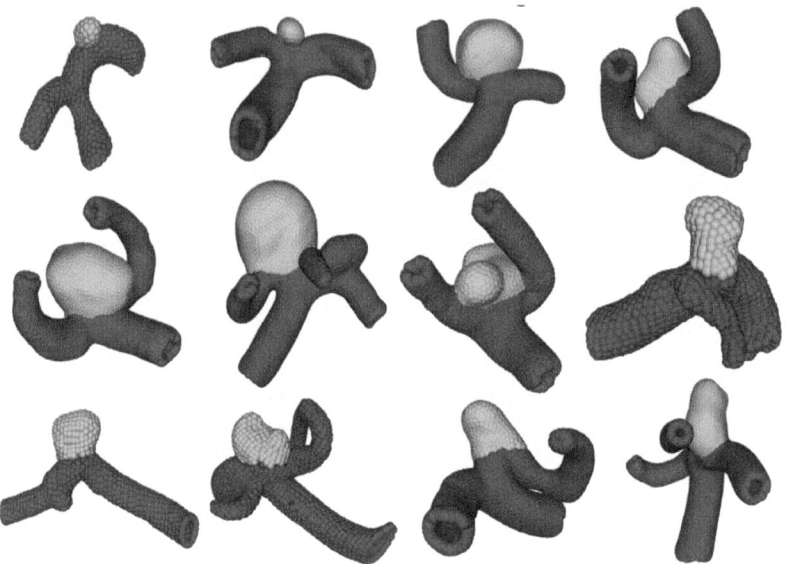

Fig. 6. The visualization of the IntrA dataset.

In point cloud segmentation tasks, Intersection over Union (IoU) is a commonly used evaluation metric used to measure the similarity between predicted segmentation results and actual segmentation results. It is achieved by calculating the ratio of the intersection area between the predicted segmentation result and the actual segmentation result to their union area. Therefore, the calculation of IoU for class i is shown in Eq. (8), and mIoU is used to calculate the average IoU for each category, as shown in Eq. (9), where C represents the number of categories.

$$\text{IoU}_i = \frac{TP_i}{TP_i + FP_i + FN_i} \tag{8}$$

$$\text{mIoU} = \frac{1}{C} \sum_{i=1}^{C} \frac{TP}{TP + FP + FN} \tag{9}$$

In the evaluation of component segmentation tasks in the ShapeNetPart public point cloud dataset, two standard metrics are usually used: average instance intersection to union ratio (Instance mIoU, Instance. mIoU) and average class intersection to union ratio (Class mIoU, Cls. mIoU). Inst.mIoU calculates the average intersection to union ratio of all individual shape instances, while Cls.mIoU calculates the average intersection to union ratio of all shapes divided by category, thus measuring the comprehensive performance of the model on each category. For the segmentation task of medical point cloud IntrA dataset, especially when dealing with structures such as blood vessels and tumors, DSC (Dice Similarity Coefficient) is often an important evaluation indicator, also known as Dice coefficient. It is similar to IoU, but the calculation method is slightly different. The calculation formula is shown in (10).

$$DSC = \frac{2 \times |X \cap Y|}{|X| + |Y|} \tag{10}$$

Among them, $|X \cap Y|$ is the intersection size of the predicted segmentation and the true segmentation, which is the common number of positive predicted points between the two, $|X|$ is the total number of predicted segmentation points, and $|Y|$ is the total number of true segmentation points.

4.2 ShapeNetPart Experimental Results and Analysis

On the ShapeNetPart dataset, this section quantitatively analyzed the comparison results between our method and other methods. The experimental results are shown in Table 2. Our method achieved 86.6% on Inst.mIoU and 85.2% on Cls.mIoU.

Compared to PointNet + +, which has insufficient local feature extraction, the method in this chapter has improved by 1.5% and 3.3% on the two evaluation metrics of Inst.mIoU and Cls.mIoU, respectively. Compared with DGCNN, the method in this chapter has improved the evaluation metrics of Inst.mIoU and Cls.mIoU by 1.4% and 2.3%, respectively. This is because edge convolution may lose edge geometric information when downsampling points in the middle layer, especially when a large acceptance domain is set. Compared to position adaptive convolution (PAConv), the method in this chapter has improved by 0.5% and 0.6% on the two evaluation metrics of Inst.mIoU and Cls.mIoU, respectively. Compared to the latest method PointNorm, the method in this chapter has improved by 0.4% and 0.5% on the two evaluation indicators of Inst.mIoU and Cls.mIoU, respectively.

The experimental results show that compared to existing segmentation methods, this chapter's method achieves higher segmentation accuracy in the segmentation task on the ShapeNetPart dataset.

In order to verify the actual segmentation effect of the method in this chapter, different segmentation objects were selected on the ShapeNetPart dataset and different methods were used to test the segmentation results. Figure 6 shows the visual comparison of the segmentation results between our method and other methods in this chapter. The first column of Figs. 4,5,6,7 represents the source shape of the segmented object, the second column corresponds to the segmentation results obtained using DGCNN, the third column represents the segmentation results obtained using PointNorm, and the last column represents the segmentation results of the method proposed in this paper. Through the black square markings in the figure, it can be seen that different segmentation methods have the problem of incomplete segmentation at the markings.

Table 2. Experimental results of ShapeNetPart.

Methods	PointNet + + [8]	DGCNN [9]	PAConv [28]	PointNorm [30]	DGCBA
Inst.mIoU(%)	85.1	85.2	86.1	86.2	**86.6**
Cls.mIoU(%)	81.9	82.3	84.6	84.7	**85.2**
airplane	82.4	84.0	84.3	82.7	85.1
bag	79.0	83.4	85.0	84.9	86.2
cap	87.7	86.7	90.4	88.9	89.6
car	77.3	77.8	79.7	79.8	80.0
chair	90.8	90.6	90.6	90.2	90.5
areophone	71.8	74.7	80.8	81.9	78.7
guitar	91.0	91.2	92.0	91.6	91.7
knife	85.9	87.5	88.7	87.4	88.2
lamp	83.7	82.8	82.2	82.9	82.5
laptop	95.3	95.7	95.9	95.8	96.3
motorbike	71.6	66.3	73.9	78.4	78.7
mug	94.1	94.9	94.7	95.5	95.2
pistol	81.3	81.1	84.7	84.5	85.6
rocket	58.7	63.5	65.9	65.6	66.7
skateboard	76.4	74.5	81.4	81.4	83.5
table	82.6	82.6	84.0	83.8	83.9

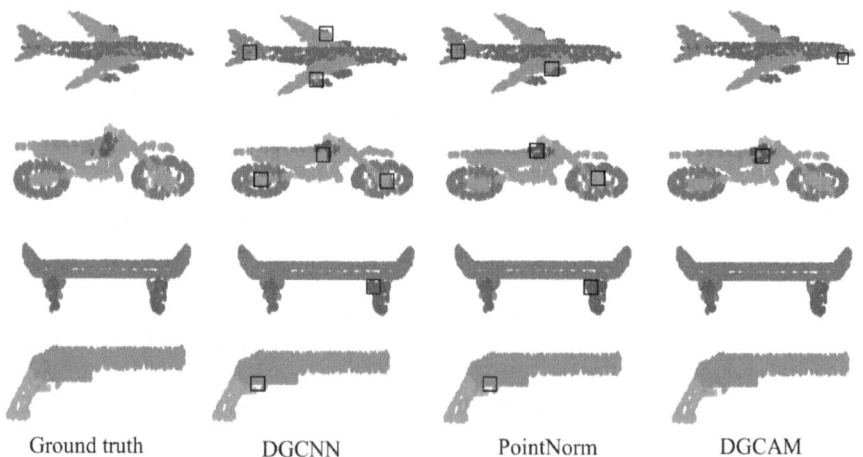

Ground truth   DGCNN   PointNorm   DGCAM

Fig. 7. Comparison of ShapeNetPart segmentation effects.

4.3 Intra Experimental Results and Analysis

In order to verify the segmentation performance of our method in medical point cloud segmentation tasks, this section quantitatively analyzed the comparison results between our method and other methods on the medical point cloud IntrA dataset. The experimental results are shown in Table 3, where V represents the main vascular segment and A represents the aneurysm segment.

Table 3. Experimental results of IntrA.

Methods	Input	IoU(%)		DSC(%)	
		V	A	V	A
PointNet + + [8]	512	93.41	76.34	96.47	83.91
	1024	93.35	76.28	96.48	84.62
	2048	93.24	76.30	96.40	84.50
DGCNN [9]	512	93.26	77.94	96.21	86.89
	1024	93.53	77.86	96.34	86.74
	2048	93.47	78.23	96.26	87.25
PointConv [29]	512	94.16	79.42	96.94	86.02
	1024	94.59	79.15	97.05	86.03
	2048	94.64	79.56	96.98	86.69
PCT [10]	512	92.49	78.09	96.04	86.21
	1024	92.05	78.23	96.25	86.44
	2048	91.74	77.20	95.43	86.14
PointNorm [30]	512	94.31	81.34	96.92	89.31
	1024	94.54	81.72	96.21	89.94
	2048	94.25	81.47	96.13	89.21
DGCBA	512	95.14	82.34	96.43	90.13
	1024	**95.20**	**83.12**	**97.37**	**90.31**
	2048	94.78	83.09	96.71	90.24

From Table 3, it can be seen that our method in this chapter outperforms other methods in both IoU and DSC evaluation metrics when the number of input points is 512, 1024 and 2048. The highest IoU and DSC values were obtained in the main vessel segment segmentation with 1024 input points, which are 95.20% and 97.37%, respectively. Meanwhile, the method in this chapter also performed the best in the segmentation of arterial aneurysm segments, with IoU and DSC values of 83.12% and 90.31%, respectively. Due to the fact that point cloud methods in the general field focus on the design of complex feature extractors and neglect boundary segmentation, the segmentation performance difference between this chapter and the ShapeNetPart dataset is not very significant.

However, on the medical point cloud IntrA dataset, the segmentation accuracy of this chapter's method is more accurate than the previously proposed method .Fig. 8

In order to visually demonstrate the effectiveness of the method in medical point cloud segmentation tasks in this chapter, Fig. 7 shows segmentation results of different shapes. The first column shows the source shape of the segmentation object for intracranial aneurysms, and the second to fifth columns are PointNet + +, DGCNN, PointNorm, and the segmentation results corresponding to the method in this chapter, respectively. From the visual segmentation results, it can be seen that the segmentation performance of the method in this chapter on medical point cloud datasets is higher than the currently popular point cloud segmentation methods. When dealing with challenging boundary areas between aneurysms and healthy blood vessels, the experimental visualization results clearly demonstrate the excellent ability of the method in capturing medical point cloud boundary segmentation .

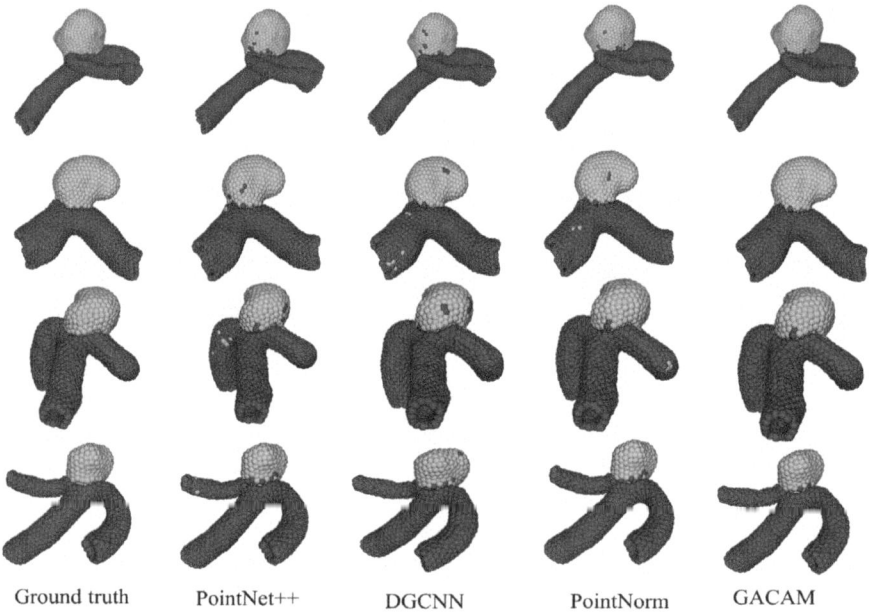

Ground truth   PointNet++   DGCNN   PointNorm   GACAM

Fig. 8. Comparison of IntrA segmentation effects.

4.4 Ablation Study

This section conducts ablation experiments on the medical point cloud IntrA dataset, using intersection to union ratio and average intersection to union ratio as evaluation indicators to study the impact of the proposed dilation graph convolution (DGConv) layers, learnable difference pooling, and Multi-Graph Boundary-Aware (MGBA) modules on medical point cloud segmentation tasks. Firstly, the influence of DGConv layers on local feature extraction of point clouds was explored to verify its impact on segmentation performance. Subsequently, in order to further explore the impact of different pooling strategies on segmentation performance, maximum pooling and average pooling were introduced as references to observe the impact of different pooling strategies on segmentation performance. Finally, this section also explores the effectiveness of the multi-graph boundary-aware module in improving the accuracy of medical point cloud segmentation.

(1) The influence of DGConv layers on segmentation results.

Under the same additional experimental conditions, this section explores the effect of DGConv layers on the performance of the network model. The experimental results are shown in Table 4. The initial setting of the experiment is two layers of DGConv, and the number of layers of DGConv is increased in sequence. When the number of layers of DGConv reaches four, IoU reaches its maximum. However, as the number of DGConv layers increases, it actually leads to a decrease in segmentation accuracy. Therefore, it can be concluded that an excessive number of DGConv layers is not conducive to improving model performance, but rather leads to an increase in model complexity due to an increase in model depth.

Table 4. The influence of different DGConv layers on segmentation results.

DGConv Layers	IoU(%)		
	V	A	mIoU
DGConv-2	91.36	79.77	85.57
DGConv-3	93.74	82.48	88.11
DGConv-4	**95.20**	**83.12**	**89.16**
DGConv-5	95.04	83.01	89.02
DGConv-6	94.92	82.18	88.55

(2) The impact of different pooling strategies on segmentation results.

Under the same additional experimental conditions, this section explores in detail the impact of different pooling strategies on segmentation results, comparing the specific effects of Max Pooling, Avg Pooling, and the learnable differential pooling function (LDP) proposed in this chapter on segmentation performance. The relevant experimental results are detailed in Table 5. Experiments have shown that in the task of segmenting healthy main vessel segments (V) and aneurysms (A), the method of using learnable differential pooling functions significantly outperforms the method of using maximum pooling and average pooling functions on the IoU metric. This result validates the significant advantage of learnable difference pooling in improving the segmentation performance of medical point clouds.

(3) The impact of MGBA on segmentation results.

Under the same additional experimental conditions, this section investigates the specific impact of MGBA on the segmentation performance of medical point clouds. The

Table 5. The influence of pooling strategies on segmentation results.

Pooling Strategies	IoU(%)		
	V	A	Mean
Max-Pooling	94.36	82.17	88.27
Avg-Pooling	94.23	81.74	87.99
LDP	**95.20**	**83.12**	**89.16**

experimental results are summarized in Table 6. Specifically, compared to the baseline model that did not use multi-graph boundary-aware, the intersection to union ratio improved by 1.46%. Therefore, the results of the ablation experiment demonstrate the effectiveness of GIFF in improving the segmentation performance of medical point clouds.

Table 6. The influence of GIFF on segmentation results.

MGBA	IoU(%)		
	V	A	Mean
—	93.71	81.68	87.70
√	**95.20**	**83.12**	**89.16**

5   Conclusion

In this paper, we propose a novel neural network approach for medical point cloud segmentation. We design dilated graph convolution for local feature extraction on medical point clouds to effectively capture complex local geometric structures. Additionally, we introduce a learnable differential pooling function that synthesizes feature information from all points, including features from non-maximum points, thereby significantly reducing information loss. Finally, to enrich the expressive power of medical geometric features at a deeper level, we propose a multi-graph boundary-aware module to enhance the continuity of boundary regions, thereby promoting deep fusion between local and global features, and ultimately optimizing the performance of medical point cloud segmentation tasks.

Acknowledgments. This study was supported by the National Natural Science Foundation of China (61373160), Hebei Province Central Guidance Local Science and Technology Development Fund Project (No. 236Z0306G), the Natural Science Foundation of Hebei Province (F2021210003), Hebei Provincial Science and Technology Plan Project (22567636H).

Disclosure of Interests. The authors declare no conflicts of interest.

References

1. Wei, Y., et al.: Surroundocc: multi-camera 3d occupancy prediction for au-tonomous driving. In: Proceedings of the IEEE/CVF International Conference on Computer Vision, 21729–21740 (2023)
2. Wu, C.Y., et al.: Multiview compressive coding for 3D reconstruc-tion. In: Proceedings of the IEEE/CVF Conference on Computer Vision and Pattern Recognition, 9065–9075 (2023)
3. Beery, S., et al.: The auto arborist dataset: a large-scale benchmark for multiview urban forest monitoring under domain shift. In: Proceedings of the IEEE/CVF Con-ference on Computer Vision and Pattern Recognition, 21294–21307 (2022)
4. Bassier, M., Yousefzadeh, M., Vergauwen, M.: Comparison of 2D and 3D wall reconstruc-tion algorithms from point cloud data for as-built BIM[J]. J. Inf. Technol. Constr. **25**, 173–192 (2020)
5. Li, P., Gao, C., Liu, F., et al.: THISNet: tooth instance segmentation on 3D dental models via highlighting tooth regions[J]. IEEE Trans. Circuits Syst. Video Technol. **31**(1), 1–13 (2023)
6. Yang, X., Xia, D., Kin, T., Igarashi, T.: Intra:3d intracranial aneurysm dataset for deep learning. In Proceedings of CVPR, pp. 2656–2666 (2020)
7. Qi, C.R., et al.: Pointnet: deep learning on point sets for 3d classification and segmentation. In: Proceedings of the IEEE Conference on Computer Vision and Pattern Recognition, 652–660 (2017)
8. Qi, C.R., et al.: Pointnet++: deep hierarchical feature learning on point sets in a metric space[J]. Adv. Neural Inf. Proce. Syst. **30** (2017)
9. Wang, Y., Sun, Y., Liu, Z., et al.: Dynamic graph cnn for learning on point clouds[J]. Acm Trans. Graphics (tog) **38**(5), 1–12 (2019)
10. Guo, M.H., Cai, J.X., Liu, Z.N., et al.: Pct: point cloud transformer[J]. Comput. Visual Media **7**, 187–199 (2021)
11. Gupta, D., Anand, R.S.: A hybrid edge-based segmentation approach for ultra-sound medical images. Biomed. Signal Process. Control **31**, 116–126 (2017)
12. Dong, Y., Yang, W., Wang, J., et al.: An improved supervoxel 3D region growing method based on PET/CT multimodal data for segmentation and reconstruction of GGNs[J]. Multimedia Tools Appl. **79**, 2309–2338 (2020)
13. Chen, Q., Jin, X., Zhu, H., et al.: Classification of pit and fissure for caries risk based on 3D sur-face morphology analysis of tooth[C]//Lasers in Dentistry XXVI. SPIE **11217**, 45–52 (2020)
14. Zhao, H., et al.: Pointweb: enhancing local neighborhood features for point cloud processing. In: Proceedings of the IEEE/CVF Conference on Computer Vision and Pat-Tern Recognition, 5565–5573 (2019)
15. Qian, G., Li, Y., Peng, H., et al.: Pointnext: revisiting pointnet++ with improved training and scaling strategies[J]. Adv. Neural. Inf. Process. Syst. **35**, 23192–23204 (2022)
16. Zhao, W., Jia, L., Zhai, H., et al.: PointSGLN: a novel point cloud classification network based on sampling grouping and local point normalization[J]. Multimedia Syst. **30**(2), 1–13 (2024)
17. Ma X, Qin C, You H, et al. Rethinking network design and local geometry in point cloud: A simple residual MLP framework[J]. arXiv preprint arXiv:2202.07123 (2022)
18. Zhang, S., Tong, H., Xu, J., et al.: Graph convolutional networks: a comprehensive review[J]. Comput. Soc. Netw. **6**(1), 1–23 (2019)
19. Niu, Z., Zhong, G., Yu, H.: A review on the attention mechanism of deep learning[J]. Neurocomput. **452**, 48–62 (2021)
20. Chen, C., Fragonara, L.Z., Tsourdos, A.: GAPointNet: graph attention based point neural network for exploiting local feature of point cloud[J].Neurocomputing **438**, 122–132 (2021)

21. Wang, L., et al.: Graph attention convolution for point cloud se-mantic segmentation. In: Proceedings of the IEEE/CVF Conference on Computer Vision and Pattern Recognition, pp. 10296–10305 (2019)
22. Zanjani, F.G., et al.: Deep learning approach to semantic segmentation in 3D point cloud intra-oral scans of teeth. In: International Conference on Medical Imaging with Deep Learning. PMLR, pp. 557–571 (2019)
23. Li, Y., et al.: Pointcnn: Convolution on x-transformed points[J]. Adv. Neural Inf. Proce. Syst., **31** (2018)
24. Xi, C., Lu, G., Yan, J.: Multimodal sentiment analysis based on multi-head attention mechanism. In: Proceedings of the 4th international conference on machine learning and soft computing, 34–39 (2020)
25. Chen, Y., Rohrbach, M., Yan, Z., Shuicheng, Y., Feng, J., Kalantidis, Y.: Graph-based global reasoning networks. In: Proceedings of CVPR, pp. 433–442 (2019)
26. Ma, Y., Guo, Y., Liu, H., Lei, Y., Wen, G.: Global context reasoning for semantic segmentation of 3d point clouds. In: The IEEE Winter Conference on Applications of Computer Vision (2020)
27. Yi, L., Kim, V.G., Ceylan, D., et al.: A scalable active framework for region annotation in 3d shape collections[J]. ACM Trans. Graph. (ToG) **35**(6), 1–12 (2016)
28. Xu, M., et al.: Paconv: position adaptive convolution with dynamic kernel assembling on point clouds. In: Proceedings of the IEEE/CVF Conference on Computer Vision and Pattern Recognition, pp. 3173–3182 (2021)
29. Wu, W., Qi, Z., Fuxin, L.: KP: deep convolutional networks on 3d point clouds. In: Proceedings of the IEEE/CVF Conference on Computer Vision and Pattern Recognition, pp. 9621–9630 (2019)
30. Zheng, S., et al.: Pointnorm: dual normalization is all you need for point cloud analysis. In: 2023 International Joint Conference on Neural Networks (IJCNN). IEEE, pp. 1–8 (2023)

Question-Guided Hybrid Learning and Knowledge Embedding for Visual Question-Answering

Wenbin Zhao[1](✉), Hanlei Zhai[1], Pingsheng Dai[2], Haoxin Jin[1],
Haoyang Zhao[1,3], and Chaocheng Zhang[1]

[1] School of Information Science and Technology, Shijiazhuang Tiedao University,
Shijiazhuang, Hebei, China
`zhaowb2013@stdu.edu.cn`
[2] Beijing Jingxing Huidong Technology Co., Ltd., Beijing 10000, China
[3] Hebei Science and Technology Information Processing Laboratory, Hebei, China

Abstract. Visual question-answering is an important application of the fusion of vision and language in multimodal learning. Its basic task is to understand the input 2D image or 3D point cloud and answer text questions based on it. At present, the development of visual question and answering technology is facing some problems. In some fields, it is difficult to obtain accurate answers by inferring only the information contained in images and problems. The data imbalance in practical application scenarios also limits the ability of models to handle rare or complex problems. In this article, we propose an image question-answering model based on problem-guided hybrid learning and knowledge embedding. This model queries relevant knowledge in the knowledge graph and integrates the queried knowledge with the problem text to form new textual information. During the training process, tuples of similar problem types (v, q, a) are mixed to generate new data samples. Then, feature extraction and cross modal fusion are performed on the new samples, and the samples are fed into the answer prediction network to obtain the answers. The optimization model narrowed the gap between predicted answers and mixed answers, ultimately enabling the model to generate more accurate answers. A series of comparative experiments conducted on the OKVQA dataset and the SLAKE dataset has verified that the model can effectively improve the accuracy of image question-answering.

Keywords: Visual question-answering · Medical visual question-answering · Feature extractor · Knowledge graph · Data augmentation

1 Introduction

Visual Question-Answering is a multidisciplinary problem that combines computer vision and natural language [1]. Its basic idea is to give an image and

© The Author(s), under exclusive license to Springer Nature Singapore Pte Ltd. 2024
H. Yu et al. (Eds.): CCF NCCA 2024, CCIS 2274, pp. 293–311, 2024.
https://doi.org/10.1007/978-981-97-9671-7_19

a text question related to the image, and the VQA model provides answers based on the image and the question. The continuous development of visual question-answering algorithms has promoted a deep understanding of visual content by computers, enabling them to not only recognize objects in images, but also understand their properties, states, and relationships. In certain professional fields, such as medical image analysis [2], military reconnaissance, etc., visual question-answering algorithms can help professionals quickly extract key information from images and assist in making decisions.

With the application of visual question-answering in various fields, researchers have encountered many challenges. Some answers to questions go beyond the information provided by the image itself. For example, when facing questions about animal dietary habits, relying solely on image data cannot provide sufficient answers, and additional knowledge bases need to be introduced to answer such questions. This dependence on external knowledge not only increases the difficulty of searching and integrating relevant information but also requires efficient query mechanisms to ensure that the acquired knowledge can be accurately used to answer specific questions. In visual Q&A tasks in professional fields, the imbalance of datasets is also an issue that cannot be ignored. Due to the high specificity of professional fields, the scale of related image and question-answering datasets is often limited, which poses challenges to the training and generalization ability of models.

In response to these issues, this article introduces an external knowledge base based on the visual question-answering model, and increases the text information representation of the model to better answer some questions that require professional knowledge. Simultaneously introducing hybrid learning based on question types to provide the model with more diverse training data and improve its ability to answer new image-related questions.

2   Related Work

2.1   Visual Question-Answering

In early visual question-answering models, joint embedding methods were mainly used. Malinowski et al. [3] first proposed a joint embedding model, which uses CNN to extract image features, then uses the LSTM network to extract features from the problem, and then combines them to pass on to the sequence generation network, finally predicting the answer. However, the method based on joint embedding has important shortcomings. The image representation of the model is all the features of the image and does not correspond the image region with the problem keywords. This makes it difficult to obtain the final answer generation when the two features are jointly embedded due to interference from useless information. The model proposed by Feng et al. [4] introduces multi-scale feature methods to characterize and fuse the features of both, improving the accuracy of visual question-answering. Asri et al. [5] utilized dependency-based multimodal fusion blocks to enhance the association between keywords and key regions and generate more effective answers. Li et al. [6] proposed a context-aware multi-level problem embedding fusion model, whose core is to understand and handle multi-level problems through context-aware mechanisms. It utilizes advanced embedding techniques to integrate word, phrase, and sentence-level features together, thereby capturing multi-dimensional semantic information of the problem. Zhang et al. [7] introduced the concept of visual information entropy in a small sample visual question-answering model, measuring the spatial distribution of visual features affected by relevant problems, reducing redundant data refining multimodal features, and improving the model's inference ability. Lu et al. [8] proposed a modal fusion collaborative algorithm to address language bias in visual question-answering. This algorithm promotes mutual modeling between different modalities through collaborative training methods, achieving effective feature fusion. Zhang et al. [9] proposed an efficient model to capture fine-grained semantic details when studying text video cross modal retrieval, which can effectively capture fine-grained features from videos and texts for cross modal retrieval.

When the attention mechanism in deep learning is applied in natural language, it can attract more attention to important language parts and improve the accuracy of the model. Lubna et al. [10] used a visual question-answering system to analyze individual blood samples. This method used convolutional neural networks for problem classification and object detectors with attention mechanisms for visual understanding. In the experiment, two attention mechanisms were used: convolutional block attention module and squeeze excitation network, achieving fast and reliable answer prediction. Yan et al. [11] designed a common attention model that combines scene text features in images. The model extracts text information from images using OCR technology utilizes a common attention mechanism to enhance the interaction between problem text and image content, significantly improving the model's understanding and answering ability toward text in images. In the field of visual question-answering, researchers are constantly exploring methods of combining image content with knowledge bases to improve the accuracy and application scope of generated answers. The work of Chen et al. [12] is to introduce an attention optimization module into the traditional encoder-decoder architecture to deeply optimize the obtained attention weights to clarify the correlation between queries and attention results and to generate a more accurate weighted average for each query. In the stage of multimodal fusion, the model can dynamically decide on the focus of visual or textual information according to the specific needs of the current task. The visual language pre-training model proposed by Zeng et al. [13] based on the multi granularity attention mechanism achieves more fine-grained modal coordination through the reconstruction of the dataset and the alignment of images and texts. The Zero shot visual question-answering algorithm proposed by Chen et al. [14] transforms visual question-answering tasks from classification to mapping-based alignment tasks, predicts answers that do not appear in the training set, and demonstrates a new approach to external knowledge integration. Sun et al. [15] proposed a dual stream attention multi-hop inference architecture, constructing two different attention streams to reduce unnecessary knowledge facts, while collecting necessary knowledge by learning implicit correlations between knowledge facts and problems. They also designed a hypergraph knowledge extraction module to extract the best knowledge facts by evaluating the correlation between each knowledge fact and the problem. Lerner et al. [16] used a multimodal dual encoder to study visual Q&A on named entity recognition, and used cross-modal retrieval to make up for the semantic differences between entities and their descriptions, and combined with unimodal retrieval to verify the effectiveness of the method. Shao et al. [17] proposed a Prophet framework to augment the input of the GPT-3 language model by using answer heuristics, including candidate answers and answer-aware examples, to more efficiently extract and understand relevant knowledge from visual data.

2.2   Medical Visual Question-Answering

The field of medical visual Q&A has rapidly developed since its first introduction in the Image CLEF competition in 2018. This task drives the universal visual question -answering model to meet specific medical needs. In 2018, Peng et al. [18] strengthened the connections between different modalities by incorporating collaborative attention mechanisms into multimodal decomposition bilinear pooling methods, winning the championship that year. In 2019, Yan et al. [19] combined the VGG network and Bert model to apply the MFB method in the modal fusion stage, once again winning the championship. At the same time, the launch of datasets specifically designed for medical visual question-answering, such as VQA-RAD [20] and SLAKE [21], further promoted research progress.

In the subsequent development, Ajay et al. [22] proposed the RepsNet model, which can not only generate medical reports but also parse medical images. It integrates an improved ResNeXt-101 [23] and a pre trained Bert [24] model, marks text using Word Piece technology, and fuses image and problem features using BAN [25] technology. Bidirectional contrastive learning helps ensure consistency between images and text, while an improved version of the GPT-2 language decoder based on image features and prior knowledge is responsible for generating answers. The model developed by Moon et al. [26] combines ResNet-50 trained on ImageNet with a text embedding layer based on Bert to extract and fuse visual and text features. This model utilizes cross-modal representation to enhance the spatial and sequential information of input data and is pre-trained and finetuned on a medical image report pairing dataset to optimize its performance for medical visual question-answering and report generation tasks. The model proposed by Eslami et al. [27] is a visual question-answering system based on the CLIP model, which combines multiple CLIP visual encoders and advanced natural language processing technologies. By calculating the cosine similarity between images and text, fusing the output of image denoising autoencoder and GloVe word embedding, and using bilinear attention for feature fusion, medical visual question-answering is achieved.

3   Method

The model overview is shown in Fig. 1. The model consists of four parts: image feature embedding, text feature embedding, cross modal feature fusion network, and answer prediction network. Among them, the image feature extraction part is composed of mixed learning of images and image encoders, the text feature embedding is composed of mixed learning of text information and text encoders, and the answer prediction module is composed of an answer prediction network and a mixed part of answer labels, and the generated answer is closer to the correct answer through loss function optimization model.

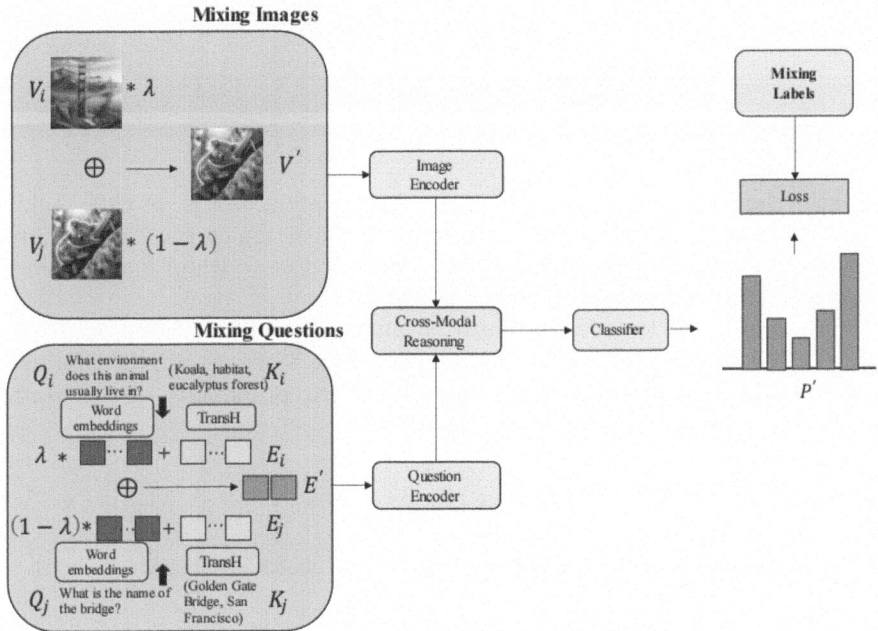

Fig. 1. Overall Architecture.

3.1   Question Encoder

Text Embedding. For the computer to process the text information in the image question and answer, the text information needs to be transformed, which is usually unstructured and difficult to process directly. Therefore, this paper uses the GloVe model for word vector embedding to map text information to feature space.

The core idea of GloVe is to learn the vector representation of words through global statistics. It uses the global word-word co-occurrence matrix to record the co-occurrence frequency of each pair of words in a given context window and performs singular value decomposition to this co-occurrence matrix, which can effectively capture the semantic relationship between words.

Table 1. Co-occurrence matrix.

Label	Central word	Context
0	What	What sort of
1	sort	What sort of vehicle
2	of	What sort of vehicle uses
3	vehicle	sort of vehicle uses this
4	uses	of vehicle uses this item
5	this	vehicle uses this item
6	item	uses this item

Firstly, the co-occurrence matrix of the statistical dataset is set to X. Taking the text problem in the dataset as an example, the existing corpus "What sort of vehicle uses this item?". Assuming a statistical window width of 5 and a length of 2, a table can be obtained as shown in Table 1.

Use the window to traverse the entire corpus once to obtain the co-occurrence matrix X. The elements of the co-occurrence matrix X are set to $X_{i,j}$, and then based on the obtained co-occurrence matrix X, the probability of the word k appearing in another word i is solved as follows:

$$P_{i,k} = \frac{X_{i,k}}{X_i} \tag{1}$$

then calculate the correlation between word k and word i, using the formula:

$$F = \frac{P_{i,k}}{P_{j,k}} \tag{2}$$

by constructing an approximate relationship between word vectors and co-occurrence matrices based on correlation, the word vector representation of text problems can be obtained.

Knowledge Embedding. Next, semantic analysis of the problem text is carried out to identify the key entities (people, places, objects, etc.) and potential relationships in the problem. This step is done using named entity recognition technology [28], which can identify specific categories of entities in the text. After entity identification, further analysis of the potential relationship between entities, such as "located", "owned" and so on. This helps model clause structures to extract verbs or prepositional phrases, clarify relationships between entities, and then classify questions according to the type of question, such as true or false questions, multiple choice questions, or open-ended questions.

Based on the obtained entity relationships, the natural language problem is transformed into a knowledge-based query language, which is used in SPARQL in this chapter. This transformation process can map the identified entities and relationships to the corresponding entities and attributes in the knowledge base, and construct multiple query statements to cover different knowledge domains for complex problems. Query results are returned in the form of triples (subject, predicate, object) that represent pieces of knowledge related to the problem.

This chapter adopts the TransH [29] method for knowledge feature learning. The process of TransH feature learning mainly involves mapping entities and relations in knowledge graph to low-dimensional vector Spaces, while considering the diversity and complexity of relations. Here are the detailed steps for learning TransH features.

First, we assign an initial vector e to each entity in the knowledge graph, and two vectors to each relation: one is the vector r of the relation, and the other is the normal vector w_r that defines the hyperplane of the relation. These vectors are represented in a lower-dimensional vector space. For each triplet (h, r, t), where h is the head entity, r is the relation, and t is the tail entity, first compute the projection of the vectors of the head and tail entities to the hyperplane corresponding to the relation r:

$$h_\perp = h - w_r^T h w_r \tag{3}$$

$$t_\perp = t - w_r^T t w_r \tag{4}$$

The score for each triplet is calculated using projected entity vectors and relational vectors. TransH uses the distance measure as a scoring function, usually L1 or L2 distance. The score function is defined as:

$$f_r(h, t) = \|h_\perp + y - t_\perp\| \tag{5}$$

A lower score means the triples are more likely to be correct. Using the obtained feature representation, the retrieved knowledge triplet is encoded and converted into a continuous vector representation and the problem text vector is fused with the selected knowledge vector to generate an enhanced text representation.

Text Mixing. Let $S = \{V, Q, A\}$ represent A sample in the VQA data set, where V represents the input image, Q represents the enhanced text representation, and A represents the corresponding answer, encoding the answer into a unique thermal code Y. First, two different samples S_i and S_j are selected and enhanced to generate a new sample S'.

The mixing coefficient λ is first obtained from the Beta distribution $Beta(\alpha, \alpha)$, where α is the hyperparameter. Since the input space of the text is discontinuous, it is necessary to adopt a linear combination of two input texts in the embedded space to obtain an embedded representation of the mixed text E':

$$E' = \lambda E_i + (1 - \lambda) E_j \tag{6}$$

where E' is the embedded representation of the mixed problem and E_i and E_j are the integrated representation after knowledge embedding and problem text embedding link. Finally, it is sent to LSTM to obtain the language feature, and the text feature represents f_q.

3.2   Image Encoder

In this article, the two images corresponding to the text representation are linearly mixed to obtain a new mixed image:

$$V' = \lambda V_i + (1 - \lambda)V_j \tag{7}$$

The mixed images are fed into the Enhanced Visual Feature Mixing (MEVF) component [30] for feature extraction, which is initialized by MAML and CDAE training weights, and then fine-tuned end-to-end on different visual question-answer datasets. Figure 2 shows the structure of Enhanced Visual Feature Blending (MEVF). (1) The MAML classification model is represented by parameterized function f_θ with meta parameter θ. When adapting to the new task T_i, the model parameter θ becomes θ_i'. Collection of data used in this chapter for $D = \{x_i, y_i\}_{i=1}^{N}$, where N as the sample size, $\{x_i, y_i\}$ for the $image(x_i)$ and its categories TAB (y_i). MAML the task is defined as "k cyber-shot n - way" classification problem, and each task of the data set is defined as $D' = \{x_i', y_i'\}_{i=1}^{N'}$. The samples in D' come from n different classes, which are subsets of the classes in D. The task data set D' is evenly divided into two groups: D^{tr} - training set and D^{val} - verification set; In D^{tr}, each class contains k training images. In each iteration, h, m tasks are generated to form a meta-batch of MAML training. The

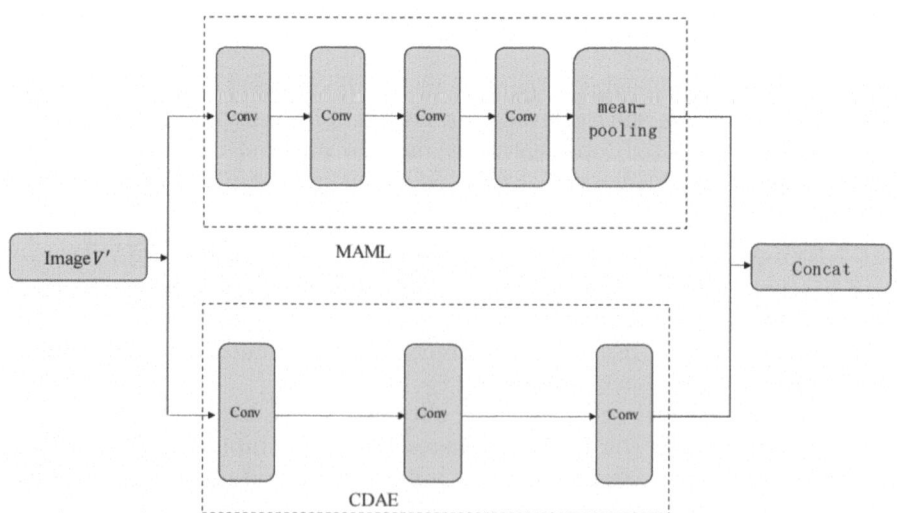

Fig. 2. Enhanced visual feature hybrid component structure.

MAML model designed in this chapter includes four 3×3 convolution layers with a step size of 2 and ends with an average pooling layer. Each convolutional layer has 64 filters, followed by ReLU layers, and the weight of the metamodel is used for subsequent fine-tuning. (2) CDAE is a convolutional denoising autoencoder that maps the noisy version x' of the original image x to a potential representation z that retains useful information. The decoder converts z to output y. The training algorithm minimizes the reconstruction error between y and the original image x, as follows:

$$L_{rec} = \parallel x - y \parallel_2^2 \tag{8}$$

In the design of this chapter, the encoder is a bunch of convolutional layers; Each layer is followed by a maximum pooling layer. The decoder is a bunch of deconvolution and convolution layers. The noise version x' is achieved by adding Gaussian noise to the original image x, and the training weights of the encoder and decoder are used for subsequent fine-tuning.

3.3   Cross-Modal Reasoning

In the cross-modal fusion stage, this chapter uses the bilinear attention network BAN model to carry out feature fusion of text features and image features.

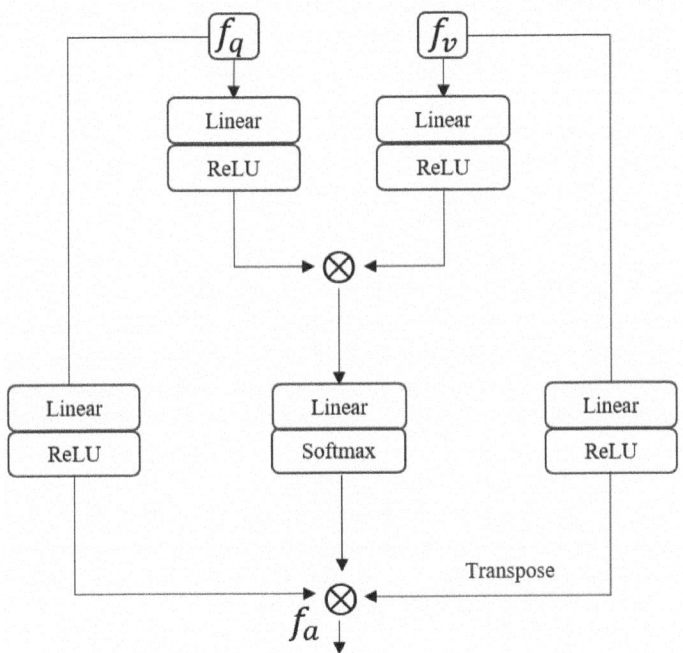

Fig. 3. BAN network architecture.

It introduces the transpose matrix and applies attention to the processing of the two modes, which greatly enhances the correlation between the two modes of learning. Figure 3 shows the network structure of BAN.

3.4   Classifier

Question Guided Answer Tag Mix. When mixed in two different samples, there are four conditions including $\{V_i, Q_i\}, \{V_j, Q_j\}, \{V_i, Q_j\}$ and $\{V_j, Q_i\}$. $\{V_i, Q_i\}, \{V_j, Q_i\}$, the answer is A_i and A_j, and $\{V_i, Q_j\}$ and $\{V_j, Q_i\}$, the answer is missing, the answer is set to A_k and A_l, answer as Y_i respectively, and the label Y_j, Y_k, Y_l. The mixing process of labels can be expressed as:

$$Y' = \lambda^2 Y_i + (1 - \lambda)^2 Y_j + \lambda(1 - \lambda) Y_k + \lambda(1 - \lambda) Y_l \tag{9}$$

The coefficients for each label represent the probability distribution of the corresponding image-problem pair. Since labels are missing, there are meaningless situations in mixing and some methods are needed to constrain the learning of the supervised model. The first thing you can do is simply discard these labels, representing them as Y':

$$Y' = \lambda^2 Y_i + (1 - \lambda)^2 Y_j \tag{10}$$

However, there are some problems in this case, some information may be lost, so this chapter mixes (v, q, a) based on the question types in the data set. In this case, the answer is more closely related to the question in the hidden space, because the type of question can often be a direct indication of the type of answer. This means that by mixing questions and answers of the same class, the generated mixed labels are logically consistent, providing meaningful learning goals for the model; And the images in the dataset involve multiple modes, including different scenes, objects, and activities. Images of different modes are easy to distinguish in terms of features. By mixing images of different modes, the visual diversity of training data can be effectively improved, which helps the model to learn more generalized visual feature representation. Later applications may involve medical images of specific models and organs. By limiting mixing operations between images from the same area of expertise (for example, images of the same model and organ), uncertainty in the training process can be reduced. This binding hybrid strategy helps reduce the risk of overfitting the model on these domain samples, thereby improving the model's performance when dealing with specialized or subdivision problems. Label the answer obtained by this method as Y'', and use Y'' to calculate the loss to ignore the unknown existence of the answer to reduce the effect of noise. Using this strategy, the binary cross entropy loss between the prediction fraction S' and the noise label Y'' is calculated to train the visual question-answering model:

$$\text{Loss} = \frac{1}{C} \sum_{c=1}^{C} [Y_c'' \log(S_c') + (1 - Y_c'') \log(1 - S_c')] \tag{11}$$

where C is the number of answers in the candidate set. Therefore, let B be the batch size, then the final loss of a training batch is:

$$L_B = \sum_{b=1}^{B} L_b \tag{12}$$

Answer Generation. In the answer prediction stage, a two-layer perceptron is used, connected by ReLU activation function, and the fusion feature f is obtained to calculate the answer result. The formula is as follows:

$$H = \mathrm{ReLu}\,(f_a W_h + b_h) \tag{13}$$

where H stands for the hidden layer, ReLU stands for activation function, and W stands for weight, H stands for number of hidden units, b stands for deviation parameter, OUT stands for output, o stands for number of output units. The classified answer set is a pre-defined answer classification based on the data set, and the model will select the answer with the highest prediction score in OUT as the answer to the question. The predicted answer is compared with the mixed answer, the loss function is calculated, and the gap between the predicted answer and the correct answer is narrowed.

4   Experiments and Results

4.1   Datasets and Evaluation Indicators

OKVQA Dataset. The OKVQA dataset is one of the most interesting data sets in the field of visual question answering, and it is also one of the largest public data sets labeled by professionals, providing researchers with a wealth of visual information and external knowledge base support. This dataset not only contains rich image information but also integrates the knowledge graph labeled by experts, which promotes the development of a visual question-answering field. The OKVQA dataset contains different scenes and objects, as well as image-related annotations, annotating rich object information and scene descriptions. The OKVQA dataset also contains many question texts and corresponding answers, as well as an external knowledge base containing rich triples of knowledge covering a variety of domains, providing rich textual information and semantic knowledge for the visual question-answering model.

SLAKE Dataset. The newly released SLAKE dataset is the largest public dataset in the field of medical visual answer research to date and contains an external knowledge base manually labeled by medical visual answer research professionals, including 12 diseases and 39 organs. SLAKE's knowledge base is that the author constructs the tuple <head, relation, tail> based on Wikipedia, where head and tail both represent the whole, such as disease or organ, while relation represents the relationship between the two wholes, such as function or treatment. Then it is transformed into dk feature representation with dimension by using the presentation learning method in this paper.

When evaluating the accuracy of the visual question-answering model, it is usually analyzed from three perspectives, namely, the Overall accuracy, the Open accuracy, and the Closed accuracy. Open is defined as a question type with No fixed answer, which changes as the question changes and Closed is defined as a question type with a fixed answer, that is, no matter how the question changes, the answer type is always the same, such as Yes/No. Previous studies have shown that Open questions often require more and more detailed feature information than Closed questions.

4.2   OKVQA Results and Analysis

The OKVQA data set is classified according to each question type, the accuracy of each category is calculated separately, and the results of all classes are averaged, the results are marked as Open, and the average results of the types of questions with the answer to yes are marked as Close. In this chapter, existing mainstream models evaluated on the same OKVQA data set with the same evaluation criteria are selected for comparison, and the results are shown in Table 2.

Table 2. Comparison of Model Experiment Results.

Modal	Open(%)	Close(%)	Overall(%)
Article Net	6.53	8.64	5.28
MUTAN+AN	24.57	34.56	26.41
BAN+AN	25.75	41.66	25.17
XNM Net	23.43	42.17	25.61
ViLBERT	24.56	43.33	31.35
Ours	**30.44**	**43.45**	**35.42**

AN(ArticleNet) [31]: This model is the official benchmark comparison model for the OKVQA dataset, which retrieves articles from Wikipedia based on question image pairs, and then trains a network to predict if and where answers to basic facts will appear in the article and each sentence.

MUTAN [32]: This model is based on a tensor bilinear interaction model for multi-modal fusion.

BAN: This model is multimodal fusion based on collaborative attention.

XNM Net [33]: This model uses scene graph objects as nodes and pair relationships as edges for interpretable and explicit reasoning with structured knowledge.

ViLBERT [34]: This model proposes that image and text are extracted in their respective data streams for feature extraction, and then the two streams are fused through the cooperative attention mechanism layer.

As can be seen from the table, the method in this chapter performs best when dealing with "Open" class problems, reaching AN accuracy rate of 30.44%, which is significantly higher than 24.56% of ViLBERT, 23.43% of XNM Net and 6.53% of MUTAN+AN. This demonstrates the significant advantages of this chapter approach in understanding and answering complex questions that require extensive knowledge. For "Close" questions, those that usually have a fixed answer, the chapter's approach was also ahead, achieving 43.45% accuracy. This result was better than ViLBERT's 43.33% and well above the performance of XNM Net and MUTAN+AN. This further demonstrates the high efficiency of this chapter's approach in answering specific and clear questions. In terms of overall performance, the method in this chapter leads with an overall accuracy of 35.42% and an accuracy of 31.35%. These results show that while all models vary in their performance when dealing with these two types of problems, the approach presented in this chapter provides the most balanced and the strongest performance overall.

4.3 SLAKE Results and Analysis

This paper selects existing mainstream models evaluated on the same SLAKE dataset with the same evaluation criteria for comparison. Table 3 shows the comparative experimental results.

Table 3. Comparison of Model Experiment Results.

Modal	Overall(%)	Open(%)	Close(%)
VGG+SAN	72.7	70.3	76.1
$VGG_{seg} + SAN$	75.3	72.2	79.8
MAML	75.5	74.1	77.5
PubMed CLIP	78.0	76.5	80.4
Ours	**79.0**	**78.0**	**81.2**

VGG+SAN [35]: This model uses VGG16 to extract image features and designs a bilingual tokenizer to create a bilingual word embedding. Then, a 1024-dimensional LSTM is used to extract text semantic features, and SAN is used for cross-modal feature fusion.

VGGseg+SAN [35]: This model pre-trains the VGG feature extraction network in it.

MEVF-MAML [30] In the stage of image feature extraction, the process of encoding and decoding is introduced, and the idea of meta-learning is introduced. In the training process, the encoder is iteratively trained at the same time, so that the model can better extract features.

MEVF-PubMedCLIP: The PubMed CLIP is a model that Sedigheh Eslami et al. used medical image text to fine-tune the original CLIP model. The images used were ROCO data set, including various types of radiological medical images, and the text information in the data set were collected from PubMed articles related to medicine. Provides a wealth of information about the content of the image.

In medical visual question answering, the pre-training of the feature extraction network can significantly improve the image feature extraction ability of the model and improve the answer accuracy of the model, compared with the unpre-trained VGG+SAN. MEVF enhances the model's learning ability of image key features through the iterative training of meta-learning but lacks the breadth and inclusiveness of features. PubMed CLIP utilizes large-scale external data for pre-training, improving the model's extensive feature extraction capabilities, but also introduces noisy information unrelated to medical visual question-and-answer.

4.4 Visualization of Model Results

As shown in Fig. 4 and Fig. 5, the results were visualized for some datasets. In the figure, the area where the model focuses on the image is the colored area, and the area with a weak correlation with the answer keywords is the blue area. It can be seen that the model accurately focuses on the image area corresponding to the question keywords and answers the correct answer.

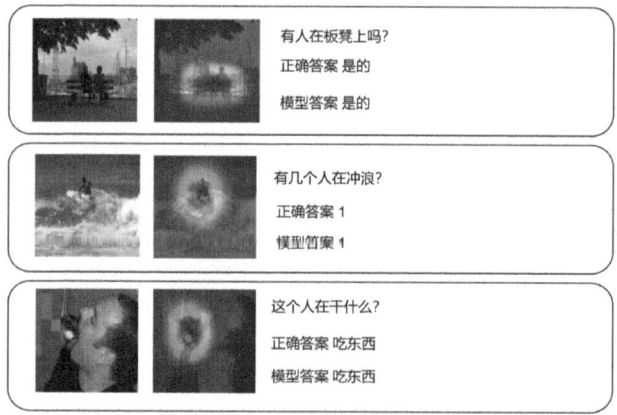

Fig. 4. SLAKE visualization example.

4.5 Ablation Study

To verify the validity of the proposed method and each module, a series of ablation experiments were conducted on the OKVQA data set and SLAKE data set. Among them, -KGE means removing the embedded part of the knowledge

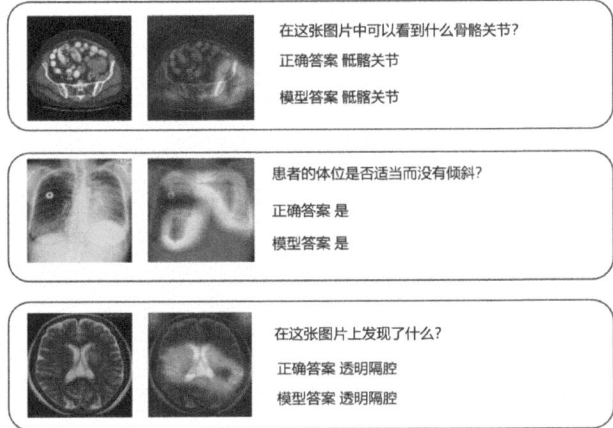

Fig. 5. OKVQA visualization example.

graph based on the method in this chapter, and -Mix means removing the data mixing part based on the method in this chapter. The experimental results are shown in Table 4:

Table 4. Results of ablation experiment.

Modal	OKVQA			SLAKE		
	Open(%)	Close(%)	Overall(%)	Open(%)	Close(%)	Overall(%)
-KGE	27.45	41.52	33.64	77.2	80.5	78.3
-Mix	28.69	42.64	32.15	76.9	80.9	78.6
Ours	**30.44**	**43.45**	**35.42**	**78.0**	**81.2**	**79.0**

To investigate the effect of blended learning on performance, models with different mixing rates are also evaluated. Sample the mixing rate from the Beta distribution Beta (α, α), with the hyperparameter α ranging from 0.2 to 2 and a step size of 0.2. Figure 6 shows the results. For all α values considered, the best performance occurs when $\alpha = 1.0$.

The experimental results show that integrating external knowledge into the visual question-answering system can enrich the semantic information of the text and strengthen the model's understanding of the problem text and image content. By using the entity, relation, and attribute information in the knowledge graph, the model can grasp the relation between semantics and context more accurately, and improve the accuracy of generating answers.

In the process of model training, the mixed learning strategy is adopted to mix different data set tuples based on problem types to form new training samples, reduce the negative impact of label noise in training data on model performance, and enhance the generalization ability of the model in the face of new data sets.

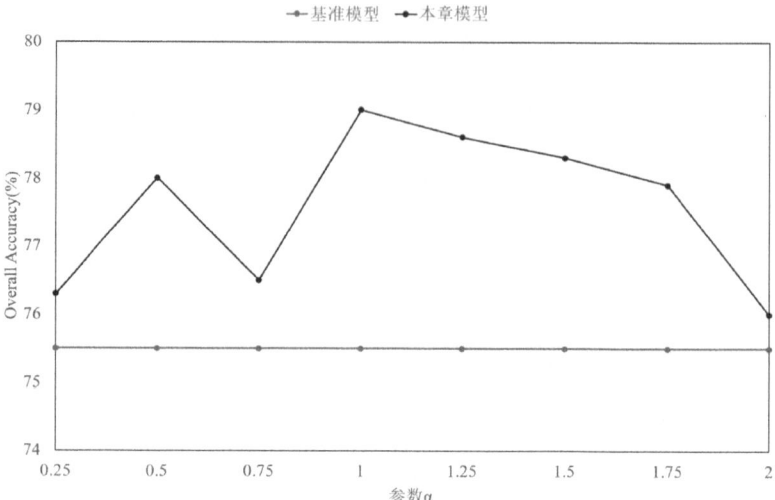

Fig. 6. Parameter α variation table.

5   Conclusion

This paper explores and studies the problems in image visual question answering. Firstly, external knowledge is integrated into the model to improve its reasoning ability. Then, problem-type-based hybrid learning is introduced to combine two training samples with a random coefficient to improve the diversity of training data. Specifically, the first step is to use the missing label strategy to learn, roughly discarding missing answers; On this basis, using prior knowledge of the problem categories as conditional mixed constraints, a learning method for conditional mixed labels was further established, making the labels meaningful and alleviating the inherent problem of missing answers and meaningless answers caused by the combination of (v,q,a) tuples. Finally, a comparative experiment was conducted on the model in this chapter, and the experimental results showed that the method effectively improved the accuracy of visual question answering, inspiring future research work. With the continuous deepening of research and the gradual widespread application of visual question answering, the next step of this article will focus on the fusion of multimodal features and the inference ability of answers in the subsequent part. In addition, optimizing the evaluation methods for visual question-answering tasks to make them more in line with artificial evaluation standards is also a meaningful direction.

Acknowledgement. This study was supported by the National Natural Science Foundation of China (61373160), Hebei Province Central Guidance Local Science and Technology Development Fund Project (No. 236Z0306G), the Natural Science Foundation of Hebei Province (F2021210003), Hebei Provincial Science and Technology Plan Project (22567636H).

Disclosure of Interests. The authors have no competing interests.

References

1. Anderson, P., He, X., Buehler, C., et al.: Bottom-up and top-down attention for image captioning and visual question answering. In: Proceedings of the IEEE Conference on Computer Vision and Pattern Recognition, pp. 6077–6086 (2018)
2. Yang, H., Chen, J., Chi, Y., Xie, X., Hua, X.: Discriminative coronary artery tracking via 3D CNN in cardiac CT angiography. In: Shen, D., et al. (eds.) MICCAI 2019. LNCS, vol. 11765, pp. 468–476. Springer, Cham (2019). https://doi.org/10.1007/978-3-030-32245-8_52
3. Malinowski, M., Fritz, M.: A multi-world approach to question answering about real-world scenes based on uncertain input. In: Proceedings of the 27th International Conference on Neural Information Processing Systems-Volume 1, pp. 1682–1690 (2014)
4. Lu, S., Ding, Y., Liu, M., et al.: Multiscale feature extraction and fusion of image and text in VQA. Int. J. Comput. Intell. Syst. **16**(1), 54 (2023)
5. Asri, H.S., Safabakhsh, R.: Advanced visual and textual co-context aware attention network with dependent multimodal fusion block for visual question answering. Multimedia Tools Appl. **20**, 1–28 (2024)
6. Li, S., Gong, C., Zhu, Y., et al.: Context-aware multi-level question embedding fusion for visual question answering. Inf. Fusion **102**, 102000 (2024)
7. Zhang, J., Liu, X., Chen, M., et al.: Cross-modal feature distribution calibration for few-shot visual question answering. In: Proceedings of the AAAI Conference on Artificial Intelligence, vol. 38, no. 7, pp. 7151–7159 (2024)
8. Lu, Q., Chen, S., Zhu, X.: Collaborative modality fusion for mitigating language bias in visual question answering. J. Imaging **10**(3), 56 (2024)
9. Zhang, S., Mu, H., Li, Q., et al.: Fine-grained features alignment and fusion for text-video cross-modal retrieval. In: ICASSP 2024-2024 IEEE International Conference on Acoustics, Speech and Signal Processing (ICASSP), pp. 3325–3329. IEEE (2024)
10. Lubna, A., Kalady, S., Lijiya, A.: Visual question answering on blood smear images using convolutional block attention module powered object detection. Vis. Comput. **10**, 1–19 (2024)
11. Yan, F., Silamu, W., Chai, Y., et al.: OECA-Net: a co-attention network for visual question answering based on OCR scene text feature enhancement. Multimedia Tools Appl. **83**(3), 7085–7096 (2024)
12. Chen, X., Chen, C., Tian, X., et al.: DBAN: an improved dual branch attention network combined with serum Raman spectroscopy for diagnosis of diabetic kidney disease. Talanta **266**, 125052 (2024)
13. Zeng, Y., Zhang, X., Li, H.: Multi-grained vision language pre-training: aligning texts with visual concepts. In: International Conference on Machine Learning, pp. 25994–26009. PMLR (2022)
14. Chen, Z., Chen, J., Geng, Y., Pan, J.Z., Yuan, Z., Chen, H.: Zero-shot visual question answering using knowledge graph. In: Hotho, A., et al. (eds.) ISWC 2021. LNCS, vol. 12922, pp. 146–162. Springer, Cham (2021). https://doi.org/10.1007/978-3-030-88361-4_9
15. Sun, Y., Zhu, Z., Zuo, Z., et al.: DSAMR: dual-stream attention multi-hop reasoning for knowledge-based visual question answering. Expert Syst. Appl. **245**, 123092 (2024)

16. Lerner, P., Ferret, O., Guinaudeau, C.: Cross-modal retrieval for knowledge-based visual question answering. In: European Conference on Information Retrieval, pp. 421–438. Springer, Cham (2024)

17. Shao, Z., Yu, Z., Wang, M., et al.: Prompting large language models with answer heuristics for knowledge-based visual question answering. In: Proceedings of the IEEE/CVF Conference on Computer Vision and Pattern Recognition, pp. 14974–14983 (2023)

18. Peng, Y., Liu, F., Rosen, M.P.: UMass at image CLEF medical visual question answering (Med- VQA) 2018 Task. In: CLEF (working notes), pp. 1–9 (2018)

19. Al-Sadi, A., Talafha, B., Al-Ayyoub, M., et al.: JUST at image CLEF 2019 visual question answering in the medical domain. In: CLEF (working notes) (2019)

20. Lau, J.J., Gayen, S., Ben Abacha, A., et al.: A dataset of clinically generated visual questions and answers about radiology images. Sci. Data 5(1), 1–10 (2018)

21. Chen, Y., Xing, X.: Slake: facilitating slab manipulation for exploiting vulnerabilities in the linux kernel. In: Proceedings of the 2019 ACM SIGSAC Conference on Computer and Communications Security, pp. 1707–1722 (2019)

22. Tanwani, A.K., Barral, J., Freedman, D.: Repsnet: combining vision with language for automated medical reports. In: International Conference on Medical Image Computing and Computer Assisted Intervention, pp. 714–724. Springer, Cham (2022)

23. Xie, S., Girshick, R., Dollár, P., et al.: Aggregated residual transformations for deep neural networks. In: Proceedings of the IEEE Conference on Computer Vision and Pattern Recognition, pp. 1492–1500 (2017)

24. Kenton, J.D.M.W.C., Toutanova, L.K.: BERT: pre-training of deep bidirectional transformers for language understanding. In: Proceedings of NAACL-HLT, pp. 4171–4186 (2019)

25. Kim, J.H., Jun, J., Zhang, B.T.: Bilinear attention networks. In: Proceedings of the 32nd International Conference on Neural Information Processing Systems, pp. 1571–1581 (2018)

26. Moon, J.H., Lee, H., Shin, W., et al.: Multi-modal understanding and generation for medical images and text via vision-language pre-training. IEEE J. Biomed. Health Inform. 26(12), 6070–6080 (2022)

27. Eslami, S., Meinel, C., De Melo, G.: Pubmed clip: how much does clip benefit visual question answering in the medical domain? In: Findings of the Association for Computational Linguistics: EACL 2023, pp. 1181–1193 (2023)

28. Chen, Y., Wu, C., Qi, T., et al.: Named entity recognition in multi-level contexts. In: Proceedings of the 1st Conference of the Asia-Pacific Chapter of the Association for Computational Linguistics and the 10th International Joint Conference on Natural Language Processing, pp. 181–190 (2020)

29. Wang, Y., Wumaier, A., Sun, W., et al.: TransH-RA: a learning model of knowledge representation by hyperplane projection and relational attributes. IEEE Access 11(29510–295), 20 (2023)

30. Nguyen, B.D., Do, T.-T., Nguyen, B.X., Do, T., Tjiputra, E., Tran, Q.D.: Overcoming data limitation in medical visual question answering. In: Shen, D., et al. (eds.) MICCAI 2019. LNCS, vol. 11767, pp. 522–530. Springer, Cham (2019). https://doi.org/10.1007/978-3-030-32251-9_57

31. Marino, K., Rastegari, M., Farhadi, A., et al.: OK-VQA: a visual question answering benchmark requiring external knowledge. In: Proceedings of the IEEE/CVF Conference on Computer Vision and Pattern Recognition, pp. 3195–3204 (2019)

32. Ben-Younes, H., Cadene, R., Cord, M., et al.: Mutan: multimodal tucker fusion for visual question answering. In: Proceedings of the IEEE International Conference on Computer Vision, pp. 2612–2620 (2017)
33. Shi, J., Zhang, H., Li, J.: Explainable and explicit visual reasoning over scene graphs. In: Proceedings of the IEEE/CVF Conference on Computer Vision and Pattern Recognition, pp. 8376–8384 (2019)
34. Lu, J., Batra, D., Parikh, D., et al.: ViLBERT: pretraining task-agnostic visiolinguistic representations for vision-and-language tasks. In: Proceedings of the 33rd International Conference on Neural Information Processing Systems, pp. 13–23 (2019)
35. Liu, B., Zhan, L.M., Xu, L., et al.: Slake: a semantically-labeled knowledge-enhanced dataset for medical visual question answering. In: 2021 IEEE 18th International Symposium on Biomedical Imaging (ISBI), pp. 1650–1654. IEEE (2021)

Integrating Image Super-Resolution Network and Semantic Segmentation for 3D Reconstruction of Medical Sequence Image

Wenbin Zhao[1]([📧]), Shuhang Chai[1], Haoyang Zhao[1], Jianmin Wang[1], and Pingsheng Dai[2]

[1] Shijiazhuang Tiedao University, Shijiazhuang 050043, Hebei, China
Zhaowb2013@stdu.edu.cn
[2] Beijing Jingxing Huidong Technology Co., Ltd., Beijing 10000, China

Abstract. The three-dimensional model of medical sequence image can display human tissues and organs in stereo, and effectively overcome the insufficiency of two-dimensional examination results for disease diagnosis. In this paper, the three-dimensional reconstruction of sequential medical images is studied from the perspective of image super resolution and semantic segmentation. In view of the problem of ignoring global information association in existing image super-resolution reconstruction methods and the problem of long distance dependence due to lack of information in existing image semantic segmentation methods. In this paper, the method of fusion of global and local features is introduced into the super-resolution reconstruction task, and the method of cross-level fusion compensation is introduced into the semantic segmentation task of medical images, so as to realize the three-dimensional reconstruction of sequential medical images. Finally, the reliability of the proposed method is verified by experiments. In the experiment of 3D reconstruction of medical sequence images, we can observe that some details are preserved completely and the clarity is improved significantly after reconstruction.

Keywords: Image super resolution · Image semantic segmentation · Subpixel convolution technology · 3D reconstruction of medical image

1 Introduction

For a long time in the past, doctors had to determine the types of diseases according to the results of two-dimensional plane examinations such as CT and MRI, which required doctors to have a lot of clinical diagnosis and treatment experience. Different patients' tissues and organs were very different, and the identification of lesions was mainly based on the subjective judgment of doctors, which was very easy to cause errors. For rare and difficult diseases that were difficult to distinguish according to experience, The applicability of flat two-dimensional images is greatly reduced. The three-dimensional reconstruction technology of medical sequence images can enable the multi-angle three-dimensional display of human tissues and organs. Compared with two-dimensional plane detection, its clear and intuitive results are more conducive to the diagnosis of complex and rare diseases, greatly reducing the subjectivity of doctors in diagnosing diseases, and allowing disease discrimination to have a starting point. The medical image 3D reconstruction technology aims to restore the 3D model of human tissues and organs from a series of 2D tomography images, and use the visualization technology of computer imaging to display the 3D model in holographic stereo. However, in the final analysis, three-dimensional reconstruction relies on two-dimensional tomography images, and the collection of two-dimensional medical images mostly uses X-rays, which inevitably exposes the body of the collector to radiation, increasing the risk of cancer. Therefore, in order to reduce the impact of two-dimensional tomography image acquisition on the body of the collector, the method of reducing the accuracy of the instrument is often used to reduce the radiation to the patient. However, the results of three-dimensional reconstruction of two-dimensional tomography images obtained by this method may appear fuzzy artifacts, which will affect the diagnosis of diseases by medical workers. Therefore, it is necessary to use medical image super-resolution reconstruction technology to reconstruct medical two-dimensional tomography images in high resolution.

Based on the three-dimensional reconstruction of sequential medical images, this paper conducts super-resolution, segmentation and three-dimensional reconstruction of sequential medical images, aiming at the problems such as insufficient utilization and low efficiency of effective image information in existing super-resolution methods, loss of high-frequency information in the process of feature extraction, and long distance dependence due to lack of feature information in existing segmentation methods. In this paper, a new method of medical image super-resolution reconstruction based on global and local features and a new method of 3D reconstruction based on semantic segmentation are proposed to realize the super-resolution, segmentation and 3D reconstruction of medical image sequence.

2 Related Work

2.1 Image Super Resolution

Image Super-Resolution (SR) technology aims to restore low-quality images from small sizes to large-size high-quality images. According to different research methods, this technique can be divided into traditional super resolution method and deep learning-based super resolution method. The traditional super resolution method is based on interpolation, and the image super resolution method based on deep learning aims to realize the reconstruction from low resolution image to high resolution image by learning the relationship between a series of high and low resolution image pairs. With Transformer's success in natural language processing, it has attracted the attention of researchers in the field of computer vision. Chen et al. [1] proposed a skeleton network model IPT based on transformer, which introduced transformer into the field of image recovery for the first time and opened the application of transformer in the low-level vision field. SwinIR [15] is an application of Transformer technology in the field of image super-resolution reconstruction. This model uses the moving window mechanism to effectively solve problems such as patch boundary artifacts and pixel information loss, and maintains high network performance while using fewer parameters. SwinFSR [3] is an improvement based on SwinIR. The network model introduces fast Fourier convolution to address the problems such as noise and artifacts in SwinIR reconstruction. ELAN [4] is an effective remote attention network that uses a lightweight design concept to improve the efficiency of super-resolution reconstruction by utilizing forward-looking information to predict future states. HAT [5] is a super-resolution reconstruction model using attention mechanism, which introduces attention mechanism in different links of feature extraction to improve network performance. SRFormer [6] adopted the displacement strategy to improve the self-attention mechanism, so as to extract more abundant image features and then reconstruct more perfect image details.

2.2 Image Semantic Segmentation

Image semantic segmentation refers to dividing the image into several different regions according to the different color, spatial structure and geometric characteristics of the image, that is, separating the object to be segmented from the image background. According to different segmentation methods, image semantic segmentation can be divided into traditional image semantic segmentation and image semantic segmentation based on deep learning. The traditional image semantic segmentation methods are mainly realized by using related knowledge such as mathematics and topology. As an image semantic segmentation method based on deep learning, U-Net [7] consists of a symmetric encoder (contracting path) and a decoder (expanding path) expanding path. The encoder consists of convolution, pooling and other operations to extract image features, and the expanding path recovers image information through upsampling. The combination of the two methods can achieve higher image segmentation accuracy with less training data. Because U-Net has the characteristics of strong adaptability and high scalability, many subsequent network models are based on the improvement of this model. TransUNet [8] is an improvement on the UNet model, which consists of an encoder and decoder and has significant advantages in medical applications such as multi-organ segmentation and heart segmentation. U^2-Net [9] is a fusion of UNet and ResNet. This model can accept image context information of different scales, improve the extraction ability of image features, and reduce the segmentation error rate. Scunet [10] is a combination of Swin Transformer and UNet. This network model introduces the self-attention mechanism of Transforemer into UNet network, which improves the ability of image feature extraction and thus improves the accuracy and efficiency of medical image segmentation.

2.3 Three-Dimensional Reconstruction

The aim of medical image 3D reconstruction technology is to reconstruct medical 2D tomography image into 3D model, that is, to put 2D image data into 3D data field for analysis and processing, so as to obtain three-dimensional 3D model. 3D reconstruction methods for sequential medical images can be divided into surface rendering and volume rendering according to different data processing methods. Surface rendering technology refers to the method of using the geometric characteristics of the object, constructing geometric elements and splicing them together, so as to reconstruct the surface of the object. This method has a fast rendering speed, but the accuracy of the reconstruction of the object is relatively low, and is suitable for the three-dimensional reconstruction of tissues and organs with simple structure. Volume rendering technique [11] Through interpolation calculation and synthesis of the color or opacity of each voxel, volume data of the two-dimensional plane scan image in the three-dimensional data field is obtained, and then the three-dimensional model is reconstructed. The main representative methods are three-dimensional Texture Mapping Algorithm, ray casting method and Shear Warp Algorithm and so on.

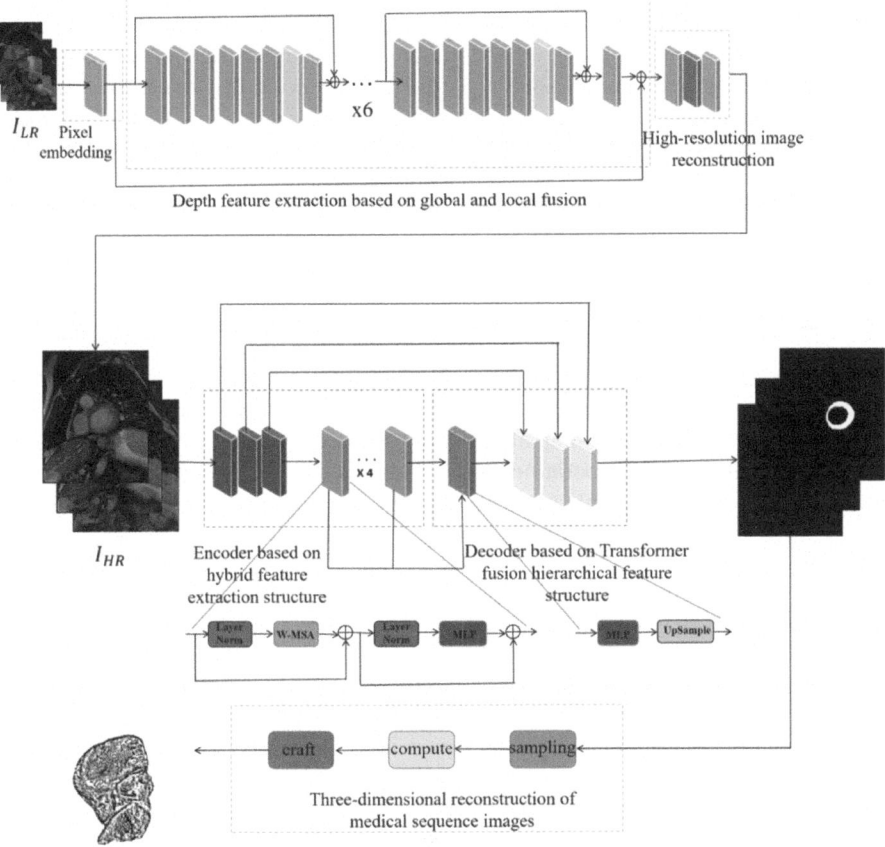

Fig. 1. Overall Architecture.

3 Methodology

Our network model, as shown in Fig. 1, consists of a super-resolution reconstruction of the image combining global and local features and a three-dimensional reconstruction of the sequential medical image based on semantic segmentation. The medical image super-resolution reconstruction with global and local features is composed of pixel embedding layer, depth feature extraction layer with global and local fusion and high-resolution image reconstruction layer. The 3D reconstruction of medical sequence image based on semantic segmentation is composed of encoder layer based on hybrid feature extraction structure, decoder layer based on Transformer fusion hierarchical feature structure and 3D reconstruction layer of medical sequence image. For input low resolution CT images $I_{LR} \in R^{W \times H \times 1}$, Shallow features are extracted through a convolution layer with a convolution kernel size of 3×3, as shown in formula (1):

$$F_{BL} = H_{IBL}(I_{LR}) \tag{1}$$

where, H_{IBL} represents the pixel embedding layer, F_{BL} represents the image output result after convolution. The mapping from low dimensional space to high dimensional space is realized, which provides stable guarantee for the following depth feature extraction. Then, the feature map is passed into the depth feature extraction layer of global and local fusion for depth feature extraction, which contains n residual skip DFL modules and a 3×3 convolution, as shown in formula (2) and formula (3):

$$F_{DL,n} = H_{DFL,n}(F_{DL,n-1}) + F_{DL,n-1} \tag{2}$$

$$H_{DFL,n}(F_{DL,n-1}) = H_{DFL,n}(...H_{DFL,1}(F_{BL}) + F_{BL}...) \tag{3}$$

where, $H_{DFL,n}$ represents the NTH DFL module, $F_{DL,n}$ and $F_{DL,n-1}$ is the output feature map of the corresponding layer, in order to ensure better integration of shallow and deep features. The feature map after depth feature extraction is passed into a 3×3 convolution for convolution operation, and it is connected with shallow feature by jump connection, as shown in formula (4):

$$F_{OL} = H_{COV}(F_{DL,n}) + F_{BL} \tag{4}$$

where, H_{COV} represents the convolution operation, F_{OL} represents the final output feature map. Then it is transferred to the high-resolution image reconstruction layer for high-resolution image reconstruction. In this part, the feature image first goes through a 3×3 convolution, then to the sub-pixel convolution module for upsampling, and finally to a 3×3 convolution layer for high-resolution image output, as shown in formula (5):

$$I_{HR} = H_{COV}(H_{SCL}(H_{COV}(F_{OL}))) \tag{5}$$

where, H_{SCL} denotes subpixel convolution operation, I_{HR} indicates that high-resolution graphics are finally reconstructed. An encoder layer based on the hybrid feature extraction structure is then passed to encode it into a high-level feature representation for downstream decoding and reconstruction tasks. As shown in formula (6):

$$I_{HX} = H_{HFE}(I_X) \tag{6}$$

where H and W respectively represent the width and height of the input image, I_{HX} represent the advanced features obtained by the encoder, H_{HFE} represent the encoder coding operation. Then, it is passed into the decoder layer based on the Transformer fusion hierarchical feature structure, which represents the resolution space of the decoding back to the input image size, as shown in formula (7):

$$I_{LD} = H_{THD}(I_{HX}) \tag{7}$$

where, I_{LD} represents the final output image through the decoder, H_{THD} represents the decoder decoding operation, and finally obtains a pixel-level label map with a size of $H \times W$.

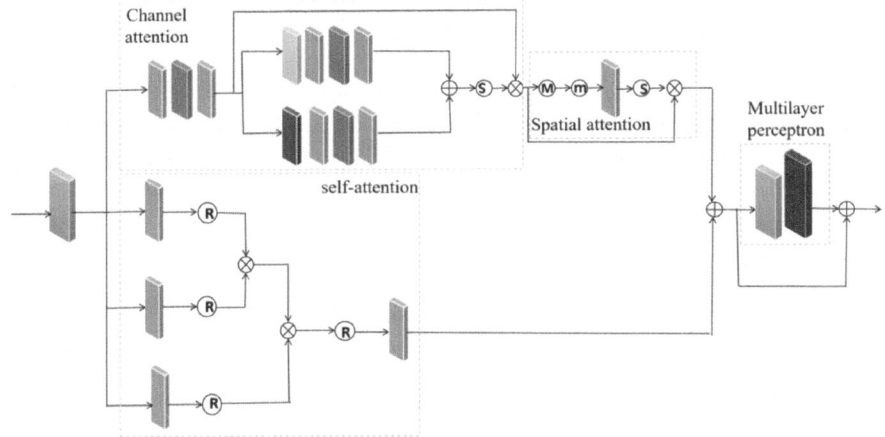

Fig. 2. Attention feature extraction based on series-concatenated combination structure.

3.1 Image Super Resolution

This part is composed of two parts: feature extraction based on series-parallel combination structure and deep cross attention based on pixel. The part of attention feature extraction based on the series-parallel combination structure is composed of the convolution based channel attention and spatial attention mechanism in series, and the standard multi-head self-attention mechanism based on the original transformer in parallel, so as to realize the integration of local features and global features, as shown in Fig. 2. Channel attention and spatial attention based on convolution are global attention mechanisms in feature images and maintain the relevance of global information. According to [23], average pooling highlights the background information of image features, and maximum pooling highlights the texture details of image features, and the combined effect of the two can achieve better results. Therefore, this paper adopts channel attention based on average pooling and maximum pooling at the same time, and adds spatial attention after channel attention to achieve a better representation of global information. Assuming X as the input feature, the global feature enhancement formula is shown in (8):

$$X_{OS} = SA(C_M(X) + C_A(X)) \tag{8}$$

where, C_M denotes channel attention based on maximum pooling, C_A denotes channel attention based on average pooling, SA denotes spatial attention. Based on transformer's multi-head self-attention mechanism, feature images of size $H \times W \times C$ is first divided into HW/M^2 Windows size of $M \times M$, and local self-attention is calculated in each window $X \in R^{M^2 \times C}$.. The formula for window local self-attention is shown in (9):

$$\text{Attention}(Q, \ K, \ V) = \text{SoftMax}(QK^T/\sqrt{d} + B)V \qquad (9)$$

where, Q is the query matrix, K is the key matrix, and V is the value matrix, which are obtained by the product of the window feature X and the projection matrix, d is the ratio of the dimension of the query matrix to the dimension of the key matrix, and B is the learnable relative position coding. In order to establish a relationship between adjacent Windows that perform local self-attention, we also use the moving window mechanism to recalculate the window self-attention by moving half of the window size each time. For further feature transformation, more high-frequency information can be recovered easily. After the attention operation, we use a multi-layer perceptron with two fully connected layers, with deep convolution added. Convolution can help transformer achieve better visual effects. The multi-layer perceptron is mainly composed of two linear layers enclosing a deep convolutional layer, and GELU nonlinearity exists between the two linear layers. We also add the LN layer before the attention and multi-layer perceptron, and the total process for the input feature X is shown in formulas (10) and (11):

$$X_A = GA(LN(X)) + LA(LN(X)) + X \qquad (10)$$

$$X_O = DMLP(LN(X_A)) + X_A \qquad (11)$$

GA and LA represent global attention and local attention respectively, X_A is the intermediate feature, X_O is the final output feature of the module. In order to alleviate the feature splitting problem caused by window self-attention mechanism, more feature pixels are involved in the calculation. We used a pixel-based deep intersectional attention module, as shown in Fig. 3.

The module is similar to the standard multi-head self-attention in transformer, consisting mainly of a cross self-attention layer and a multi-layer perceptron layer based on deep convolution. For the input feature X, the cross self-attention layer also computs the query matrix Q, the key matrix K, and the value matrix V for feature X, but the K and V matrices are computed from a larger window than the Q matrix. In order to ensure the consistency of the number of Windows, the size of the window should meet the formula (12):

$$M_{LW} = M_{SW} + \alpha M_{SM} \qquad (12)$$

where M_{LW} is the size of the key value matrix window, M_{SW} is the size of the query matrix window, α is the variable that controls the size of the overlap area. And the features of the computed key matrix should be zero-filled according to the filling size of $\alpha M_{SM}/2$. Then it passes through the multi-layer perceptron layer based on depth convolution and outputs.

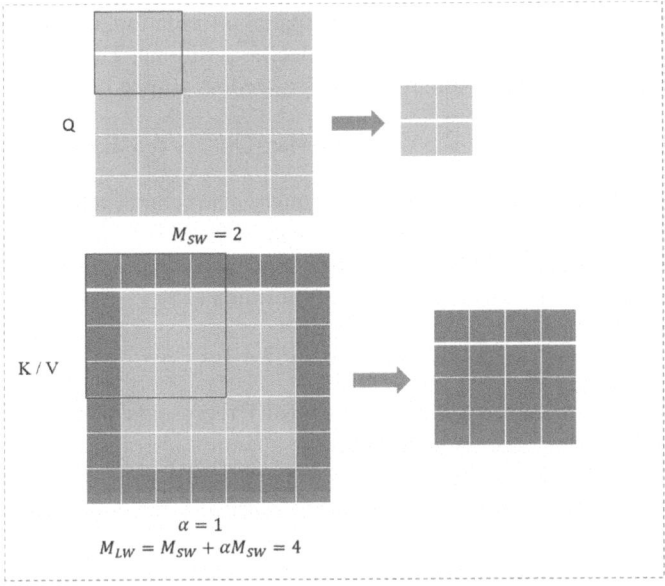

Fig. 3. Pixel-based deep cross attention.

3.2   Image Super Resolution

The main task of the encoder based on hybrid feature extraction structure is to convert the input image to feature representation, which includes shallow low-level features and deep global features. The encoder consists of a dense low-level feature coding layer and a global high frequency information coding layer, which act on the image successively for feature extraction. The dense low-level feature coding layer consists of three convolution layers with convolution kernel size 3×3. The global high-frequency information coding layer consists of four Transformer-based deep feature extraction layers, which are connected in series for feature extraction. Since the mixed feature extraction structure of convolution and Transformer is better than the pure Transformer structure as an encoder, the convolution operation can retain low-resolution information and be used to compensate for the corresponding upsampling process of the decoder, so as to better recover image boundary, contour and other details. For input image I_X, low-level features are extracted and retained through convolution operation, as shown in formula (13):

$$I_{LF} = H_{CON}(I_X) \tag{13}$$

Among them, I_{LF} represents the feature map generated after the convolution operation, H_{CON} represents the convolution operation, and then the resulting image feature $I_{LF} \in R^{H_1 \times W_1 \times C}$ is input to the Tranformer layer for advanced feature extraction. The important components of this layer are mainly composed of multi-head self-attention (MSA) and multi-layer perceptron (MLP). First, the feature size of $H_1 \times W_1 \times C_1$ is segmented into $H_1 W_1 / M^2$ images size of $M \times M$, and this is reshaped into a sequence length of $H_1 W_1/M^2$. In order to better encode spatial information, we map the sequence to the D-dimensional space and embed it in specific positions. The image sequence is normalized and multi-head self-attention is calculated, as shown in formula (14):

$$\text{Attention}(Q, \ K, \ V) = \text{SoftMax}(QK^T/\sqrt{d} + B)V \qquad (14)$$

where, K, Q and V are respectively key matrix, query matrix and value matrix, which are the product of image sequence and projection matrix, and d is the ratio of query matrix to key value matrix dimension. Then the output sequence is normalized and multi-layer perceptron operation is performed, as shown in formula (15):

$$I_L = \text{MLP}(\text{LN}(I_{SA})) + I_{SA} \qquad (15)$$

where, LN represents the normalized operation, MLP represents the multi-layer perceptron operation, I_{SA} is the image feature after the attention operation, and I_L is the final output, which outputs the final image feature of the module.

The main task of the decoder based on Transformer fusion hierarchical feature structure is to represent the features of the target image as pixel-level segmentation prediction. Corresponding to the encoder part, it includes two parts: cascade feature fusion layer and Transformer upsampling segmentation prediction layer. The main function of the cascade feature fusion layer is to expand the effective acceptance field of the decoder operation. We unify the features from the Transformer layer in the encoder into the same channel dimension through a hierarchical feature fusion structure, and then upsample the features to the size of 1/8 and fuse them together. For the feature calculation received by the decoder, as shown in formulas (16) and (17):

$$I_{OAF} = H_{UC}(I_L) \qquad (16)$$

$$I_{PFF} = H_{UP}(I_{OAF}) \qquad (17)$$

Among them, H_{UC} is the unified channel dimension operation, H_{UP} is the upsampling operation, and I_{OAF} and I_{PFF} are the corresponding operation output feature maps, respectively. This makes effective use of different levels of local and non-local attention, enabling more powerful expression of features. Then, the fused results are combined with the features saved by the convolution operation at the corresponding encoder level respectively in the way of residual skip, and sent to the patch extension layer based on Transformer for upsampling, to obtain a higher resolution feature map, and the feature dimension will be half of the original. For the calculation of feature $I_{PFF} \in R^{H/8 \times W/8 \times C_2}$, as shown in formula (18):

$$I_{MFM} = H_{PE}(I_{PFF}) \qquad (18)$$

H_{PE} indicates the up-sampling operation of patch expansion, $I_{MFM} \in R^{H/4 \times W/4 \times C_2/2}$ indicates the feature result of the output upper-sampling of the middle layer. After several up-sampling operations, the final segmentation image is obtained.

The cascade feature fusion layer, consisting of an MLP layer and an upper sampling layer, is a relatively lightweight decoder, and features from the Transformer encoder have a larger effective receiving field, making simple decoding possible. The features from the Transformer layer in the encoder are first unified to the same channel dimension through the MLP layer, then up-sampled to 1/8 and connected together, as shown in formulas (19) and (20):

$$I_0 = H_{MLP}(I_{IN}) \qquad (19)$$

$$I_K = H_{UP}(I_O) \qquad (20)$$

Among them, H_{MLP} represents the MLP operation, H_{UP} represents the upsampling operation, and I_{IN}, I_O, and I_K represent the image features of the input and output of the corresponding layer.

The upsampling segmentation prediction layer based on Transformer mainly reconstructs feature maps from adjacent dimensions in the cascade feature fusion layer combined with the encoder dense low-level feature coding layer into higher-resolution feature maps, and reduces the dimension number of feature maps to half of the original dimension number. For a series of features that need to be operated on, such as $I_{L,1} \in R^{\frac{H}{16} \times \frac{W}{16} \times 4C}$, the number of dimensions of the feature graph is doubled through a linear layer, as shown in formula (21):

$$I_{L,2} = H_{LM}(I_{L,1}) \qquad (21)$$

where H_{LM} represents a linear operation, $I_{L,2} \in R^{\frac{H}{16} \times \frac{W}{16} \times 8C}$ is obtained, using the rearrangement operation to expand the resolution of the feature map to 2 times the original, and the number of dimensions to a quarter of the original, as shown in formula (22):

$$I_{L,3} = H_{CP}(I_{L,2}) \qquad (22)$$

where H_{CP} represents the rearrangement operation, you get $I_{L,3} \in R^{\frac{H}{8} \times \frac{W}{8} \times 2C}$. The 3D reconstruction layer of medical sequence image mainly uses ray projection algorithm to reconstruct the 3D model of medical sequence image.

4 Experiments

4.1 Data Sets and Evaluation Indicators

The super-resolution reconstruction part of the image combining global and local features is trained by using the public data set DIV2K [12] as the original training data set. Evaluation and verification were performed on Set14, Set5, BSDS100, Urban100, Manga109 and two medical image data sets of chest CT image and abdominal CT image. PSNR and SSIM were used to evaluate the performance of super-resolution reconstruction. The three-dimensional reconstruction of sequential medical images based on semantic segmentation was trained on the public dataset SkyScapes and tested on the public datasets Cityscapes and MICCAI 2015.

4.2 Experimental Details

In the super-resolution part design, the pixel embedding module has only one convolution layer; The depth feature extraction module is composed of 6 depth feature extraction modules based on global and local fusion, and each module has 6 attention feature extraction layers based on series-linked combination structure and 1 pixel based deep cross attention layer. The number of heads for multi-head self-attention is set to 6, the window size is set to 16 × 16, and the channel number of the intermediate feature is set to 180. The parameter α, which controls the size of the overlapping region in the pixel-based deep cross attention layer, is set to 1, the patch size used for training is 64 × 64, the initial learning rate is 2×10^{-4}, the value of batch is 1, and the minimum absolute deviation is used as the loss function. In our segmentation reconstruction model, the convolution kernel size and number of convolution layers of the encoder based on the hybrid feature extraction structure are 3 × 3 and 3, and the number of transformer layers is 4. The number of Transformer layers in the decoder based on transformer fusion hierarchical feature structure is 3, the input image size is 224 × 224, the patch size used for training is 16 × 16, the batch size is 1, and the initial learning rate is 0.001. Average cross-linked mIoU is used to evaluate the performance of our model. The model was tested under a TITAN Xp GUP with 100,000 iterations.

4.3 Compare with Other Methods

Table 1. Comparison results of model experiments on medical image dataset.

Method	Scale	Training Dataset	Thorax		Abdomen	
			PSNR	SSIM	PSNR	SSIM
SwinIR	×4	DIV2K	24.48	0.8787	25.29	0.8404
HAT	×4	DIV2K	28.67	0.9369	29.09	0.8903
Our	×4	DIV2K	29.12	0.9408	29.39	0.8925

The super resolution reconstruction part is compared with the super resolution methods based on CNN and Transformer. These advanced methods include: LapSRN [13], DRRN [14], MemN [15], IDN [16], CARN [17], SwinIR [2], HAT [5]. As you can see, our method significantly outperforms other methods on all baseline datasets. It is also significantly better than other methods on two medical datasets. Since more pixels are involved in the reconstruction process to a certain extent, our method performs better in the baseline datasets Urban100 and Manga109, which contain more structured and self-repeating patterns, as well as in the medical image datasets chest CT images and abdominal CT images, as shown in Tables 1, 2, 3 and 4.

The three-dimensional reconstruction of sequential medical images based on semantic segmentation is compared with a series of advanced methods, including ICNet [18], PsPNet [19], DeepLabV3 + [20], FCN [22], EncNet [21], TransUNet [8]. The comparison results are shown in Table 5.

Table 2. Comparison results on the baseline data set with 2x scaling factor.

Method	Scale	Training Dataset	Set5		Set14		BSDS100	
			PSNR	SSIM	PSNR	SSIM	PSNR	SSIM
Bicubic	×2	-	33.66	0.9299	30.24	0.8688	29.56	0.8431
LapSRN	×2	DIV2K	37.52	0.9591	32.99	0.9124	31.80	0.8952
DRRN	×2	DIV2K	37.74	0.9591	33.23	0.9136	32.05	0.8973
MemNet	×2	DIV2K	37.78	0.9597	33.28	0.9142	32.08	0.8978
IDN	×2	DIV2K	37.83	0.9600	33.30	0.9148	32.08	0.8985
CARN	×2	DIV2K	37.76	0.9590	33.52	0.9166	32.09	0.8978
HAT	×2	DIV2K	37.84	0.9606	33.46	0.9173	32.07	0.8985
Our	×2	**DIV2K**	**37.88**	**0.9608**	**33.51**	**0.9176**	**32.11**	**0.8990**

Table 3. Comparison results on the baseline data set with 4x scaling factor.

Method	Scale	Training Dataset	Manga		Urban	
			PSNR	SSIM	PSNR	SSIM
Bicubic	×4	-	24.89	0.786	23.14	0.657
SwinIR	×4	DIV2K	26.58	0.833	24.08	0.704
LapSRN	×4	DIV2K	29.09	0.890	25.21	0.756
DRRN	×4	DIV2K	29.45	0.894	25.44	0.763
MemN	×4	DIV2K	29.42	0.894	25.50	0.763
HAT	×4	DIV2K	29.45	0.894	25.51	0.767
IDN	×4	DIV2K	29.41	0.894	25.41	0.763
Our	×4	**DIV2K**	**30.23**	**0.904**	**25.99**	**0.783**

4.4 Experimental Effect

In order to express part of the effects of super-resolution reconstruction more clearly, we randomly selected an image from the chest CT image and abdominal CT image under sampling for 4 times, and compared the experimental effects with our method and the above advanced methods, as shown in Fig. 4.

In order to further verify the validity of the segmentation proposed in this paper, we carried out further test experiments in medical data MICCAI 2015, and randomly selected 11 continuous medical images from the data set for segmentation. The segmentation effect is shown in the Fig. 5.

The purpose of selecting multiple consecutive medical images is to prepare for subsequent 3D reconstruction. It can be seen from Fig. 5 that the combined encoder based on hybrid feature extraction structure and decoder based on Transformer fusion hierarchical feature structure are effective in the field of medical image segmentation.

Table 4. Comparison results on the baseline data set with 4x scaling factor.

Method	Scale	Training Dataset	Set5		Set14		BSDS	
			PSNR	SSIM	PSNR	SSIM	PSNR	SSIM
Bicubic	×4	-	28.42	0.810	26.00	0.702	25.96	0.667
SwinIR	×4	DIV2K	30.16	0.855	27.23	0.746	26.68	0.704
LapSRN	×4	DIV2K	31.54	0.885	28.09	0.770	27.32	0.727
DRRN	×4	DIV2K	31.68	0.888	28.21	0.772	27.38	0.728
MemN	×4	DIV2K	31.74	0.889	28.26	0.772	27.40	0.728
HAT	×4	DIV2K	31.74	0.889	28.31	0.775	27.37	0.730
IDN	×4	DIV2K	31.82	0.890	28.25	0.773	27.41	0.729
Our	×4	**DIV2K**	**32.06**	**0.893**	**28.55**	**0.781**	**27.55**	**0.736**

Table 5. Comparison of model experiments on Cityscapes dataset.

Method	Dateset	Encoder	mIoU
ICNet	Cityscapes	MobileNetV2	67.7
PsPNet	Cityscapes	MobileNetV2	70.2
DeepLabV3 +	Cityscapes	MobileNetV2	75.2
FCN	Cityscapes	ResNet-101	76.6
EncNet	Cityscapes	ResNet-101	76.9
TransUNet	Cityscapes	R50-ViT	78.7
Our	**Cityscapes**	**Our-Encoder**	**79.4**

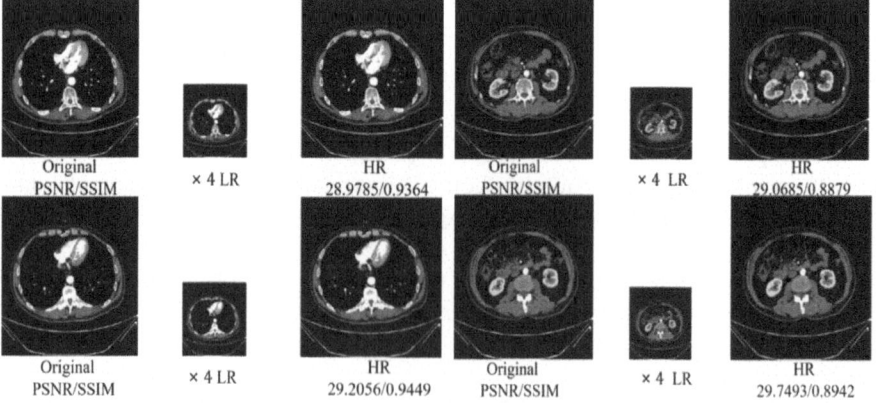

Fig. 4. The super resolution part compares the experimental effect with other methods.

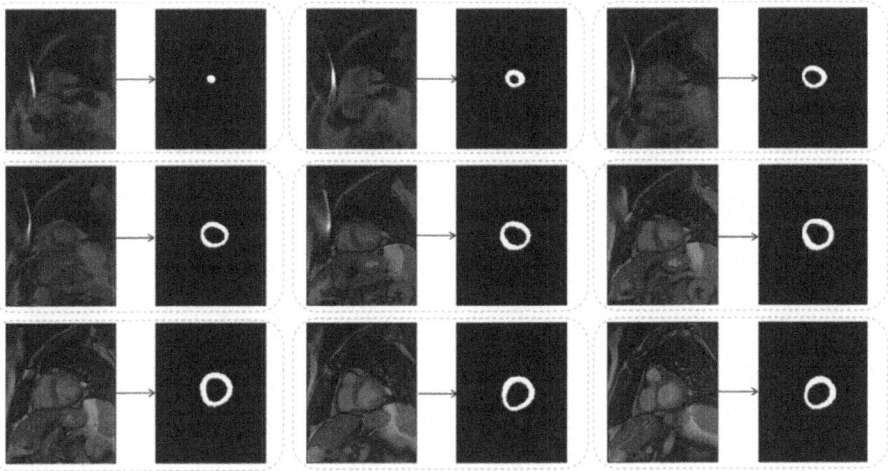

Fig. 5. Medical image segmentation effect.

The segmented 11 sequential medical images were reconstructed in 3D, and the 3D reconstruction results were shown in Fig. 6.

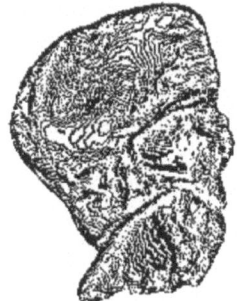

Fig. 6. T3D reconstruction effect of sequence medical image.

As can be seen from Figs. 4 and 3, 3D reconstruction of multiple consecutive segmentation medical images can clearly restore the original appearance of organs and tissues and obtain 3D models of medical organs and tissues with relatively good quality.

5   Conclusion

In order to help medical staff to better diagnose the disease and establish the clinical treatment plan, this paper reconstructs the two-dimensional medical tomography image into a three-dimensional model. The super-resolution reconstruction of medical image and the three-dimensional reconstruction of medical image based on semantic segmentation are studied, and the three-dimensional reconstruction of medical image sequence is realized. A large number of experiments show that the proposed model is effective and has achieved good results in medical application.

Acknowledgments. This study was supported by the National Natural Science Foundation of China (61373160), Hebei Province Central Guidance Local Science and Technology Development Fund Project (No. 236Z0306G), the Natural Science Foundation of Hebei Province (F2021210003), Hebei Provincial Science and Technology Plan Project (22567636H).Disclosure of Interests. The authors declare no conflicts of interest.

References

1. Chen, H., et al.: Pre-trained image processing transformer. In: IEEE Conference on Computer Vision and Pattern Recognition, pp. 12299–12310 (2021)
2. Liang, J., Cao, J., Sun, G., et al.: Swinir: image restoration using swin transformer. In: Proceedings of the IEEE/CVF International Conference on Computer Vision, pp. 1833–1844 (2021)
3. Chen, K., Li, L., Liu, H., et al. :Swinfsr: stereo image super-resolution using swinir and frequency domain knowledge. In: Proceedings of the IEEE/CVF Conference on Computer Vision and Pattern Recognition, pp. 1764–1774 (2023)
4. Zhang, X., Zeng, H., Guo, S., et al.: Efficient long-range attention network for image super-resolution. In: European Conference on Computer Vision., pp. 649–667. Springer Nature Switzerland, Cham (2022)
5. Chen, X., Wang, X., Zhou, J., et al.: Activating more pixels in image super-resolution transformer. In: Proceedings of the IEEE/CVF Conference on Computer Vision and Pattern Recognition, pp. 22367–22377 (2023)
6. Zhou, Y., Li, Z., Guo, C.L., et al.: Srformer: permuted self-attention for single image super-resolution. In: Proceedings of the IEEE/CVF International Conference on Computer Vision, pp. 12780–12791 (2023)
7. Ronneberger, O., Fischer, P., Brox, T., U-net: convolutional networks for biomedical image segmentation. In: 18th International Conference on Medical image computing and computer-assisted intervention–MICCAI, Munich, Germany, 5–9 October 2015, proceedings, part III 18, pp.234 241. Springer International Publishing (2015) https://doi.org/10.1007/978-3-319-24574-4_28
8. Wang, N., Lin, S., Li, X., et al.: MISSU: 3D medical image segmentation via self-distilling TransUNet. IEEE Transactions on Medical Imaging (2023)
9. Qin, X., Zhang, Z., Huang, C., et al.: U2-Net: Going deeper with nested U-structure for salient object detection. Pattern Recogn. **106**, 107404 (2020)
10. Chen, Y., Zou, B., Guo, Z., et al.: Scunet++: swin-unet and cnn bottleneck hybrid architecture with multi-fusion dense skip connection for pulmonary embolism ct image segmentation. In: Proceedings of the IEEE/CVF Winter Conference on Applications of Computer Vision, pp. 7759–7767 (2024)
11. Burns, M., Klawe, J., Rusinkiewicz, S., et al.: Line drawings from volume data. ACM Trans. Graph. (TOG) **24**(3), 512–518 (2005)
12. Lim, B., Son, S., Kim, H., Nah, S., Mu Lee, K.: Enhanced deep residual networks for single image super-resolution. In: Proceedings of the IEEE Conference on Computer Vision and Pattern Recognition Workshops, pp. 136–144 (2017)
13. Lai, W.S,. Huang, J.B., Ahuja, N, et al.: Deep laplacian pyramid networks for fast and accurate super-resolution. In: Proceedings of the IEEE Conference on Computer Vision and Pattern Recognition, pp. 624–632 (2017)

14. Tai, Y., Yang, J., Liu, X.: Image super-resolution via deep recursive residual network. In: Proceedings of the IEEE Conference on Computer Vision and Pattern Recognition, pp. 3147–3155 (2017)
15. Tai, Y., Yang, J., Liu, X., et al.: Memnet: a persistent memory network for image restoration. In: Proceedings of the IEEE International Conference on Computer Vision, pp. 4539–4547 (2017)
16. Zeyde, R., Elad, M., Protter, M.: On single image scale-up using sparse-representations. In: Boissonnat, J.D. (ed.) et al. Curves and Surfaces. Curves and Surfaces 2010. LNCS, vol. 6920. Springer, Berlin, (2012). https://doi.org/10.1007/978-3-642-27413-8_47
17. Ahn, N., Kang, B., Sohn, K.-A.: Fast, accurate, and lightweight super-resolution with cascading residual network. In: Proceedings of the European conference on computer vision (ECCV), pp. 252–268 (2018)
18. Zhao, H., Qi, X., Shen, X., et al.: Icnet for real-time semantic segmentation on high-resolution images. In: Proceedings of the European Conference on Computer Vision (ECCV), pp. 405–420 (2018)
19. Zhu, X., Cheng, Z., Wang, S., et al.: Coronary angiography image segmentation based on PSPNet. Comput. Methods Programs Biomed. **200**, 105897 (2021)
20. Azad, R., Asadi-Aghbolaghi, M., Fathy, M., et al.: Attention deeplabv3+: Multi-level context attention mechanism for skin lesion segmentation. In: European Conference on Computer Vision, pp. 251–266. Springer International Publishing, Cham (2020)
21. Zhang, H., Dana, K., Shi, J., et al.: Context encoding for semantic segmentation. In: Proceedings of the IEEE conference on Computer Vision and Pattern Recognition, pp. 7151–7160 (2018)
22. Dai, J., Li, Y., He, K., et al.: R-fcn: object detection via region-based fully convolutional networks. Adv. Neural Inform. Process. Syst. **29** (2016)
23. Woo, S., Park, J., Lee, J.-Y., Kweon, I.-S.: CBAM: convolutional block attention module. In: European Conference on Computer Vision (2018)

Data Science and Technology

Data Sources and Fast Preprocessing of Shoreline and Bathymetric Data from the Coastal Ocean Numerical Model: Examples of the Pearl River Estuary

Yanqiang Wang[1,3] , Tianyu Zhang[2(✉)], and Wensheng Jiang[4]

[1] National Marine Environmental Forecasting Cerner, Beijing 100081, China
[2] Key Laboratory of Climate, Resources, and Environment in Continental Shelf Sea and Deep Sea of the Department of Education of Guangdong Province, College of Ocean and Meteorology, Guangdong Ocean University, Zhanjiang 524088, Guangdong, China
zhaangty@sina.com
[3] Ocean University of China College of Oceanic and Atmospheric Science, Qingdao, China
[4] The Key Laboratory of Marine Environment and Ecology, Ministry of Education, Qingdao 266100, Shandong, China

Abstract. This paper collects data from the National Fundamental Geographic Information System, Electronic Navigational Charts, etc., which solve the problem of insufficient accuracy of international public data in the Pearl River Estuary area. A technical method of extracting and converting complex shoreline and bathymetric data for coastal ocean numerical models can improve the preprocessing efficiency. This technical method can meet the needs of high-precision complex shorelines and bathymetric data due to the development of high-resolution numerical models and changes in coastlines. This method can provide technical support for the establishment of a marine forecast system and the study of marine dynamic processes.

Keyword: Oceanic numerical model · Pearl River Estuary · Shoreline extraction · Bathymetric data

1 Introduction

Ocean numerical modeling is an important tool for the study of multiscale dynamic processes in the ocean and the establishment of ocean forecasting systems. The development trend of regional ocean numerical models is high-resolution refinement, thus creating the need to produce complex and refined shoreline and bathymetry data. In the preprocessing process of ocean numerical models, shoreline and bathymetry data are often prepared [1]. A considerable amount of time is spent in the preparation of

The original version of the chapter has been revised. Affiliations are correctly numbered. A correction to this chapter can be found at https://doi.org/10.1007/978-981-97-9671-7_24

H. Yu et al. (Eds.): CCF NCCA 2024, CCIS 2274, pp. 331–343, 2024.
https://doi.org/10.1007/978-981-97-9671-7_21

shoreline and bathymetry data. Different data formats and data specifications for different periods are inconsistent, requiring complex data conversion. The marine numerical model uses discrete Reynolds-averaged Navier–Stokes equations with finite difference, finite element, and finite volume methods. The input mesh of the numerical model of the finite difference method is usually a structural mesh, and the nested approach is generally used to improve the model resolution of the key areas of concern to improve the degree of model refinement. The input mesh of the finite element model and the finite volume method model is a nonstructural mesh such as triangles, and the nonstructural mesh can be encrypted directly to the key areas of concern to improve the local resolution of the model. Therefore, for the characteristics of different refined marine numerical models, it is necessary to quickly process the input files of models such as shoreline bathymetry.

The shoreline and bathymetry provide important basic information in producing the computational grid for numerical modeling, and they are also important data sources for the computation of ocean numerical models, which are related to the accuracy of the models. Hou et al. [2] Li et al. [3] For example, Fan et al. [4] provided an overview of the commonly used topographic data for ocean numerical modeling and listed the commonly used public bathymetric data, but for numerical models with higher resolution requirements, it is necessary to use a combination of international public data sources and measured data, such as the use of onsite measured chart data. [5] The following are some examples of the most commonly used public bathymetry data. Changes in the shoreline due to coastal evolution, offshore engineering, reclamation, etc., in estuaries will also produce changes in offshore ocean circulation and ocean tides. Dong et al. [6] Li et al. [7] Wang et al. [8] Teng et al. [9] The precise shoreline bathymetry is related to the accuracy of numerical ocean modeling.

The Pearl River Estuary is located in the intersection area of the inland-delta-river network-estuarine bay-oceanic area, with crisscrossing water systems, complex topography, and numerous harbors. It is difficult and time-consuming to acquire and process data by digitizing paper materials and manually pointing online maps. In this work, more than forty electronic navigation charts near the Pearl River Estuary are collected for the construction of a refined pearl River Estuary marine numerical model. Taking the Pearl River Estuary as an example, it lists the commonly used shoreline bathymetry data for offshore marine numerical models and summarizes the preprocessing techniques of offshore marine numerical models in response to the processing and conversion needs involved in a large amount of measured data, which can provide a method for the rapid establishment of offshore marine numerical models.

2 Data

In this section, offshore bathymetric shoreline data at home and abroad are sorted by taking the Pearl River Estuary region as an example. The global high-resolution shoreline and bathymetry data are listed first, but the accuracy of these data in the Pearl River Estuary region cannot meet the needs of offshore high-resolution numerical modeling. Regional data such as shoreline data from the National Fundamental Geographic Information System (NFGIS) and shoreline data from electronic navigational charts (ENCs) are further listed, and the problem of insufficient data ranging from global open data in the Pearl River Estuary is supplemented by the above data.

2.1 Global High-Resolution Shoreline Data

The GSHHG (Global Self-consistent, Hierarchical, High-resolution Geography Database) (https://www.ngdc.noaa.gov/mgg/shorelines/gshhs.html) is a global high-resolution hierarchical geographic dataset in which the coastline consists entirely of closed polygons arranged in layers. The data are categorized into five levels, namely, a) full-resolution b) high-resolution c) intermediate-resolution d) low-resolution e) crude-resolution, and for each lower level, the data are reduced by approximately 80% compared with the previous level. Taking the Pearl River Estuary as an example, a comparison of the shoreline at different GSSHG shoreline resolutions is shown in Fig. 1. The graded data can satisfy the needs of different resolution modes for the shoreline, but the riverbank data are not accurate and need to be further supplemented with other data.

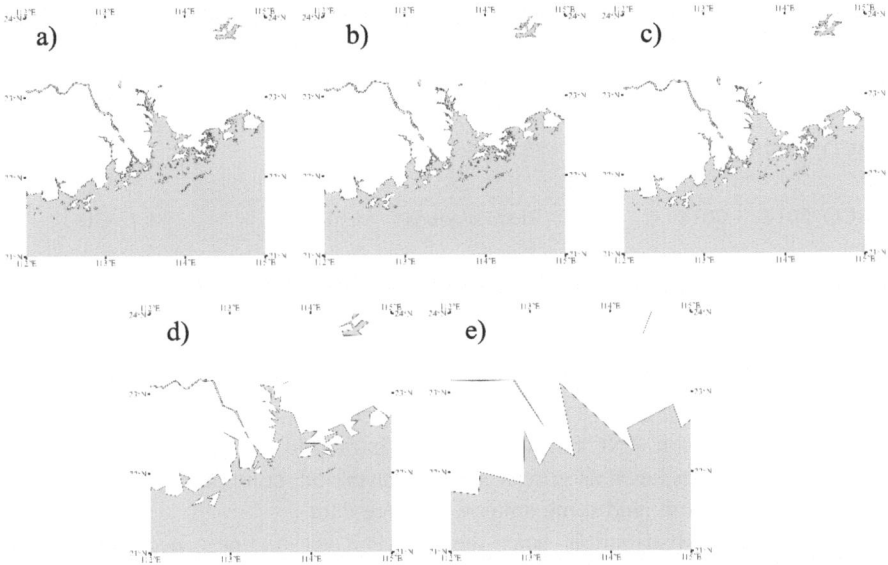

Fig. 1. GSHHS shoreline near the Pearl River Estuary with different hierarchical a) full-resolution, b) high-resolution, c) intermediate-resolution, d) low-resolution, e) crude-resolution.

2.2 Global High-Resolution Bathymetric Topographic Data

Commonly used global bathymetric topographic maps include the General Bathymetric Chart of the Oceans, GEBCO (https://www.gebco.net/.) [10] and ETOPO1 Gridded Global Relief Data [11]. The data of ETOPO1 in the Chinese offshore area are derived from GEBCO data, and the historical releases of GEBCO data are shown in Table 1. There are six different versions of GEBCO data, 2008, 2014, 2019, 2020, 2021, and 2023, and the resolution has increased from 1 arcminute to 15 arcseconds. TIFF (Tagged Image File Format) images and American Standard Code for Information Interchange

(ASCII) code raster data have been added to the traditional network Common Data Form (netCDF) format since 2020, expanding the types of geographic data. The amount of data is 15 times greater than that of the first data.

Table 1. GEBCO data information for different versions

Data name	Year of release	resolution (of a photo)	data format	Data size after decompression
GEBCO_2023 Grid	2023	15 arcseconds	netCDF Geo Tiff Esri ASCII raster	8 GB 6.95 GB 18.2 GB
GEBCO_2021 Grid	2021	15 arcseconds	netCDF Geo Tiff Esri ASCII raster	8 GB 6.95 GB 18.2 GB
GEBCO_2020 Grid	2020	15 arcseconds	netCDF Geo Tiff Esri ASCII raster	8 GB 6.95 GB 18.2 GB
GEBCO_2019 Grid	2019	15 arcseconds	netCDF	10.9 GB
GEBCO_2014 Grid	2014	30 arc seconds	netCDF	1.73 GB
GEBCO One Minute Grid	2008	1 arcmin	netCDF	445 MB

The main sources of GEBCO data are categorized into direct measurements, such as single-beam, multibeam, ENC bathymetry point extraction, and optical measurements; indirect measurements, such as gravity satellite inversion, computer interpolation, and ENC contour extraction; and some unknown source data.

Figure 2 data distribution near the Pearl River Estuary according to the GEBCO_2020_TID data identifier, most of its bathymetry data come from gravity satellite inversion data, which cannot meet the needs of the offshore high-resolution regional numerical model, and its land area marking information is not consistent with the actual shoreline, which needs to be supplemented with other data information.

2.3 National Fundamental Geographic Information System

NFGIS (National Fundamental Geographic Information System) National 1:1,000,000 public version, 1:250,000 public version [12] (https://www.webmap.cn/) The basic geographic data cover the land area of the whole country and the major islands, including Taiwan Island, Hainan Island, Diaoyu Island, South China Sea Islands, and their adjacent sea areas. The geospatial reference adopts the CGCS2000 (China Geodetic Coordinate System 2000) latitude and longitude coordinates and the 1985 national elevation datum. The layer content contains the water system (point, line, surface layer) and other element

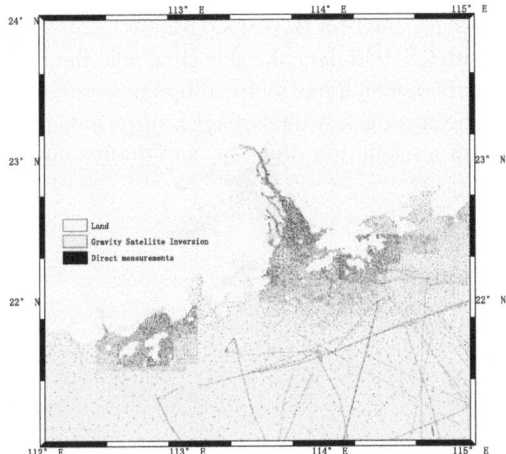

Fig. 2. The source data identify GEBCO near the Pearl River Estuary

layers. The data format adopted by the NFGIS is shapefile files or geo database data. Table 2 lists the geographic information elements in the NFGIS database geographic data used in the ocean numerical model, and the various water system elements are labeled with GB attribute values for oceans, rivers, lakes, etc., which can be used to extract different types of elements.

Table 2. Frequently used basic geographic information features for ocean numerical models

Classification of elements	data layering	geometric type		Main elements of content	data sources
Hydro Systems	Hydro Systems (Areas)	HYDA	Areas	Lakes, reservoirs, two-lane rivers, etc	1: 250,000 1: 1 million
	Hydro Systems (Lines)	HYDL	Lines	Single-line rivers, ditches, river structure lines, etc	
	Hydro Systems (Points)	HYDP	Points	Springs, wells, etc	
Boundaries	Boundaries (Areas)	BOUA	Areas	various administrative boundary areas	1: 1 million
	Boundaries (Lines)	BOUL	Lines	various administrative boundary lines	
	Boundaries (Points)	BOUP	Points	territorial sea base	

Shoreline using National Center for Basic Geography 1:250,000 and 1:1 million data Fig. 3a) b), Compared with GSHHS data, the shoreline near the Pearl River Estuary can be more accurately described, which has more complete water system information and higher resolution shoreline accuracy. A total of 1:1 million data points can be processed as a rectangular grid with a resolution of 1 km, and the resolution of 1:250,000 data points can reach 100 m.

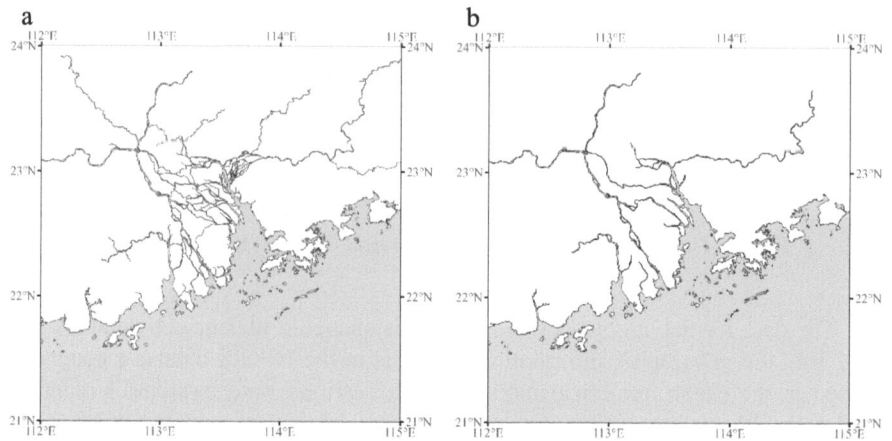

Fig. 3. Extraction of the Pearl River Estuary Shorelines from the National Fundamental Geographic Information System a) 1:250,000 b) 1:1,000,000.

2.4 Electronic Navigational Charts Shoreline and Bathymetric Data

Electronic navigational charts (ENCs) are commonly used in the S-57 format, which is a standard format for the transmission of digital hydrographic data promulgated by the International Hydrographic Organization (IHO) and is used mainly for the exchange of hydrographic data/ENCs between countries around the world and the distribution of ENCs by end-users. The standard is mainly used for the exchange of hydrographic data/electronic ENCs between countries around the world and the distribution of ENCs by original equipment manufacturers and end users. The S-57 data model defines an object as a feature and a spatial object. The feature object describes the type, nature, and characteristics of the entity, whereas the spatial object describes the spatial location characteristics of the entity. In the process of numerical modeling, data such as shoreline and water depth data can be differentiated according to various features.

On the basis of ocean numerical modeling needs,

Table 3 lists the geographic landmarks and abbreviations of nautical charts needed for numerical modeling, among which land, sea, seabed, and other landmarks can extract sea and land scales; coastlines, riverbanks, and other landmarks can be used to extract shore boundaries; and bathymetry and isobaths can be used to extract bathymetric data.

The ENCs of China's sea areas are divided into six classes according to scale, generally including general charts, general charts, coastal charts, nearshore charts, harbor charts, berth charts, and so on. According to the needs of offshore marine numerical modeling, the appropriate scale of chart shoreline and bathymetry data can be selected according to the modeling area.

Table 3. Frequently used S57 ENCs in ocean numerical models

Object label	Abridge
Coast line	COALNE
Depth area	DEPARE
Depth contour	DEPCNT
Land area	LNDARE
River	RIVERS
River bank	RIVEBNK
Sea area	SEAARE
Seabed	SBDARE
Sounding	SOUNDG

Compared with the GSSHG shoreline data Fig. 1, the NFGIS data Fig. 3. The GSSHG shoreline data are relatively complete in terms of shoreline preservation, but the full-resolution data of the rivers can distinguish only the data of the Xijiang River mainstream, the Modaomen waterway, and the Yamen waterway, and most of the data of the Xijiang River, the Beijiang River, and the Dongjiang River are missing. The 1:1,000,000 NFGIS shoreline data can distinguish the data of the first-class river channels in the Pearl River Basin, the 1:250,000 NFGIS shoreline data can distinguish the data of the first-class to fourth-class river channels in the Pearl River Basin, and the data are relatively complete. The 1:1,000,000 NFGIS shoreline data can distinguish the data of the first- to fourth-class river channels in the Pearl River Basin, and the data are relatively complete.

3 Data Processing

Shoreline and bathymetry data come from a wide range of sources and are not in a uniform format; thus, quickly processing and converting them into input data for offshore ocean numerical modeling is a very important but not easily solved problem in the establishment of ocean models. In this part, which combines the commonly used shoreline bathymetry data, third-party software or native software is used to unify the shoreline and bathymetry files with different sources and formats into the shapefile file format, which is the most commonly used geospatial format. In the process of conversion, data coordinates are converted, data datums are revised, merged, and finally converted into the required type of numerical mode input file.

3.1 Shoreline Data Processing - Shoreline Fusion

The requirement of the marine numerical model preprocessing program for the shoreline is that the shoreline should be composed of closed polygons. First, the data from different sources, such as rivers and oceans, are merged and fused according to the element attributes to obtain the hydrological fusion data of the modeling area. The complete modeling area data are then used to erase the modeling area hydrological fusion data to obtain the modeling area land data. Furthermore, through the land data element to the line, the element folding point is used as a point to generate a complete, continuous shoreline point dataset, which is used to build the model input data. When the data in the modeling area are erased from the hydrological data to obtain the land data, it is necessary to specify the appropriate XY tolerance according to the scale of the elements, i.e., the minimum distance between the coordinates of the elements, to ensure that the data obtained from the erasure reach the resolution requirements of the modeling. The flow chart of shoreline extraction via this method is shown in Fig. 4.

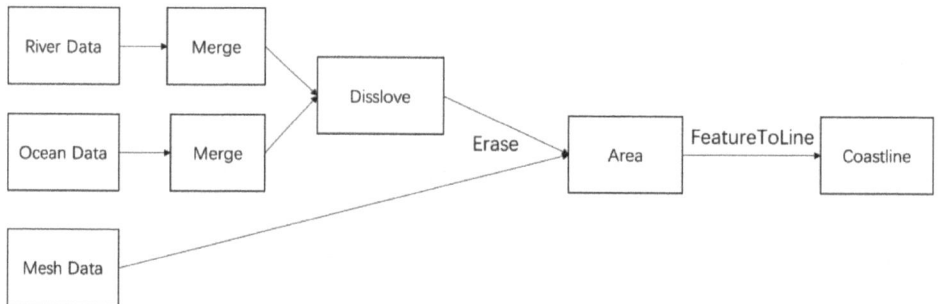

Fig. 4. Data fusion and conversion process for shorelines

3.2 Bathymetric Data Processing

Batch Conversion of Bathymetric Data Coordinates of Electronic Navigation Charts

The original data format of the electronic nautical chart is S57, and the data information that needs to be converted when building the numerical model is as follows Fig. 5. Most of the data can be directly converted via third-party software. However, it should be noted that the bathymetric data extracted from the SOUNDG object and the original bathymetric data are three-dimensional points, and the bathymetric values are lost when they are directly converted to two-dimensional data. Therefore, it is necessary to design a special process to ensure that no information is lost when converting the data. In the process, we first determine whether the data are 3D data; if so, we need to increase the value of the Z attribute and write the Z value of the 3D data into the Z attribute of the output shapefile file. The conversion of two-dimensional vector point data via this method ensures that the ENCs data conversion is complete.

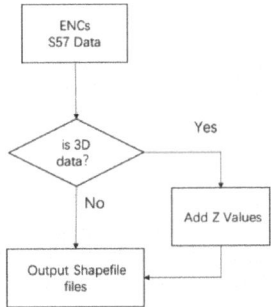

Fig. 5. The S57 ENCs data conversion process

Quality Control of Bathymetric Data

In the mapping data, due to manual recording and other reasons, some symbols are marked incorrectly, the size of the note markers is not uniform, the contour markers are unreasonable, the data null value exceeds the limit value, and other quality problems exist. Moreover, the collected seawater depth value also deviates from the actual water depth when ocean mapping. In marine surveying and mapping, abnormal data can be automatically identified by computers, usually via the least squares fitting method, to eliminate abnormal values. The more commonly used methods are categorized into two types: one is the statistical test of error based on classical statistical theory, and the other is the comparative discrimination method based on function or statistical imputation. Huang [13] summarized related methods and proposed the application of modern anti-differential estimation theory to improve statistical tests of marine measurement anomaly data.

The ENCs are labeled with publication dates and update dates, the charts with different publication dates generally have large changes, and the charts with the same publication date but different update dates are usually labeled with small corrections. The numerical modeling study should be based on the time range of the model validation data and select the ENCs with a better present situation to ensure the accuracy of the model.

Vertical Datum Correction of Bathymetric Data

There are different sources of bathymetry data for ocean numerical modeling, such as the Chart Theory Lowest Tide Surface Datum, 1985 Elevation Datum, Pearl River Elevation Datum, and Mean Sea Surface Datum Fig. 6. The modeled water level in the numerical model calculation is relative to the mean sea level. In the numerical model calculation, the model water level is related to the mean sea level rise and fall, so in the numerical calculation, it is necessary to revise the bathymetry data from various sources to the local mean sea level uniformly.

To ensure that the instantaneous water depth at any moment can be less than the charted water depth data, the theoretical lowest tide surface is usually used as the starting surface of the water depth. China's hydrographic norms use the lowest possible tide level calculated by the superposition of 13 tidal variables as the theoretical depth datum. The elevation datum of the current ENCs is the 1985 National Elevation Datum, which is a

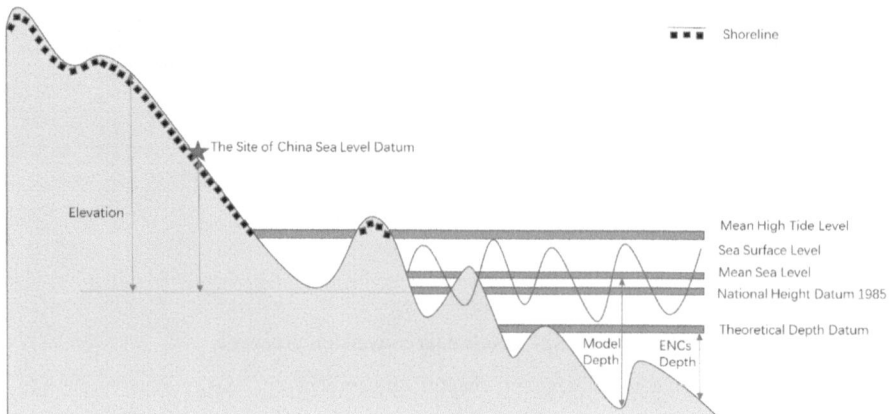

Fig. 6. Schematic diagram of datum corrections for ocean depth

multiyear mean sea level calculated from the tidal observation data of Qingdao for 19 of the 27 years from 1952--1979.

Usually, the base surface can be converted according to the value marked in the tide table of paper charts, and the water depth data of ENCs can be revised to the mean sea surface by using the value of water depth plus the value of the mean sea surface marked in the Electronic Navigational Charts. The difference between the 1985 National Elevation and the local mean sea level needs to be calculated according to the two sources of charts based on the 1985 National Elevation and the mean sea level, and the difference can be calculated by using the value of the large tide rise minus the value of the mean sea level of each set of charts and then calculating the difference between the two sets of charts. Sea surface value from each chart and then calculate the difference between the two sets of these charts to be [14]. The difference between the two sets of charts is then calculated. According to the revised method, the elevation can be revised to the mean sea level uniformly.

3.3   Ocean Numerical Model Grid Production

The tools for producing structured grid input files are generally MATALB's toolboxes, such as Seagrid, Roms_Tools, and GridBuilder. The tools for producing unstructured grid input files are generally specialized software, such as SMS (surface water model system) software. According to the requirements of different toolkits/software formats, the input files for shoreline and bathymetry need to be further produced.

Bathymetric data usually consists of an x-coordinate, y-coordinate, and z-coordinate, which can be converted from a shapefile file to an ASCII code file and then converted to the format required by the preprocessing software.

Shoreline data converted by step 3.1 usually consists of three or more attributes, of which the key attributes are the x, y, and id, respectively; the x coordinate, y coordinate, and id information in step Fig. 7 a; the same shoreline id is the same, such as in the Fig. 7 MATLAB-based toolkit; the data of the shoreline format are usually required to be continuously closed; each line is represented by the x coordinate and y coordinate of the shoreline, respectively; and the different shorelines are divided by not a number (NaN), as shown in Fig. 7b.

SMS software to generate the grid needs to use the shoreline file for the.cst extension of the file. The basic requirements of the file are as follows: the first line of the file identification information is a shoreline file; the second line is the total number of shorelines in the file; the third line of the total number of nodes in a single shoreline and whether the shoreline is closed identification information, from the beginning of the fourth line to the end of the shoreline for each node's coordinate value, the end of the shoreline for the next shoreline, the total number of nodes and shoreline information, and so on. After the end of this shoreline, the total number of nodes and shoreline information of the next shoreline are included. As shown in Fig. 7c.

The orgi_id information of different shorelines after shapefile file conversion is different; the MATLAB-based toolkit can add NaN NaN segmentation according to the id information and then convert it to the format required by the preprocessing software. For the format required by SMS software, the number of nodes, node information, and shoreline information of each independent shoreline can be determined according to the converted id of the shapefile file and converted into a **cst** file.

```
a                                  b  line1x1   line1y1        c   COAST
                                      line1x2   line1y2            m 0.0
   line1x1    line1y1   id1            ...                         n1 0
   line1x2    line1y2   id1           line1xn1  line1yn            line1x1   line1y1   0.0
   ...                                NaN       NaN                line1x2   line1y2   0.0
   line1xn1   line1yn   id1           line2x1   line2y1            ...
   line2x1    line2y1   id2           line2x2   line2y2            line1xn1  line1yn1  0.0
   line2x2    line2y2   id2           ...                          n2 0
   ...                                line2xn2  line2yn2          line2x1   line2y1   0.0
   line2xn2   line2yn2  id2           NaN       NaN                line2x2   line2y2   0.0
   ...                                ...                          ...
   ...                                NaN       NaN                line2xn2  line2yn2  0.0
   linenx1    lineny1   idm           linenx1   lineny1            nm 0
   linenx2    lineny2   idm           linenx2   lineny2            line2x1   line2y1   0.0
   ...                                ...                          line2x2   line2y2   0.0
   linenxnm   linenynm  idm           linenxnm  linenynm           ...
                                                                   line2xnm  line2ynm  0.0
```

Fig. 7. Shoreline format description a) Shapefile format b) MATLAB format c) cst format

4 Conclusion

In this work, the National Fundamental Geographic Information System data and Electronic Navigational Charts data are used as the basic data to address the problems of insufficient data accuracy and resolution of international public data such as GEBCO and GSHHS in the Pearl River Estuary. The GIS method is used to construct a fast data extraction and conversion model to efficiently realize the preprocessing of the marine numerical model.

In the process of shoreline processing, the spatial surface elements are merged, fused, and topologized to generate a complete and closed shoreline, which provides the basis for numerical model input. In the S57 file ENCs, bathymetry data conversion solves the problem of three-dimensional data conversion to two-dimensional data information loss during the conversion process from the S57 format data to the shapefile file and can replace the traditional method usually used in the digitization of paper charts, which can greatly improve the efficiency of the production of bathymetry. When dealing with the fusion of bathymetric data from different sources, the problem of how to revise different base surfaces is solved. The research method in this paper is based on data near the Pearl River Estuary, which can also be applied to other regional marine numerical models and can provide technical support for the rapid establishment of high-resolution numerical computational models.

Acknowledgment. This research was jointly funded by the National Natural Science Foundation of China (41976200); GHfund B(202407025349); Innovative Team Plan for Department of Education of Guangdong Province (No. 2023KCXTD015); Guangdong Science and Technology Plan Project (Observation of Tropical marine environment in Yuexi); Guangdong Ocean University Scientific Research Program (Grant number 060302032106). We acknowledge the comments of anonymous reviewers.

References

1. Song Z., Liu W., Liu X., Su T., Liu H., Liu X.: Research Progress and perspective of the key technologies for ocean numerical model driven by the mass data. Adv. Marine Sci. **37**, 161 (2019)
2. Hou, J., Gao, Y., Li, T.: Review of topography and bathymetry data used in ocean model. Marine Forecasts **29**, 44 (2012)
3. Li, S., Guo, J., Jiang, X., Bai, Z., Song, J., Wang, Y.: Sources and analysis of multitemporal-spatial scale marine hydrometeorology data. Marine Sci. Bull. **39**, 24 (2020)
4. Fan, D., Li, S., Shan, J., Zhang, J., Huang, Y., Wang, A.: Comparing and analyzing several global seafloor topography models. Hydrographic Survey. Charting **41**, 20 (2021)
5. Zhu, M., Yang, B., Wang, M., You, B., Liu, G.: The present situation and analysis of the comprehensive study of water depth in digital chart. Hydrographic Surv. Charting **38**, 11 (2018)
6. Dong, J., Chen, Y., Liu, C., Ma, H., Liu, C.: Numerical simulation of the hydrodynamics affected by coastline and bathymetry changes in the Bohai Sea. Adv. Marine Sci. **4**, 676 (2020)
7. Li, L., Cao, J., He, Z., Yao, Y.: Impacts of coastline changes of Hangzhou Bay-Changjiang Estuary on tidal dynamics in Hangzhou Bay. J. Zhejiang Univ. (Eng. Sci.) **52**, 1605 (2018)
8. Wang, Z., et al.: The influence of coastline changes on tidal dynamics in the pearl river estuary. Sci. Technol. Eng. **20**, 1171 (2020)
9. Teng, F., Li, S., Wang, G., Xu, T.: Synthesis and evaluation of indonesian seas bathymetry data based on tidal simulation. Adv. Marine Sci. **39**, 210 (2021)
10. Li, S.: Development and enlightenment of general bathymetric chart of the oceans (GEBCO). Marine Inform. **1** (2015)
11. ETOPO1 1 Arc-Minute Global Relief Model. https://www.ngdc.noaa.gov/mgg/global/global.html,
12. Jiang J.: The design and application research for national fundamental geographic information system China 1:250,000 database. Remote Sensing Inform. **14** (1999)

13. Huang, M., Zhai, G., Wang, R., Ouyang, Y., Guan, Z.: The detection of abnormal data in marine survey. Acta Geodaetica et Cartographica Sinica **28**, 269 (1999)
14. Xu, J., Shen, J., Miao, S., Huang, C.: Establishment and transformation of the vertical datum in hydrographic surveying and charting. Hydrographic Surv. Charting **31**, 4 (2011)

GPR-STA: A Style Transfer Algorithm for Enhancing GPR Data in Airport Runway Structural Defect Detection

Haifeng Li, Boyu Wang, Sensen Liu, and Nansha Li[✉]

College of Computer Science and Technology, Civil Aviation University of China,
Tianjin 300300, China
nsli@cauc.edu.cn

Abstract. Airport runways, essential for aircraft safety, are susceptible to subsurface defects like voids, settlement, and cracks due to continuous exposure to high loads and complex environments. Traditional detection methods struggle with the identification of various underground anomalies and lack sufficient algorithmic generalization. This study introduces a novel style transfer algorithm, leveraging Ground Penetrating Radar (GPR) data through generative adversarial networks (GANs), to enhance the detection and generalization capabilities in underground object detection at airports. Our method integrates mask and decoder modules within the network architecture and utilizes adversarial learning to manage variations in underground backgrounds, which are common across different airport environments. This approach effectively minimizes the impact of background disparities on the detection algorithms while maintaining spatial consistency of underground objects. The proposed algorithm significantly outperforms traditional methods, as demonstrated in comparative experiments that involve pre- and post-data enhancement. It not only stabilizes image generation, achieving an average structural similarity index above 0.5, but also improves the average F1 score for multi-object detection by 2%. These advancements highlight the enhanced generalization performance of our detection model across varied airport datasets.

Keywords: Airport runway · structural defects · GANs · GPR · unsupervised · generalization

1 Introduction

The detection of underground objects on airport runways is a critical task to ensure the safety and reliability of airport operations, especially given the complex environmental conditions these structures endure. Particularly in environments exposed to precipitation infiltration, temperature fluctuations, and the repeated loading impacts from aircraft take-offs and landings, runway structures are inevitably prone to damage, leading to performance degradation. However, traditional manual detection methods are not only time-consuming and labor-intensive but also prone to human error, potentially overlooking critical issues. Even minor underground anomalies can signify serious underlying issues with airport runways, necessitating accurate and timely identification to prevent catastrophic failures. Therefore, developing robust and high-precision methods for detecting underground objects on airport runways is crucial for ensuring the longevity and safety of these critical infrastructures. This not only ensures the safe operation of airport runways and prolongs their lifespan but also helps reduce flight delays or cancellations caused by runway issues, thereby enhancing the overall operational efficiency of airports.

In the domain of object detection, the most critical challenge is the model's generalization capability, which determines its effectiveness across diverse and previously unseen datasets. An exemplary object detection model not only excels on its training data but also maintains high detection accuracy and resilience when applied to new, unseen datasets. "However, enhancing a model's generalization capability is challenging due to the significant distributional differences that can exist between datasets, leading to inconsistencies during training and prediction—a phenomenon known as 'domain shift,' where the model's learned representations may not transfer well across different data domains. To address the challenge of limited generalization in detection models for underground runway damages, this paper proposes a novel unsupervised image generation algorithm. It introduces modules for image mask coverage, unsupervised image generation, and image mask recovery, while utilizing adversarial learning to mitigate the adverse effects of background differences on algorithm generalization caused by environmental background fluctuations. The experimental section describes the experimental environment and network parameter settings, conducts comparative experiments and validates data effectiveness on GPR data, and finally demonstrates the feasibility and effectiveness of the proposed algorithm.

The organizational structure of this paper is as follows: Sect. 2 provides a comprehensive review of related work. Section 3 details the proposed model framework and network architecture. Section 4 presents the simulation setup, experimental results, and data validation analysis. Finally, Sect. 5 concludes the paper with a summary of findings and potential future research directions.

2   Related Work

GPR [1, 2] is one of the commonly used techniques for detecting subsurface anomalies. The study of detecting and evaluating underground objects using GPR was conducted by Maser K R et al. [3] Data analysis in GPR primarily utilizes B-scan images. Research on underground object detection using B-scan images can be classified into three main categories: traditional data enhancement methods, simulation software-based methods, and deep learning model-driven methods.

Traditional methods for GPR object detection often exhibit inherent limitations, as they predominantly focus on underground objects with high-frequency characteristics. This leads to the need for bespoke feature extractors for each anomaly, resulting in less generalized detection capabilities. Notably, Zhan Yi et al. [4] effectively employed continuous wavelet transform to analyze the measured responses of underground cylindrical objects, extracting distinctive waveform characteristics conducive to effective detection of such objects. Similarly, leveraging prior knowledge of the hyperbolic features generated by buried cylindrical objects in GPR imagery, Gamba P et al. [5] achieved successful pipeline detection and localization. Wang J et al. [6] introduced a novel approach that utilizes cross-correlation between adjacent A-scans to more precisely identify reflected hyperbolas, enhancing detection accuracy. Wang J et al. [7] proposed a rapid detection methodology for underground objects, which is based on analyzing the spatial distribution of echo energy using an innovative three-parameter hyperbolic model. Additionally, conventional machine learning algorithms such as Support Vector Machines (SVM) and Random Forest (RF) have been widely employed in underground anomaly detection, particularly for their robustness in handling structured data. However, their effectiveness can be limited by the complexity of the subsurface environment. Ji G et al. [8] proposed a detection methodology incorporating principal component analysis and digital image processing techniques, demonstrating robust detection capabilities for regular hyperbolic features. With the evolution of deep learning algorithms, state-of-the-art object detection approaches such as Faster R-CNN [9], RetinNet [10], SSD [11], YOLOX [12], and YOLOv8 [13] have been adapted for GPR underground anomaly detection, offering superior accuracy and robustness in complex subsurface environments compared to traditional methods. Hou F et al. [14] utilized a convolutional neural network based on masked region to introduce a novel distance-guided intersectional joint loss function for detection and segmentation. Liu Z et al. [15] utilized YOLO series models to detect hidden runway cracks and void objects underground using GPR 3D C-scan data in road detection projects. Zhang F et al. [16] proposed a method combining spatial attention and channel attention to enhance the detection of underground disease objects in GPR 3D C-scan data. Pan M et al. [17] introduced a feature extraction module with a wider receptive field (Receptive Field Block, RFB) to extend the feature extraction range of diseases. Liu Z et al. [18] integrated the YOLOv3 model with ResNet50 and deformable convolutions, employing Bayesian search and hyperparameter optimization to effectively detect crack objects in a small road hidden underground crack dataset. Li H et al. [19, 20] advanced this approach by fusing GPR B-scan and C-scan data, designing a multiscale model to accurately detect the three-dimensional coordinates of underground diseases and objects.

Deep learning has shown significant potential in interpreting GPR data; however, training effective deep models necessitates large volumes of labeled data, which are challenging to acquire due to the complex and costly nature of data collection and processing in this domain. Furthermore, discrepancies between the frequencies used in GPR data acquisition and the errors introduced during manual post-processing of B-scan data result in significant differences between the original and processed data, leading to poor performance of deep models trained on the original datasets when applied to real-world scenarios. Thus, improving the generalization and robustness of deep models by enhancing GPR data has become an urgent and critical challenge. The pressing need to overcome the inherent limitations of deep models in GPR data interpretation has catalyzed significant advancements in image generation algorithms, particularly those designed to augment data diversity and model robustness.

In 2014, Ian Goodfellow introduced Generative Adversarial Networks (GANs) [21], marking a groundbreaking advancement in machine learning and a pivotal moment in the field of data generation. Since its inception, GANs have undergone rapid theoretical and architectural evolution, finding widespread application in diverse fields such as computer vision, natural language processing, and human-computer interaction. Notably, GANs have excelled in image generation, leveraging adversarial training to progressively enhance their capacity for modeling complex data distributions, ultimately achieving unprecedented realism in synthesized images. Subsequent to GAN's introduction, DeepMind's unveiling of BigGAN [22] marked a watershed moment, significantly enhancing the precision and diversity of image generation, especially in producing high-resolution outputs from large-scale datasets like ImageNet. A key concept in image generation tasks is style transfer, which involves the transformation of stylistic attributes such as color palette, texture, and other visual elements. Addressing this challenge, CycleGAN [23] emerged as a pivotal variant within the GANs framework. Augmented with the concept of cycle consistency, CycleGAN not only facilitates the generation of stylized images but also enables domain-to-domain image translations in an unsupervised manner, rendering it instrumental in a myriad of image style transfer applications. Moreover, XiongH et al. [24] introduced a Structurally Adaptive GPR Generative Adversarial Network, specifically designed for synthesizing GPR defect data, which features dual normalization for enhanced parameter stability and convolutional output refinement. GPR-GAN, employing dual normalization for parameter stabilization and convolutional output refinement, coupled with an adaptive discriminator enhancement module to bolster training stability on modest datasets, culminates in the generation of GPR defects replete with intricate features.

To address the performance limitations of deep models in GPR data analysis for airport runways, this paper introduces a novel multi-model unsupervised image generation algorithm aimed at enhancing subsurface radar data. This method tackles the limitations of existing approaches by simultaneously generating data for multiple object classes. It employs an unsupervised image generation model comprising an encoder E, decoder D, joint generator G, and discriminator. The generated data undergoes constraints imposed on object positions through mask and mask recovery modules. Validation experiments conducted using real underground data from airport surfaces demonstrate the efficacy of the generated GPR data across various tasks. Additionally, to improve the model's generalization capability to different airport subsurface B-scan images, an innovative image generation data augmentation algorithm is proposed. Inspired by generative algorithms such as GANs, the proposed GPR-STA algorithm operates on three-dimensional GPR data, integrating two-dimensional Convolutional Neural Networks (CNNs) and three-dimensional Convolutional Neural Networks to extract salient features, thereby enhancing underground object detection capabilities. Simultaneously, the algorithm processes the backgrounds of B-scan images to mitigate the impact of background heterogeneity on generalization performance.

3 Our Method

To enhance GPR B-scan data, we propose a style transfer algorithm called GPR-STA. The overall framework is shown in Fig. 1. The network architecture is mainly composed of three modules: (1) B- scan Mask Module: a method designed to add mask information to underground objects on B-scan images. (2)Unsupervised Image-to-image transfer module: It borrowed the idea of Munit (Multi-modal unsupervised image to image translation) and conducted the transformation of B-scan images from the source domain to the object domain, completing the background migration of B-scan images. (3) B-scan Mask recovery Module: The image generated by the Unit module obtains the mask location information from the Mask module and fuses the object location information in the original B-scan image with the transferred image.

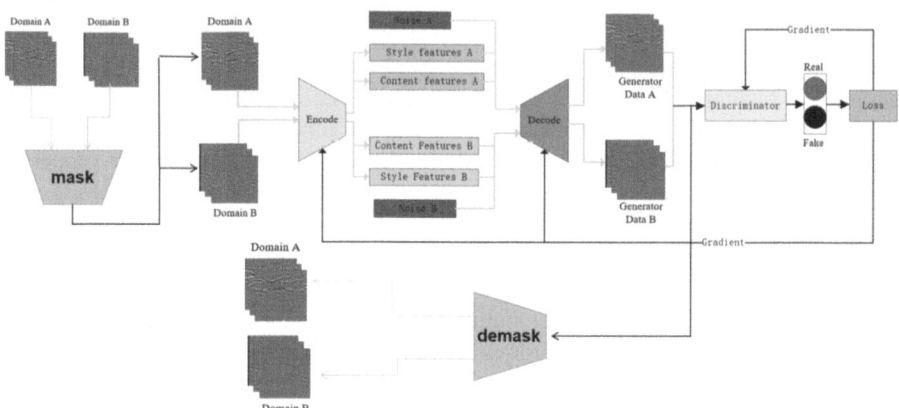

Fig. 1. GPR unsupervised image generation data enhancement algorithm structure

3.1 B-scan Mask Module

We designed an algorithm to mask the object location information, with the aim of improving the robustness and generalization of B-scan image analysis in GPR data. Considering that B-scan images collected using the same GPR may exhibit subtle background variations due to environmental factors or acquisition conditions, we accounted for these differences in our algorithm. We extracted the background threshold distribution information for each B-scan image to accurately distinguish between the background and object regions. We then conducted a statistical analysis of the background information for each individual B-scan image. Utilizing the PASCAL VOC annotation information from the B-scan dataset, we identified different objects within the images and replaced these objects with background masks to generate and retain preliminary mask information. To prevent sharp edge features from affecting subsequent feature extraction, we applied Gaussian blurring to the pixels surrounding the mask regions, thereby smoothing the transition between masked and unmasked areas.

By incorporating the annotation information from the B-scan image dataset, we generated a new dataset with added mask information, which was subsequently used as the training and testing sets for the next module. The B-scan masking algorithm, detailed in Algorithm 1, outlines the process for generating the masked B-scan images. The algorithm consists of three main steps: First, the most frequent pixel value in each B-scan image is identified and designated as the background pixel. Second, object information is extracted from the image, and the locations of these objects are processed using Gaussian blurring to soften the edges. Finally, the background pixels are employed as masks to replace the object information, effectively obscuring the object regions. These steps collectively generate a B-scan image with embedded mask information, ready for further processing in subsequent modules.

Algorithm 1 The B-scan Mask Module

Input: Image P;
Output: Image P* ;
1: initialize frequency[128];
2: **for all** pixel in P **do**
3:     frequency[pixel]+=1;
4: **end for**
5: freqnum = maxfrequency(a);
6: Get object positions A in annotations
7: **for all** position in A **do**
8:     position = Gaussian(position)
9:     **for** x,y in position.x − 2, position.y − 2 **do**
10:         position[x][y] = freqnum
11:     **end for**
12: **end for**
13: **return** ImageP*

3.2 Unsupervised Image-to-Image Background Translation Module

GANs were originally proposed by Ian Goodfellow and colleagues at Google, with the core idea being to model and fit the distribution characteristics of training set data through adversarial learning. The basic GANs framework consist of two neural networks: a generator GGG, which aims to generate realistic data samples, and a discriminator DDD, which evaluates the authenticity of the generated samples by distinguishing them from real data. When GANs are applied to GPR data augmentation, challenges arise in network training, as the generated GPR data often appears completely random, failing to generate data of specified object categories. This necessitates additional re-labeling, which further complicates the process.

Our image-to-image translation model is composed of two autoencoders, each corresponding to a different domain of interest, enabling the transformation of images from one domain to another. Each autoencoder is designed with an underlying representation that includes a content code ccc and a style code sss. This dual-code structure ensures that the translated image closely resembles the real image in the target domain. Additionally, a bi-directional reconstruction objective is employed to simultaneously reconstruct both the image and its underlying code, preserving the essential content and style during translation.

The encoder EEE decomposes the input image features into two distinct components: content encoding ccc and style encoding sss, as illustrated in Eq. 1. This decomposition facilitates the independent manipulation of content and style during the translation process.

$$(ci, si) = (E_A^c(A_i), E_A^s(A_i)) = E_A(A_i) \tag{1}$$

The image-to-image transformation is accomplished by interchanging the encoder and decoder pairs, as depicted in Fig. 2. This process allows for the effective translation of features between different domains by reconstructing the input image in a new context. For instance, to transform the background of an image $A_1 \in X_A$ to background X_B, the content encoding $c_1 = E_c^1(A_1)$ is first obtained, and a latent code s_2 is sampled from the prior distribution $q(s_2) \sim N(0,I)$. Subsequently, the decoder D is utilized to generate the resulting image X_B under background B.

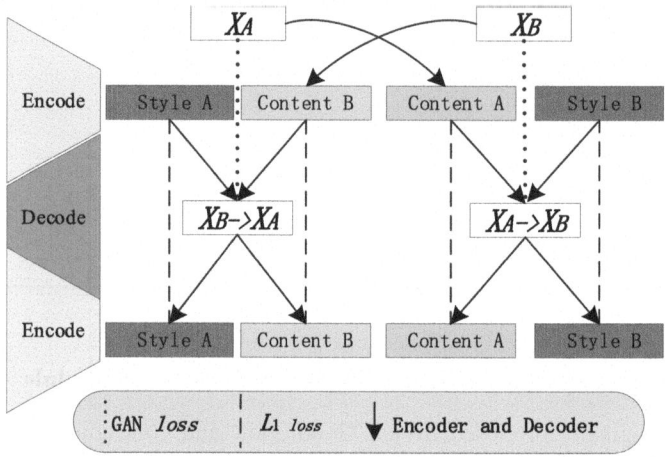

Fig. 2. GPR image-to-image translation model structure

Our loss function consists of two main components: a bidirectional reconstruction loss and an adversarial loss. The bidirectional reconstruction loss ensures that the encoders and decoders function as inverses of each other, thereby maintaining the integrity of the content and style representations during the translation process. The adversarial loss, on the other hand, aligns the distribution of the translated images with that of the target domain, ensuring that the generated images are both realistic and consistent with the expected characteristics of the object domain.

The loss function is described as follows:

Bidirectional Reconstruction Loss: To learn bidirectional encoder-decoder pairs, we utilize functions established in both the image-to-latent code-to-image and latent code-to-image-to-latent code directions for reconstruction losses..This ensures that the model can accurately reconstruct the original image from its latent representation and vice versa. These include two losses: image reconstruction loss and encoding reconstruction loss.

Image Reconstruction Loss: This loss quantifies the difference between the original image and the image reconstructed through the encoding-decoding process. The goal is to minimize this difference, ensuring that the model preserves essential image features during reconstruction. The image reconstruction loss function is mathematically defined in Eq. 2.

$$L_{recon}^{x_A} = E_{x_A \sim p(x_A)}[D_A(E_A^c(A_i), E_A^s(A_i)) - A_i] \qquad (2)$$

Encoding Reconstruction Loss: This loss ensures that both the style encoding and content encoding, which are sampled from the latent distribution during generation, can be accurately reconstructed after undergoing the decoding and re-encoding process. In other words, the model should be able to recover the original latent representations from the generated image, preserving both content and style information. The encoding reconstruction loss functions are mathematically expressed in Eqs. 3 and 4.

$$L_{recon}^{c_A} = E_{c_A \sim p(c_A), s_B \sim q(s_B)}, [E_B^c(D_B(c_A, s_B)) - c_A] \qquad (3)$$

$$L_{recon}^{s_B} = E_{c_A \sim p(c_A), s_B \sim q(s_B)}, [E_B^c(D_B(c_A, s_B)) - s_B] \qquad (4)$$

Adversarial Loss: This loss leverages the principles of GANs to align the distribution of generated images with that of the real object data. The goal is to make the generated images indistinguishable from real images, thereby enhancing the realism and authenticity of the output. The adversarial loss function, which quantifies the effectiveness of this alignment, is mathematically expressed in Eq. 5.

$$L_{GAN}^{X_B} = E_{c_A \sim p(c_A), s_B \sim q(s_B)}, [\log(1 - G_B(D_B(c_A, s_B)))] + E_{s_B \sim q(s_B)}[\log G_B(X_B)] \qquad (5)$$

where G_B is a discriminator used to discern the differences between generated images and real images, and the GANs loss for discriminator G_A and X_A is defined similarly.

Overall Loss Function: Jointly training the encoder, decoder, and discriminator to optimize the final objective, which is the weighted sum of adversarial loss and bidirectional reconstruction loss terms, as shown in Eq. 6.

$$\min_{E_A, E_B, D_A, D_B} \max_{G_A, G_B} L(E_A, E_B, D_A, D_B, G_A, G_B) = L_{GAN}^{X_A} + L_{GAN}^{X_B} +$$
$$\lambda_X L_{recon}^{x_A} + L_{recon}^{x_B} + \lambda_c(L_{recon}^{c_A} + L_{recon}^{c_B}) + \lambda_s(L_{recon}^{s_A} + L_{recon}^{s_B}) \qquad (6)$$

where λx, λc, and λs are weights controlling the importance of the reconstruction terms.

3.3 B-scan Mask Recovery Module

By applying the unsupervised image-to-image background translation module, we seamlessly convert images from domain A to domain B. We then follow a process similar to Algorithm 1, as detailed in Algorithm 2. The first step involves extracting the annotation information from the converted image, followed by applying Gaussian edge blurring to the image. Next, the object position information in the converted image is replaced with the corresponding object information from domain A, ensuring consistency in object placement. After completing this series of operations, we obtain an enhanced B-scan image with improved clarity and consistency.

Algorithm 2 The B-scan Recovery Module

Input: Image P* ;

Output: Image P;

1: Get object positions A in annotations

2: **for all** position in A **do**

3:     position = Gaussian(position)

4:       **for** x,y in position.x − 2, position.y − 2 **do**

5:         position[x][y] = freqnum

6:       **end for**

7: **end for**

8: **return** ImageP*
---

4 Experimental Result

4.1 Dataset and Implementation Details

To evaluate the generalization performance of the object detection model, we partitioned the overall data set into two distinct subsets based on the source of collection from different airports: Airport Data set A and Airport Data set B. By training the model on Data set A and evaluating its performance on Data set B, we conducted cross-domain validation of the model. Table 1 illustrates the results of the generalization comparison experiment. Comparing the results from datasets A and B, we observed a significant performance disparity, particularly in the overall F1 score for underground multi-object detection. The model exhibited a notable decrease in performance,indicating insufficient generalization capability when applied to data from different sources.

Table 1. F1 Comparison of model generalization

	GAP	PIP_L	REBAR	REBAR(PARALLEL)
A	70.83	82.35	92.02	75.86
B	44.41	51.32	5.97	29.81

The model described is implemented using Python 3.8 programming language, with the PyTorch 1.10.1 framework, running on Ubuntu 20.04.5 operating system. Training and testing are accelerated using CUDA 11.7 on NVIDIA 3090 graphics card, significantly enhancing computational efficiency. Real GPR data is utilized to augment the data set, and the results are analyzed in terms of effectiveness, diversity, and their impact on the detector's performance. To quantitatively analyze diversity, this paper employs the Structual Similarity Index (SSIM), calculated as shown in Eq. 7.

$$\text{SSIM}(I_a, I_b) = \frac{(2\mu_a\mu_b + c_a)(2\sigma_{ab} + c_b)}{(\mu_a^2 + \mu_b^2 + c_a)(\sigma_a^2 + \sigma_b^2 + c_b)} \tag{7}$$

where μ_a and μ_b are the means of images I_a and I_b respectively, σ_a and σ_b are the variances of images I_a and I_b respectively, and σ_{ab} is the covariance between I_a and

I_b. c_b is an auxiliary variable, typically with values of 1e-4 and 9e-4. Specifically, the SSIM index approaches 1 for higher-quality images, indicating a closer similarity to the original image. Conversely, an SSIM index below 0.5 suggests poorer image quality, indicating significant differences from the original image.

4.2 Validity of Generated Data

To validate the efficacy of the data, this study conducted a comparative analysis of the background time-frequency distribution between generated data and actual GPR data. As illustrated in Fig. 3, the generated GPR B-scan data closely resemble actual GPR B-scan data in terms of phase and positional distribution, despite some subtle differences in the scattering characteristics of underground heterogeneous media. Furthermore, in terms of frequency domain trends, synthetic data exhibits good consistency with actual data, indicating that the generated data effectively captures the key spectral characteristics. There are no significant differences between the two in terms of the bandwidth occupied by major frequency components. However, synthetic data appears to be richer in low-frequency components, and the positional and phase distributions of the generated data background also closely match those of real data.

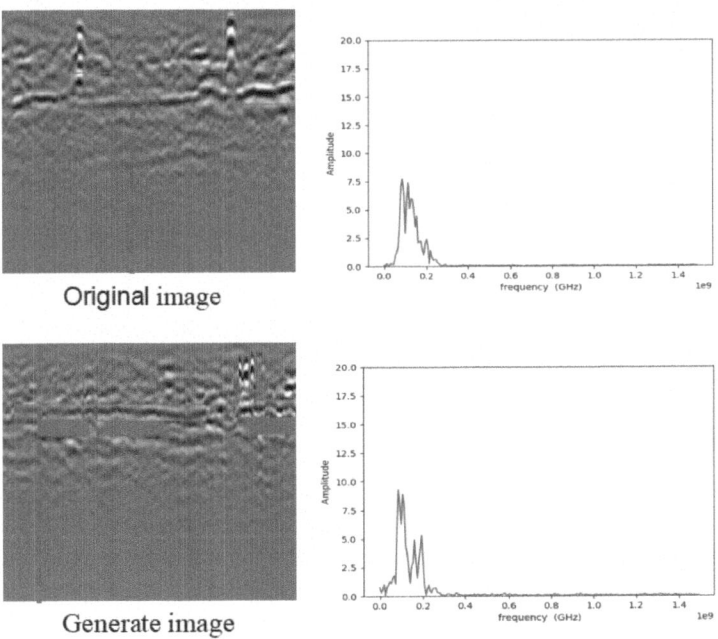

Fig. 3. Time-frequency distribution

As illustrated by the Wigner-Ville distribution in Fig. 4, the background signal distribution remains generally consistent across the time-frequency domain. Figure 5. Presents the SSIM comparison results between the original and real data. In simple backgrounds,

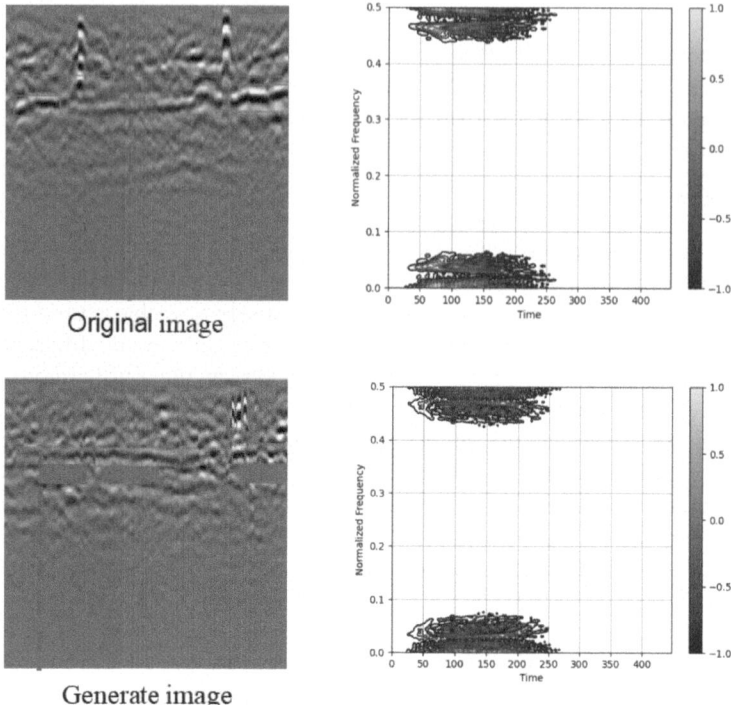

Original image

Generate image

Fig. 4. Wigner-Ville distribution

the SSIM similarity reaches up to 0.84, while in complex backgrounds, the similarity decreases to 0.55. The data generated by GPR-STA exhibits a high degree of similarity with real data in terms of background features. However, at a more detailed level, slight differences are evident in the background signals, enriching the diversity of the data. This phenomenon indicates that the proposed GPR-STA model avoids overfitting issues and generates data that is not merely a simple replication of the training data set.

4.3   Impact on Object Detection Model

To assess the efficacy of the GPR-STA algorithm in detecting subterranean objects within airport environments, it was compared with five prominent object detection methodologies widely used in the field. The experimental setup entailed partitioning 80% of the original data set along with its augmented counterpart for training purposes, while allocating the remaining 20% for testing. Subsequently, the classic Faster-RCNN object detection algorithm was employed for performance evaluation under stringent criteria, with Intersection over Union (IoU) set at 0.5 and a confidence threshold of 0.5. Two distinct sets of comparative analyses were conducted: firstly, on the unaltered domain A data set, as delineated in Table 2. The outcomes gleaned from the A data set evince a notable enhancement in the detection capabilities across varied objects post-augmentation. However, a discernible decrement in performance is noted in crack detection, attributable to

Original image   Generate image

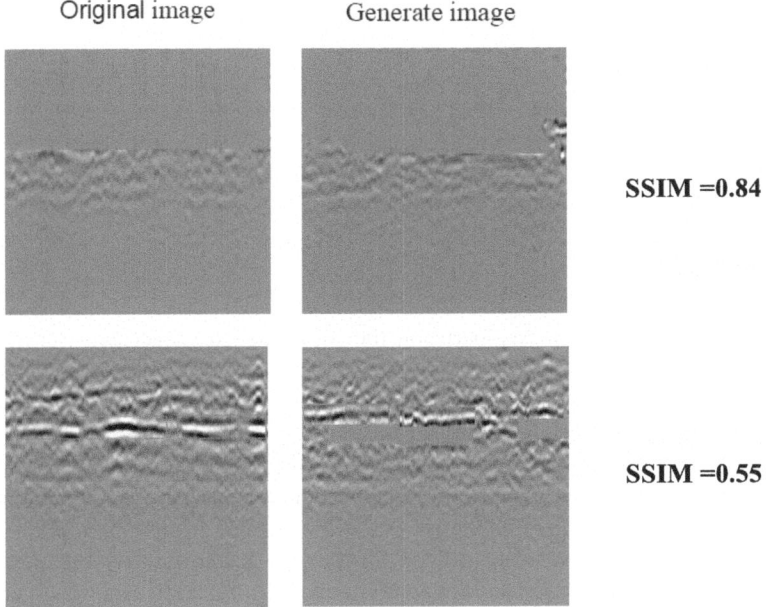

SSIM =0.84

SSIM =0.55

Fig. 5. SSIM similarity

the paucity of crack samples within the A data set. When the same model was tested on Dataset B, as shown in Table 3, a noticeable improvement in performance was observed across the four object categories. Specifically, the F1 score registers a 2% augmentation, while object accuracy demonstrates an improvement ranging from approximately 1% to 5%, indicating the ability of this method to enhance model generalization on engineering datasets.

Table 2. Comparison results of detection in data set A

	CRACK			GAP			LACUNAS		
	Prec	Rec	F1	Prec	Rec	F1	Prec	Rec	F1
RAW	61.11	65.62	63.28	63.72	79.74	70.83	26.47	32.14	29.03
ENH	59.25	50.00	54.23	**66.39**	78.70	**72.02**	**27.27**	64.28	**38.29**
	PIP_L			REBAR			REBAR(PARALLEL)		
	Prec	**Rec**	**F1**	**Prec**	**Rec**	**F1**	**Prec**	**Rec**	**F1**
RAW	87.50	77.77	82.35	87.01	97.64	92.02	68.75	84.61	75.86
ENH	87.50	77.70	82.35	**87.25**	**97.64**	**92.16**	**70.83**	**87.17**	**78.16**

Table 3. Comparison of detection in data set B

	GAP			PIP_L			REBAR			REBAR(PARALLEL)		
	Prec	*Rec*	*F1*	*Prec*	*Rec*	*F1*	*Prec*	*Rec*	*F1*	*Prec*	*Rec*	*F1*
RAW	37.27	54.88	44.41	37.17	82.85	51.32	7.69	4.87	5.97	19.67	61.53	29.81
ENH	**43.49**	48.79	**45.99**	**39.18**	80.00	**52.60**	**12.50**	**7.31**	**9.23**	**20.31**	**66.66**	**31.13**

5 Conclusion

In the current study, we initially partition the subterranean object data set of airport runways into two distinct subsets to scrutinize the model's generalization capabilities. Our observations reveal a significant trend: while the model performs well on the training data set, its efficacy declines noticeably when applied to novel engineering datasets. This decrement in performance primarily emanates from the incongruities in backgrounds attributable to the variances in subterranean environments across different airports.

In addressing this pertinent challenge, we draw insights from contemporary image generation algorithms, pioneering a GPR-STA image background substitution algorithm grounded in unsupervised learning paradigms. This innovative approach enhances the contextual background of the data and, compared to traditional data augmentation methods, creates a richer variety of background data without the need for re-labeling.

Through comprehensive analysis, we demonstrate the crucial role that the generated datasets play in enhancing the generalization capabilities of detection algorithms across diverse datasets. These advancements not only improve the precision of subterranean multi-object detection algorithms in real-world scenarios but also enhance the accuracy of underground object detection in airport runway infrastructures. As a result, this innovative approach leads to a significant improvement in detection system performance, enabling more accurate identification of subterranean objects.

Acknowledgement. The authors acknowledge the financial support from the Research Project of Tianjin Municipal Education Commission under Grant 2021KJ036.

References

1. Solla M., Perez-Gracia V., Fontul, S.: A review of GPR application on transport infrastructures: troubleshooting and best practices. Remote Sensing **13**(4), 672 (2021).https://doi.org/10.3390/rs13040672
2. Hou, F., Shi, R., Lei, W., et al.: A review of target detection algorithms for ground-penetrating radar B-scan images. J. Electr. Inform. **42**(01), 191–200 (2020)
3. Maser, K.R.: Condition assessment of transportation infrastructure using ground-penetrating radar. J. Infrastructure Syst. **2**(2), 94–101 (1996).https://doi.org/10.1061/(ASCE)1076-0342(1996)2:2(94)
4. Zhan, Y., Li, X., Liang, C., et al.: Continuous wavelet transform is used to analyze the echo signal of ground penetrating radar. Geophys. Geochem. Explorat. (06), 443–447+453 (1999)

5. Gamba, P., Lossani, S.: Neural detection of pipe signatures in ground penetrating radar images. IEEE Trans. Geosci. Remote Sensing **38**(2), 790–797 (2000).https://doi.org/10.1109/36.842008

6. Wang, J., Su Y.: Underground object detection based on cross correlation and Hough transform. In: Microwaves, Radar and Remote Sensing Symposium. IEEE (2011).https://doi.org/10.1109/mrrs.2011.6053674

7. Wang, J., Yuan, X., Li, Y., et al.: Rapid detection of ground penetrating radar targets using cross correlation and Hough transform. J. Electr. Inform. **35**(05), 1156–1162 (2013)

8. Ji, G., Gao, X., Zhang, H., Gulliver, T.A.: Subsurface object detection using UWB Ground Penetrating Radar. In: 2009 IEEE Pacific Rim Conference on Communications, Computers and Signal Processing, Victoria, BC, Canada, pp. 740–743 (2009). https://doi.org/10.1109/PACRIM.2009.5291279

9. Ren S., He K., Girshick R., et al.: Faster R-CNN: towards real-time object detection with region proposal networks. IEEE Trans. Pattern Anal. Mach. Intell. **39**(6), 1137–1149 (2017).https://doi.org/10.1109/TPAMI.2016.2577031

10. Lin, T.Y., Goyal, P., Girshick, R., et al.: Focal loss for dense object detection. IEEE Trans. Pattern Anal. Mach. Intell., PP(99), 2999–3007 (2017).https://doi.org/10.1109/TPAMI.2018.2858826

11. Berg, A.C., Fu, C.Y., Szegedy, C., et al.: SSD: Single Shot MultiBox Detector (2015).https://doi.org/10.1007/978-3-319-46448-0_2

12. Ge, Z., Liu, S., Wang, F., et al.: YOLOX: Exceeding YOLO Series in 2021 (2021).https://doi.org/10.48550/arXiv.2107.08430

13. Jacob Solawetz, F.: What is yolov8? the ultimate guide (2023). 04–30–2023. 1, 5, 8

14. Hou, F., Lei, W., Li, S., et al.: Improved Mask R-CNN with distance guided intersection over union for GPR signature detection and segmentation - ScienceDirect. Autom. Construct. 121 [2024–03–24].https://doi.org/10.1016/j.autcon.2020.103414

15. Liu, Z., Wu, W., Gu, X., et al.: Application of Combining YOLO Models and 3D GPR images in road detection and maintenance. Remote Sensing **6**(13), 1081 (2021). https://doi.org/10.3390/rs13061081

16. Li, H., Zhang, F., Piao, M., et al.: Automatic detection of underground targets on airport pavement based on access and spatial attention. Comput. Appli. **43**(03), 930–935 (2023)

17. Li, H., Pan, M., Wang, H., et al.: An algorithm of airport runway underground disease detection based on scale fusion. J. Zhengzhou Univ. Sci. Edition **55**(01), 64–70 (2023)

18. Liu, Z., Gu, X., Yang, H., Wang, L., Chen, Y., Wang, D.: Novel YOLOv3 Model with structure and hyperparameter optimization for detection of pavement concealed cracks in GPR images. IEEE Trans. Intell. Trans. Syst. **23**(11), 22258–22268 (2022). https://doi.org/10.1109/TITS.2022.3174626

19. Li, H., Li, N., Wu, R., et al.: GPR-RCNN: an algorithm of subsurface defect detection for airport runway based on GPR. IEEE Robot. Autom. Lett. PP(99) (2021).https://doi.org/10.1109/LRA.2021.3062599

20. Li, N., Wu R., Li, H., Wang. H., Gui Z., Song D.: MV-GPRNet: multi-view subsurface defect detection network for airport runway inspection based on GPR. Remote Sens **14**, 4472 (2022). https://doi.org/10.3390/rs14184472

21. Goodfellow, I., et al.: Generative adversarial nets. Adv. Neural Inform. Process. Syst. **27** (2014)

22. Andrew, B., Donahue, J., Simonyan, K.: Large scale GAN training for high fidelity natural image synthesis. arXiv preprint arXiv:1809.11096 (2018)

23. Zhu, J.Y., Park, T., Isola, P., et al.: Unpaired image-to-image translation using cycle-consistent adversarial networks. IEEE (2017). https://doi.org/10.1109/ICCV.2017.244
24. Xiong, H., Li, J., Li, Z., Zhang, Z.: GPR-GAN: a Ground-Penetrating Radar Data Generative Adversarial Network IEEE Trans. Geosci. Remote Sensing 62, 1–14 (2024), Art no. 5200114, https://doi.org/10.1109/TGRS.2023.3337172

Research on Prediction of Missing Values Based on Multiple Models

Yutang Wang[✉] and Erni Gao

Anhui Institute of Information Technology, Wuhu, Anhui, China
Ytwang30@gmail.com

Abstract. In the process of big data analysis, data integrity and consistency are often very important factors that affect the accuracy of analysis results. Therefore, before starting the analysis process, data cleaning work should be done on the collected data sources to ensure that subsequent analysis will not cause errors due to data anomalies. Therefore, maintaining data integrity is a very important task in data cleaning. One of the reasons for incomplete data is that the collected data contains missing values, which are caused by human negligence, instrument failure and other factors during the data collection process. The common methods for dealing with missing values are as follows: ignore the value group with missing values directly, or use the central tendency measurement (such as mean, median, etc.) of the missing value attribute to fill in the missing values. These methods may cause the loss of the original characteristics of the value group, which may affect the output of subsequent data analysis and application, and lead to incorrect results. To address this problem, this study uses machine learning methods to fill in missing values for a single field. We use data without missing values as training data, divide the data into multiple clusters using K-Means clustering method to capture the hidden associations between data, and build prediction models for each cluster using multiple regression and artificial neural network. For missing values that need to be predicted, we first use KNN algorithm to find the cluster to which the data belongs, and then apply the model of that cluster to calculate the predicted value. The experiment proves that the multi-model filling method proposed in this study is better than the existing filling algorithms in terms of root mean square error.

Keywords: Missing Value · Multiple Regression Analysis · Artificial Neural Network · K-means Clustering Algorithm

1 Introduction

The technology of big data analysis has made remarkable progress in the modern age where the amount of information is growing rapidly. This new technology is used to predict future behavior and events by analyzing models built from past data. With the continuous improvement of software and hardware performance, this technology has been applied in quite a few fields, such as sales forecast, customer behavior pattern forecast, etc. The data used to build the model comes from quite a variety of channels,

H. Yu et al. (Eds.): CCF NCCA 2024, CCIS 2274, pp. 359–376, 2024.
https://doi.org/10.1007/978-981-97-9671-7_23

but often due to equipment or human factors in the data collection process, data anomalies occur, which in turn reduces the accuracy of the model. Therefore, data cleaning (Data Cleansing) is a very important pre-processing in big data analysis technology to ensure the correctness of subsequent models. Generally, the collected data is presented in the form of a value group (tuple), which contains multiple values that represent various attributes of the data. For example, student data contains attributes such as "student number", "name", and "address". The main purpose of data cleaning is to maintain the consistency and integrity of the data. The basic task of maintaining consistency is to unify the format of attribute values in the data, such as gender "male" is unified as "M" value, etc., or to ensure that all collected data have the same value group structure. The maintenance of integrity mainly ensures that each attribute in the data has values and deals with the missing parts in the collected data.

The missing part of the data is called the missing value (Missing Value). The common reasons for the missing value are as follows: the negligence caused by manual input of data, the completer deliberately skipping the question when collecting the questionnaire for some reason, and the machine that collects the data. Question etc. Little and other scholars pointed out that the missing values can be divided into the following three types [7]:

1. Missing Completely At Random (MCAR), the data with missing values has nothing to do with itself or with the values of other field attributes in the data set;
2. Missing At Random (MAR), the missing value of this type may depend on the correlation of other fields;
3. Not Missing At Random (NMAR), this type of missing value may be related to the information that has been found or not found.

Missing values have a considerable impact on the correctness of subsequent data exploration and statistical analysis results. For example, algorithms such as K Nearest Neighbor (KNN) and k-Means need to calculate the distance of data point features. If there are missing values in the data, they cannot Therefore, the processing of missing values plays a very important role in the preparatory work of big data analysis. Generally, the common processing methods for missing values in pre-work are as follows:

1. Check whether there are missing values in each value group. If missing values are found, ignore the value group without analysis. Although this method is relatively simple, this method may lead to data bias, ignoring too many characteristics provided by the data itself, which will affect the analysis results. For example, for the data of the MAR missing type, deletion will cause the characteristics of specific clusters to disappear, the original characteristics cannot be provided for the deleted data during analysis.
2. Use the measure of central tendency of attributes (such as mean, median, etc.) to fill in missing values. Because only a single value of the median or mean is used for filling, the increase in the number of a certain value may change the original distribution of its characteristics and affect the subsequent analysis results.
3. Use machine learning methods (such as regression analysis, decision tree, K Nearest Neighbor, etc.) to fill in [6, 12]. This method is currently used by more people. Using machine learning to fill in the data, you can use the characteristics of the existing data

to analyze and further predict the value of the missing field. Compared with the above two methods, the machine learning method can find the characteristics of the data and predict more reasonable data for the missing data, instead of filling or deleting only one value.

If the above missing value filling methods are used to ignore or use a single model to predict missing values, it does not take into account that the attributes of the data are often not described by a single model, so the predicted results are often different from the real values. Considering the heterogeneity among data attributes, a framework is proposed in this study, which adopts the method of establishing multiple models to classify data into models of different attributes, and applies the most suitable model calculation when predicting missing values. Improve the accuracy of forecasts. In the research experiment, the multi-model prediction method proposed in this paper and other commonly used single-model prediction methods are also used to compare the root mean square error (Root Mean Squared Error, RMSE) of the predicted value and the actual value to judge the prediction. Accuracy, and the experimental results show that the prediction method proposed in this study can obviously make more correct predictions. The structure of this thesis is as follows: The second chapter is the literature in related research fields, introducing the types of missing values classified by scholars, the commonly used missing value management methods, and the algorithms used in this experiment. The third chapter is the structure and content proposed in this experiment. The fourth chapter compares the structure of this study with the methods commonly used at present. The fifth chapter is the conclusion and the limitations and future development of this experiment.

2 Literature Review

Missing value filling is very important in the pre-work of data cleaning. If only the data with missing values are ignored in this step, it will often cause errors in subsequent data analysis and produce incorrect results. Therefore, in order to improve the accuracy of subsequent data analysis and application, it is usually necessary to perform data cleaning steps before data analysis and application. The processing methods for missing values are divided into deletion methods and filling methods. Many scholars have made contributions in this field of research. The relevant research results will be discussed one by one in this chapter. In addition, several more commonly used data analysis techniques will also be introduced in this chapter, thus building the basis of this research method.

2.1 Classification and Judgment Methods of Missing Values

Missing values are a common problem in the data collection process, and the first task of dealing with missing values is to understand the types of missing values, so as to further deal with various types of missing types. Little and other scholars classified it into three types: missing completely at random (MCAR), missing at random (MAR), and missing at random (NMAR) [1], as follows:

1. Missing Completely At Random (MCAR)

The missing data has nothing to do with itself or with the values of other variables in the data set, and its missing probability is random and not necessarily related. Under this type, the missing value is randomly distributed for all data, that is, it has nothing to do with the respective attributes of the data. Take the results of the questionnaire survey in Table 1 as an example. This questionnaire is a survey of the actual age and salary of the respondents, and various types of missing values are represented by X. In the MCAR type, the missing answers of the questionnaire respondents have no relationship with other fields in the data, and are completely randomly generated missing values.

2. Missing At Random (MAR)

The generation of missing values of this type may depend on the correlation between other fields. Taking the questionnaire in Table 1 as an example, when answering the salary question, for a specific group (for example, those younger than 26 years old) who have just left the society for employment, they may be unwilling to disclose their current salary due to insufficient work experience, which will lead to Data missing.

3. Not Missing At Random (NMAR)

This type of missing value may be related to the information that has been found or not found. Taking the questionnaire in Table 1 as an example, it can be found that groups with higher age but relatively lower salaries may be reluctant to disclose their current salaries due to personal factors, resulting in missing data.

Table 1. Schematic Diagram of Missing Value Frequency Pattern

Age	Actual Salary/month	MCAR	MAR	NMAR
20	30000	30000	X	30000
24	28000	X	X	X
25	50000	50000	X	50000
25	23000	X	X	X
26	55000	X	55000	55000
26	27000	27000	27000	X
27	50000	50000	50000	50000
30	60000	60000	60000	60000
30	30000	X	50000	X
37	70000	70000	70000	70000

In the papers proposed by scholars such as Schlomer, the steps for judging the missing type are as follows [12]:

1. Judgment method for MCAR and MAR: Evaluate the relationship between collected variables and missing values based on past analysis experience, and judge by the following methods:
   1) Use dummy variable to indicate: use dummy variable 0 or 1 to indicate whether the field is missing.

2) Use statistical methods to test the relationship between this variable and other variables

A. If the dummy variable has no correlation with other variables, it can be determined that the type of the missing value is MCAR or NMAR.

B. If the dummy variable is associated with other variables, it can be judged as MAR or NMAR.

2. Using scholar Little's MCAR test, the method is as follows

    1) Use statistical tools, such as the Little test module in SPSS software, to test.

    2) When the value of p-value is not significant, the data belongs to MCAR.

    3) This method is a comprehensive test, which regards the data as a whole rather than a single variable.

When the type does not belong to MCAR or MAR, the missing can be regarded as NMAR, and finally the missing data can be classified by missing type. Among the three types of missing values mentioned above, the identification of the second and third types of missing values (MAR and NMAR) often requires the answers of experts in the field, and it is difficult to analyze the objective data model. Missing values predominate.

2.2 How to Deal with Missing Values

Many scholars are committed to exploring the methods of missing values. According to the research of Houari and other scholars, the deletion and filling methods [6] are the most valued by experts and scholars. The method is described as follows:

1. Deletion Method

Deleting is the most common way to deal with missing values, but deleting missing fields may lose most of the data, which will cause errors in the correctness of the analysis. Common deletion methods include Listwise Deletion and Pairwise Deletion. The row deletion method deletes all data rows containing missing values, and only keeps data rows with complete values for all fields.

For missing data of non-MCAR types, this method may completely delete certain types of features, resulting in the inability to provide proper information for this type of data during data analysis, resulting in deviations in the analysis results [2]. Taking the data in Table 1 as an example, if a certain group (for example, respondents under the age of 26) do not want to disclose information on specific topics in the questionnaire, if the method of row deletion is used, the information of this group may be lost. Characteristics, no data available for groups under the age of 26, possibly.

As a result, data of certain specific characteristics are deleted, and effective data cannot be provided for subsequent analysis.

The pairwise deletion method, also known as Available Case Analysis, does not directly delete the entire column of data, but uses statistical methods to perform t-test or Pearson Correlation for missing fields for correlation analysis. If the missing field is related to other fields, then use other fields as replacements during analysis. For non-time series data, this method can use non-missing data for analysis, but it will lead to serious deviation for time series data [2].

2. Imputation Method

It mainly uses the value predicted by the prediction model or statistics to fill in the missing values. At present, the more commonly used methods are mean filling, regression method for predicting filling of a single missing value, machine learning method to fill in the value after analyzing the data, and multiple filling methods used in multi-field missing. Mean filling uses the mean value of the missing field to fill in. Although it is more convenient, the data filled by the mean value may reduce the variance with its own field and reduce the variance with other fields [12]. Scholars such as Houari also pointed out in their research that the method of mean filling will cause such as the overestimation of the sample size, the reduction of the difference between the data, and the loss of attribute variance, and the correlation has a negative bias [6]. The method is also gradually paid attention to improve the accuracy.

The regression method regards the fields with missing values as dependent variables, and adjusts the parameters of the independent variables (non-missing value fields) to predict the missing fields. In the research of scholars such as Tabasi, the regression model was used to predict energy, and the missing field was used as a dependent variable, and other non-missing fields were used as independent variables for prediction, and finally the predicted value was filled [13].

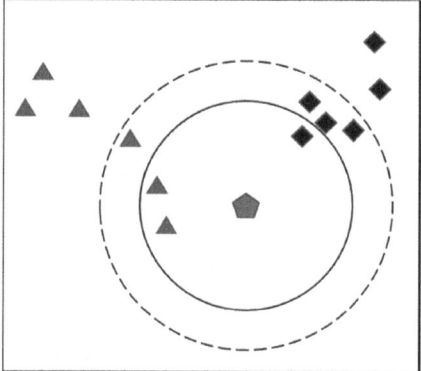

Fig. 1. Schematic diagram of KNN classification

The K-Nearest Neighbor algorithm (K-Nearest Neighbor, KNN) was proposed by T. Cover and other scholars in 1967. It mainly uses data points to map to vector space and calculates their distance for classification [4]. The main core of the KNN algorithm is based on the K closest sample labels around the predicted point. If there are more labels of a certain category among the surrounding K points, the predicted point is classified as that category. The red point in Fig. 1 is the prediction point, within the solid circle (K = 3), because the number of light blue labels is large, the prediction point will be classified as light blue; in the dotted circle (K = 3), because the number of dark blue is large, the predicted points will be classified into the category of dark blue.

Scholars such as Beretta used the KNN method to fill missing values, and pointed out that filling with KNN has the following characteristics [3]:

1. The feature of KNN filling is the value that has appeared in K points around the point to be measured, not the value generated by simulation.
2. Forecast based on information provided by other fields in the missing field, thus preserving the structure of the original data.
3. KNN is non-parametric and does not require a clear model to confirm the relationship between the dependent variable Y and the independent variable X.

Scholars such as Houari proposed a multiple filling method, and a group of m (usually between 5 and 10) reasonable replacement values were extracted from the predicted distribution to fill in missing values [6]. This method avoids single value interpolation because only one value is predicted.

It is the uncertainty caused by filling, but it is limited by the requirement that the data must be normally distributed, and the normal distribution is a common continuous probability distribution. Scholars Pedersen et al. divided multiple filling into three steps [10]:

1. Select variables that are likely to be helpful in calculating missing fields and create multiple predicted datasets, each drawn from the missing data distribution given in the observed data.
2. Using the method of choice, such as Classification and Regression Tree (CART) and Regression (Regression), etc., calculate the association with a higher degree of association in each filled data set. The standard deviation of the coefficient can be calculated as the correlation standard in the calculation data set.
3. A merge of all prediction sets is performed, which reduces the standard deviation between different prediction sets and takes into account the variation between different imputations.

Due to the concept of repeated sampling, data with multiple imputation can reduce its uncertainty compared with the method of single value imputation. For data sets with multiple missing fields, multiple imputation can fill in the missing values of multiple fields.

2.3 K-Means Grouping Algorithm

The K-means algorithm is an unsupervised grouping algorithm proposed by James Mac-Queen in 1967 [8], followed by a more efficient improved version proposed by many scholars such as Hartigan [5]. This method uses the score given by the user.

The number of groups, iterative calculations are performed repeatedly, and the average of the data points of each group is regarded as the center of the group, and finally the new surrounding points of each group are regarded as the same cluster (Cluster, or cluster). The calculation principle is that firstly, the user selects K cluster centers, calculates the distance from n data points to the cluster centers, and the data points that are closer to the cluster centers are classified into the clusters. Recalculate the position of the center (cluster mean) of the group center data point, and move the original group center to the newly calculated group center. After multiple iterations, the points near the group center are regarded as the same cluster.

2.4 Artificial Neural Network (ANN)

The neural network is composed of neurons (Neuron) simulated by humans. Its network architecture is divided into input layer (Input Layer), hidden layer (Hidden Layer), output layer (Output Layer), as shown in Fig. 2. Neurons pass the weight (Weight) and bias (Bias) connected to the neuron and through the activation function (activation function) to determine whether the neuron is activated, and pass it to the hidden layer through the feed forward (Feed Foreword) method. After the operation in the same way, it is passed to the output layer. The calculated output value is compared with the actual label, and the weight and bias in the neural network are corrected through the Backpropagation method, and finally the model is established [1, 11].

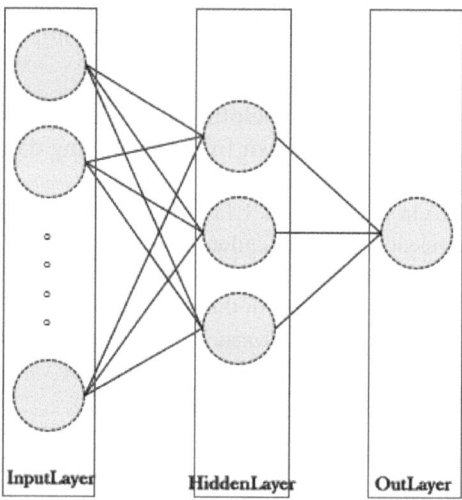

Fig. 2. Schematic diagram of a neural network

2.5 Generating Adversarial Networks

Generative Adversarial Nets (GAN) is a common method used to calculate missing values in graphics or text. After correct training, it can correctly generate missing parts. But for time series data(such as time-stamped data), GAN cannot effectively calculate missing values. Although some scholars have proposed GAN-2-Stage, a variant of GAN, which integrates GAN and recurrent neural network (RNN) to predict gaps in time-series data.

Lost value. However, GAN-2-Stage has the problems of high computational cost and error delay and amplification. In order to solve this problem, scholar Zhang et al. proposed an end-to-end GAN with RF (E2GAN-RF) model [15], integrating the Encoder Network into the GAN-2-Stage model, and the original method must have an independent step processing The cost incurred for the generated test profile. Scholars have used some public data sets, such as PhysioNet, KDD CUP 2018, Air Quality in Northern

China, etc., and other missing value prediction algorithms, such as RDA (Recurrent Denoising Autoencoder), GAIN-2-Stage, etc., to compare performance and The difference in correctness shows that the method proposed by the author has considerable advantages.

3   Research Methods

It is not easy to see whether there is a relationship between data and data in general data sets, so it is impossible to describe the attributes of data with a single model. Therefore, this study used the method of establishing multiple models to divide the data set into clusters and establish a prediction model for each cluster. When predicting data with missing values, first judge the belonging of the data cluster, and use the model of the cluster to make predictions. This study uses public data to train the model. In the experiment, the characteristics of the non-missing fields in the data are found out by the method of machine learning, and the data is divided into multiple clusters by the clustering algorithm, and a prediction model is established for each cluster.. Before predicting the test data to be tested, first find out the clusters to which the non-missing fields belong, and use the models established by different clusters to make predictions.

3.1   Assumptions

The following assumptions were made for the forecasting models consistent with the models designed for this study:

1. The data used to train the predictive model and the data required to predict missing values have the same fixed structure.
2. The values of each column in the data are quantitative values.
3. The missing value to be predicted is a single and fixed field in the data.

   For the various clustering and forecasting methods used in this research, the training data must have the same structure when building the forecasting model, while unstructured data must be converted into structured data through additional processing steps. The data that needs to predict missing values must have the same structure as the data used for model training to ensure the correctness of the prediction. In the second assumption, since various clustering and forecasting methods are mainly based on numerical values, non-numerical data can be converted into numerical types by other mapping methods to facilitate subsequent processing. In addition, the model proposed in this study is to predict the missing value of a single field in the data, and this limitation can be used to predict multiple missing values by repeatedly applying the steps of model training.

3.2   Model Building Method

Let there be n data value groups (tuple) t in the model training data set S, each data has the same structure t: $< a1, a2, ..., am >$, ai is each attribute (field) contained in the data, and both are numeric. The attribute ap is the field that needs to predict missing values, and the rest are called non-missing value fields. The way to build the model is to use the

K-means clustering method to build K clusters from the training data with non-missing field features, $C = \{C_1, C_2, ..., C_k\}$. For each cluster $Ci \in C$, a multiple regression (Multiple Regression) model and a neural network (Artificial Neural Network, ANN) model were established to evaluate the accuracy. The steps of model establishment are shown in Fig. 3.

1. Data Standardization

Due to the differences in the data values of the original data, it is necessary to standardize the data when analyzing and applying algorithms. Data standardization can scale the values proportionally. The Euclidean distance is used in the field of machine learning to calculate the distance between two points. When the value of one of the points is particularly large, it will have a great impact on the result. For this problem, the effect can be reduced by using data normalization to scale the values into a specific interval.

In this experiment, the Min-Max Normalization method was used to standardize the data, and it was also found in the experiment that the normalized data can obtain a lower root mean square error (Root Mean Square Error, RMSE). The Min-Max Normalization method is shown in formula (1), where x represents the original data value, y represents the standardized value, x_{min} is the minimum value of the field, and x_{max} is the maximum value of the field. After Min-Max Normalization calculation, the value range will fall between 0 and 1.

$$y = \frac{x - x_{min}}{x_{max} - x_{min}} \tag{1}$$

2. Data Grouping

It is often difficult to describe whether there is a correlation between the data in the overall data set with a single model. If the data are divided into multiple groups according to certain characteristics, the correlation of the data in each group will be discovered more easily. In this study, the K-Means clustering method clusters the non-missing value fields, uses these non-missing field values to map to the space vector, and uses the characteristics of each data point for clustering. By grouping the characteristics provided by the data attributes, we can find out the data with relatively consistent feature correlation in the data, which can reduce the error caused by different characteristics in the process of model building.

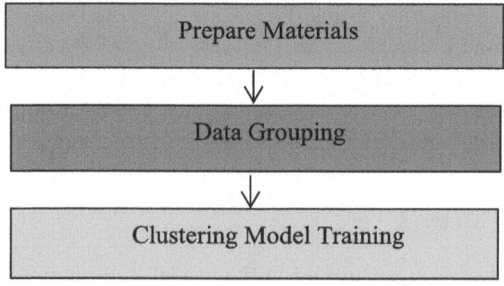

Fig. 3. Model training process

3. Grouping Model Training

After the data is grouped, the data in each group is trained and a model is established. In this stage, we establish two models to discover the correlation between the data in each group and compare the accuracy. These two models are the multiple regression model (Multiple Regression) and artificial neural network (Artificial Neural Network). Because the relationship between the independent variable and the dependent variable in the multiple regression model is a linear relationship, in order to explore the nonlinear relationship in the data, the TanH (Hyperbolic Tangent) activation function of the neural network is used to convert the nonlinear result.

Multiple regression is mainly used to explore the relationship between a dependent variable and multiple independent variables, and establish a regression equation for prediction. The regression expression is shown in formula (1).

$$Y = \beta_0 + \beta_1 X_1 + \beta_2 X_2 + \ldots + \beta_n X_n + \varepsilon \tag{2}$$

In this model, β_0 is a constant term, $\beta_1 \ldots \beta_n$ are regression coefficients, $X_1 \ldots X_n$ are independent variables in the model, that is, non-missing value fields in the data set, represents the error, and Y is the required Predicted missing value field. The selection of independent variables uses the sequential search method, and uses the F-test method to detect whether there is statistically significant (Statistically Significant) between the dependent variables and the independent variables, so that the finally selected independent variables can explain the entire regression model. The regression model through the sequential search method can reduce the amount of calculation in the process of model creation due to the reduction in the number of independent variables, and the calculation speed can be significantly improved for data with a large amount of data.

The neural network is a network that simulates the structure of the human brain. It is composed of multiple neurons and weights. The neurons determine the degree of excitation through the activation function and can be converted into a nonlinear relationship, and then predicted through the transmission of the hidden layer., and finally modify the results of prediction and actual value to achieve the effect of learning.

The training of the neural network-like model uses the non-missing fields in the data set as the input layer of the neural network, and uses the target field (the missing value field) as the label of the neural network. After calculating the error between the prediction of the neural network and the actual label value, and modifying the weight through the reverse transfer method, different neural network models are established for different clusters.

3.3 Missing Value Grouping Judgment

After grouping the training data and establishing a prediction model for each group, the missing value prediction needs to determine the group belonging of the data containing the missing value. This classification of data with missing values uses the KNN method. The non-missing value field of the training data is used as the input data of KNN, and the aforementioned clusters are used as the training labels of KNN, and finally the test data is classified. The choice of K value in KNN calculation is evaluated using the Cross-Validation [9] method to find the best K value of KNN. After determining the classification of the data containing missing values, the model of the group is used to predict the missing values. The process of complete model training and missing value prediction is shown in Fig. 4.

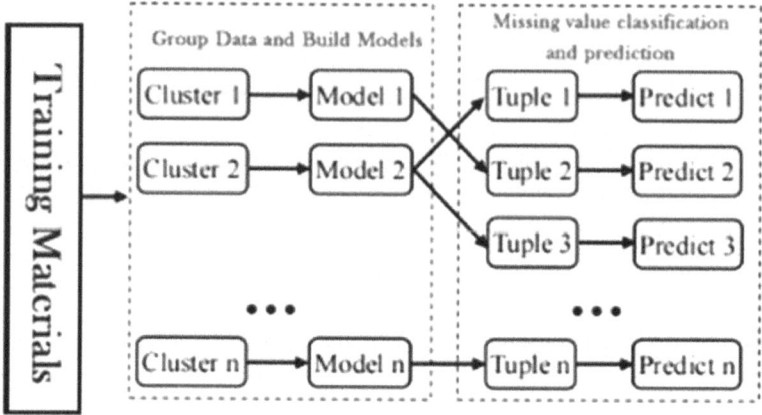

Fig. 4. Model training and missing value prediction process

4   Experimental Methods and Results

In order to verify the correctness of the modeling method proposed in this study, part of the complete data set is used as test data in this experiment, while other data are used to train the model. After each test data is grouped and classified, the missing value is predicted by the group model, and the root mean square error is calculated from the predicted value and the actual value to evaluate the accuracy of the model. The experimental steps are as follows:

1. Randomly divide the data in the data set into training data and test data
2. Use training data to establish clusters and model multiple regression and neural networks for each cluster
3. Use the clustering results of the training data as labels to classify the test data
4. Predict according to the group model after classification of test data
5. Use RMSE to compare the difference between the target value predicted by the model and the original value. The RMSE error value is calculated as shown in formula (3)

$$RMSE = \sqrt{\frac{1}{n}\sum\nolimits_{i=1}^{n}(P_i - r_i)^2} \tag{3}$$

n is the number of compared data, pi is the predicted value, and ri is the actual value. The smaller the RMSE value, the more similar it is to the predicted result.

4.1 Test Data Set

The data used in this experiment is the wine quality data provided by the UCL, Irvin [14], which contains two data sets about white wine and red wine with the same attributes. Each dataset contains 12 attributes, as shown in Table 2. There are 4898 value groups in the white wine dataset and 1599 value groups in the red wine dataset, neither dataset contains any missing values (NA). In the experiment, the target prediction field is fixed acidity, and the data is completely missing at random (MCAR) using different missing ratios to replace missing values.

Table 2. Experimental Data

EnglishName	ChineseName	EnglishName	ChineseName
Fixed Acidity	固定酸度	Total Sulfur Dioxide	总二氧化硫
Volatile Acidity	挥发性酸度	Hensity	密度
Citrie Acid	柠檬酸	pH	酸碱度
Residual Sugar	剩余糖量	Sulphates	硫酸盐
Chlondes	氯化物	Akcohol	酒精浓度
Free Sulfur Dioxide	游离二氧化硫	Quality	品质

Table 3. Test Data Split Ratio

Ratio(%)	1	3	5	10	15	20	25	30
Liquor (Number of Entries)	48	146	244	489	734	979	1224	1469
Red Wine (Number of Entries)	15	47	79	159	239	319	399	479

4.2 Ratio of Training Data and Test Data

In order to compare whether the amount of training data will affect the accuracy of the model, the experimental data set is divided into 8 groups of training data and test data according to different ratios in the experiment. The ratio of test data with missing values is shown in Table 3.

The molding method proposed in this study is compared with other commonly used fillings. The names of various comparison methods in the following statistical charts are as follows:

1. Mean filling: meanMultiple filling: mice_lr.
2. KNN padding: knn.
3. Regression model: lm_x, where lm_org means to build with all training data.

Multiple regression model, others means to divide the training data into groups and build a multiple regression model for each group, for example, lm_cluster3 means to divide the training data into 3 groups and then create a prediction method for each group.

4. Neural Network-Like Model:

ann_x, where ann_all means to build a prediction model with all training data as a neural network, and others means to divide the training data into groups and build a prediction model for each group with a neural network. For example, ann_cluster3 means to divide the training data into 3 groups. Each group builds a predictive model similar to neural networks.

4.3   Data Comparison Results of Red and White Wine

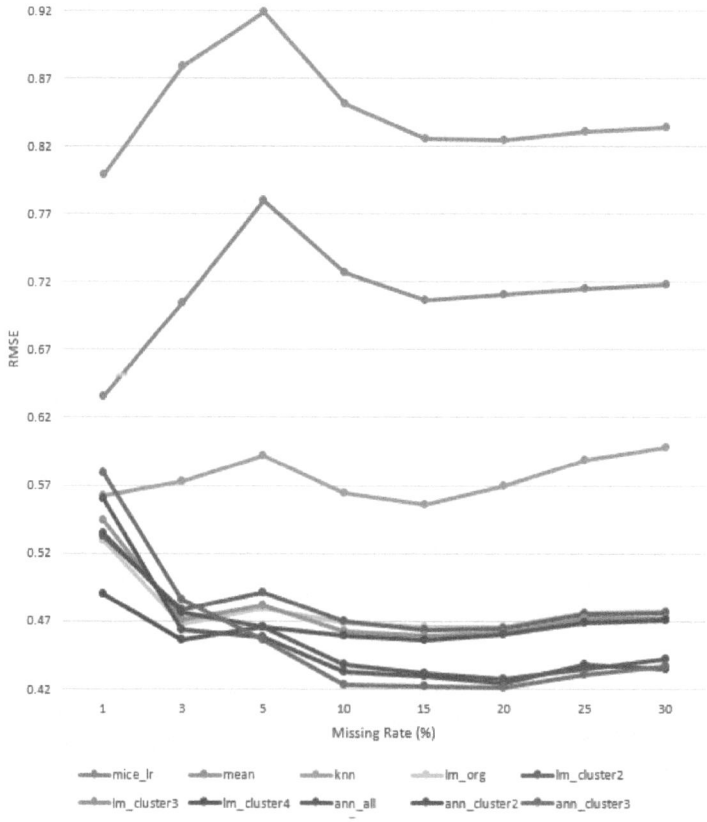

Fig. 5. RMSE comparison of liquor data under different missing rates

Figures 5 and 6 show the RMSE statistical charts of the above-mentioned prediction models based on red and white wine data. The simulation analysis results of two different data sets show that the accuracy of the traditional "mean imputation" and "multiple imputation" methods is lower than the results of other models, and KNN is used to impute data on all data. The RMSE of is also less accurate than other clustering-based modeling methods. Observing the ratio of training data to the total amount of data in the data set, it is shown in two different data sets that the accuracy of prediction will not increase due to the increase in the amount of training data. In simulated tests on both datasets, it was shown that about 90% of the training data is better than 99% of the larger training data.

The accuracy obtained by establishing a multi-prediction model by grouping is better than other prediction models. The difference between the results obtained by different groups is not large. After detailed analysis of the results, it is found that when the number of groups in the training data is 4, all A lower MSE can be obtained under the missing rate. Under the principle of building a model by grouping, the RMSE obtained by the model built with a neural network is slightly better than that of a multiple regression model in two different data sets, and the model with a neural network and a similar neural network for liquor data set The simulation results of the multiple regression model are shown in Figs. 7 and 8.

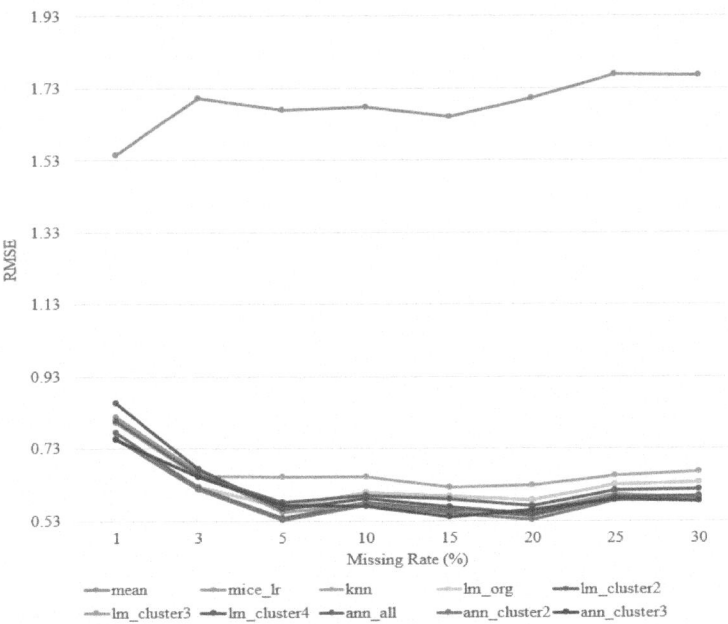

Fig. 6. RMSE comparison of red wine data under different missing rates

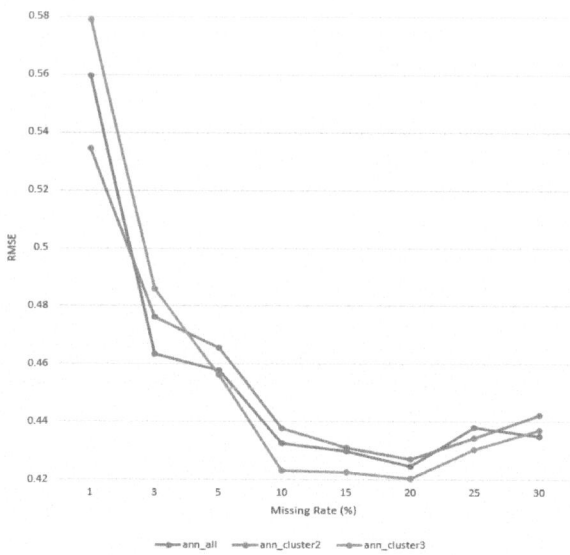

Fig. 7. RMSE Comparison of Liquor Data and Neural Model

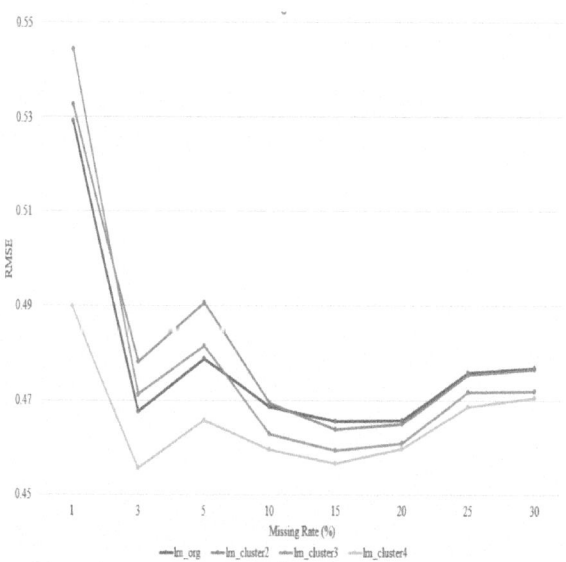

Fig. 8. RMSE comparison of liquor data in multiple regression model

5   Conclusion and Future Development

In the process of data collection, the problem of missing values is inevitable. The processing method of missing values will have a great impact on subsequent analysis and application. In this experiment, the characteristics of the attributes in the data are used for clustering, and a model is generated for each cluster. When forecasting is required, it is necessary to first determine which feature group the features in the data belong to, and then use the model established by the cluster to make predictions. In the experimental results, more accurate predictions can be obtained by using this experimental framework. The experimental results show that the model established with the neural network is slightly better than the multiple regression model in the grouping mode, so the model established with the neural network can provide higher accuracy for missing value prediction in the future Spend. In addition, the experimental results show that using different K values for K-means grouping of training data, it is also found that the model established by different K groups will also affect the results, but the error caused by the number of clusters The dataset used in the experiments is not particularly obvious compared to other methods. Therefore, the experimental results can verify that the prediction model established by the grouping method proposed in this study is superior to other traditional missing value prediction methods in terms of accuracy comparison.

In the continuation of the results of this study, future directions can improve the accuracy of the prediction model in the following directions:

In the current neural network-like model, the TanH activation function is used. In several research papers, it is proposed that the ReLU activation function has the characteristics of solving the gradient explosion problem, fast calculation speed, and fast convergence speed. In order to obtain the best prediction model, we can experiment with different incentive functions in the future.

In this experiment, it is impossible to clearly indicate the number of clusters that the data should be divided into. Instead, the experimental data is used to select the number of clusters (K value). In follow-up research, a systematic way to find a better number of clusters is needed.

In this study, the non-missing field was used to group and build a model, and a more accurate prediction can be obtained for a single missing field. However, in the case of multiple missing fields, it is impossible to predict missing values because the model of each group cannot be clearly established. In response to this problem, the method of non-supervised learning in the cluster can be further explored in the future.

References

1. Basheer, I.A., Hajmeer, M.: Artificial neural networks: fundamentals, computing, design, and application. J. Microbiol. Methods **43**(1), 3–31 (2000)
2. Bennett, D.A.: How can I deal with missingdata in my study? Aust. N. Z. J. Public Health **25**(5), 464–469 (2001)
3. Beretta, L., Santaniello, A.: Nearest neighbor imputation algorithms: a critical evaluation. BMC Med. Inform. Decis. Mak. **16**, 198–208 (2016)
4. Cover, T.M., Hart, P.E.: Nearest neighbor pattern classification. IEEE Trans. Inf. Theory **13**(1), 21–27 (1967)
5. Hartigan, J.A., Wong, M.A.: Algorithm AS 136: A k-means clustering algorithm. J. Roy. Stat. Soc. **28**(1), 100–108 (1979)
6. Houari, R., Bounceur, A., Kamel Tari, A., Tahar Kecha, M.: Handling missing data problems with sampling methods. In: IEEE International Conference on Advanced Networking Distributed Systems and Applications, Bejaia, Algeria (2014)

7. Little, R.J.A., Rubin, D.B.: Statistical Analysis with Missing Data, 3rd Edition, pp. 4–11. John Wiley & Sons Inc, NJ (2019)
8. MacQueen, J.B.: Some methods for classification and analysis of multivariate observations. In: Proceedings of 5th Berkeley Symposium on Mathematical Statistics and Probability. University of California Press, 281–297 (1967)
9. Mullin, M.D., Sukthankar, R.: Complete cross-validation for nearest neighbor classifiers. In: Proceedings of the Seventeenth International Conference on Machine Learning, pp. 639–646 (2000)
10. Pedersen, A.B., et al.: Missing data and multiple imputation in clinical epidemiological research. Clin. Epidemiol. **9**, 157–166 (2017)
11. Rumelhart, D.E., Hinton, G.E., Williams, R.J.: Learning representations by back propagating errors. Nature **323**, 533–536 (1988)
12. Schlomer, G.L., Bauman, S., Card, N.A.: Best practices for missing data management in counseling psychology. J. Couns. Psychol. **5**(1), 1–10 (2010)
13. Tabasi, S., Aslani, A., Forotan, H.: Prediction of Energy Consumption by using gression Model. Comput. Res. Progress Appli. Sci. Eng. **2**(3), 110–115 (2016)
14. Wine Quality Data Set, Machine Learning Repository, University of California, Irvine. https://archive.ics.uci.edu/ml/datasets/wine+quality, Retrieved 23 December 2019
15. Zhang, Y., Zhou, B., Cai, X., Guo, W., Ding, X., Yuan, X.: Missing value imputation in multivariate time series with end-to-end generative adversarial networks. Inf. Sci. **551**, 67–82 (2020)

Correction to: Data Sources and Fast Preprocessing of Shoreline and Bathymetric Data from the Coastal Ocean Numerical Model: Examples of the Pearl River Estuary

Yanqiang Wang⑩, Tianyu Zhang, and Wensheng Jiang

Correction to:
Chapter 21 in: H. Yu et al. (Eds.):
Computer Applications, **CCIS 2274,**
https://doi.org/10.1007/978-981-97-9671-7_21

In the originally published version of chapter 21, affiliations are renumbered incorrectly. They have now been corrected.

The updated version of this chapter can be found at
https://doi.org/10.1007/978-981-97-9671-7_21

Author Index